PENGUIN BOOKS

THE NET DELUSION

Evgeny Morozov is a contributing editor to *Foreign Policy* and runs the magazine's influential and widely quoted 'Net Effect' blog about the Internet's impact on global politics (neteffect.foreignpolicy. com). Morozov is currently a Yahoo! fellow at the Institute for the Study of Diplomacy at Georgetown University.

The Net Delusion

How Not to Liberate the World

EVGENY MOROZOV

PENGUIN BOOKS

PENGUIN BOOKS

Published by the Penguin Group
Penguin Books Ltd, 80 Strand, London WC2R ORL, England
Penguin Group (USA) Inc., 375 Hudson Street, New York, New York 10014, USA Penguin Group
(Canada), 90 Eglinton Avenue East, Suite 700, Toronto, Ontario, Canada M4P 2Y3
(a division of Pearson Penguin Canada Inc.)
Penguin Ireland, 25 St Stephen's Green, Dublin 2, Ireland (a division of Penguin Books Ltd)
Penguin Group (Australia), 250 Camberwell Road, Camberwell, Victoria 3124, Australia
(a division of Pearson Australia Group Pty Ltd)
Penguin Books India Pvt Ltd, 11 Community Centre, Panchsheel Park, New Delhi – 110 017, India
Penguin Group (NZ), 67 Apollo Drive, Rosedale, Auckland 0632, New Zealand
(a division of Pearson New Zealand Ltd)
Penguin Books (South Africa) (Pty) Ltd, Block D, Rosebank Office Park, 181 Jan Smuts Avenue,
Parktown North, Gauteng 2193, South Africa

Penguin Books Ltd, Registered Offices: 80 Strand, London WC2R ORL, England

www.penguin.com

First published in the USA by Public Affairs™, a member of the Perseus Books Group 2011
First published in Great Britain by Allen Lane 2011
Published in Penguin Books 2012

001

Copyright © Evgeny Morozov, 2011

Printed in England by Clays Ltd, St Ives plc

978-0-141-04957-1

www.greenpenguin.co.uk

MIX
Paper from
responsible sources
FSC™ C018179
www.fsc.org

Penguin Books is committed to a sustainable
future for our business, our readers and our planet.
This book is made from Forest Stewardship
Council™ certified paper.

ALWAYS LEARNING   PEARSON

To Aernout van Lynden

CONTENTS

INTRODUCTION

For anyone who wants to see democracy prevail in the most hostile and unlikely environments, the first decade of the new millennium was marked by a sense of bitter disappointment, if not utter disillusionment. The seemingly inexorable march of freedom that began in the late 1980s has not only come to a halt but may have reversed its course.

Expressions like "freedom recession" have begun to break out of the think-tank circuit and enter the public conversation. In a state of quiet desperation, a growing number of Western policymakers began to concede that the Washington Consensus—that set of dubious policies that once promised a neoliberal paradise at deep discounts—has been superseded by the Beijing Consensus, which boasts of delivering quick-and-dirty prosperity without having to bother with those pesky institutions of democracy.

The West has been slow to discover that the fight for democracy wasn't won back in 1989. For two decades it has been resting on its laurels, expecting that Starbucks, MTV, and Google will do the rest just fine. Such a laissez-faire approach to democratization has proved rather toothless against resurgent authoritarianism, which has masterfully adapted to this new, highly globalized world. Today's authoritarianism is of the hedonism- and consumerism-friendly variety, with Steve Jobs and Ashton Kutcher commanding far more respect than Mao or Che Guevara. No wonder the West appears at a loss. While the Soviets could be liberated by waving the magic wand of blue jeans, exquisite coffee machines, and cheap bubble gum, one can't pull the same trick on China. After all, this is where all those Western goods come from.

Many of the signs that promised further democratization just a few years ago never quite materialized. The so-called color revolutions that swept the former Soviet Union in the last decade produced rather ambiguous results. Ironically, it's the most authoritarian of the former Soviet republics—Russia, Azerbaijan, Kazakhstan—that found those revolutions most useful, having discovered and patched their own vulnerabilities. My own birthplace, Belarus, once singled out by Condoleezza Rice as the last outpost of tyranny in Europe, is perhaps the shrewdest of the lot; it continues its slide into a weird form of authoritarianism, where the glorification of the Soviet past by its despotic ruler is fused with a growing appreciation of fast cars, expensive holidays, and exotic cocktails by its largely carefree populace.

The wars in Iraq and Afghanistan, which were started, if anything, to spread the gospel of freedom and democracy, have lost much of their initial emancipatory potential as well, further blurring the line between "regime change" and "democracy promotion." Coupled with Washington's unnecessary abuses of human rights and rather frivolous interpretations of international law, these two wars gave democracy promotion such a bad name that anyone eager to defend it is considered a Dick Cheney acolyte, an insane idealist, or both.

It is thus easy to forget, if only for therapeutic purposes, that the West still has an obligation to stand up for democratic values, speak up about violations of human rights, and reprimand those who abuse their office and their citizens. Luckily, by the twenty-first century the case for promoting democracy no longer needs to be made; even the hardest skeptics agree that a world where Russia, China, and Iran adhere to democratic norms is a safer world.

That said, there is still very little agreement on the kind of methods and policies the West needs to pursue to be most effective in promoting democracy. As the last few decades have so aptly illustrated, good intentions are hardly enough. Even the most noble attempts may easily backfire, entrenching authoritarianism as a result. The images of horrific prisoner abuse at Abu Ghraib were the result, if only indirectly, of one particular approach to promoting democracy. It did not exactly work as advertised.

Unfortunately, as the neoconservative vision for democratizing the world got discredited, nothing viable has come to fill the vacuum. While George Bush certainly overdid it with his excessive freedom-worshiping rhetoric, his successor seems to have abandoned the rhetoric, the spirit, as well as any desire to articulate what a post-Bush "freedom agenda" might look like.

But there is more to Obama's silence than just his reasonable attempt to present himself as anti-Bush. Most likely his silence is a sign of an extremely troubling bipartisan malaise: the growing Western fatigue with the project of promoting democracy. The project suffers not just from bad publicity but also from a deeply rooted intellectual crisis. The resilience of authoritarianism in places like Belarus, China, and Iran is not for lack of trying by their Western "partners" to stir things up with an expectation of a democratic revolution. Alas, most such Western initiatives flop, boosting the appeal of many existing dictators, who excel at playing up the threat of foreign mingling in their own affairs. To say that there is no good blueprint for dealing with modern authoritarianism would be a severe understatement.

Lost in their own strategizing, Western leaders are pining for something that has guaranteed effectiveness. Many of them look back to the most impressive and most unambiguous triumph of democracy in the last few decades: the peaceful dissolution of the Soviet Union. Not surprisingly—and who can blame them for seeking to bolster their own self-confidence?—they tend to exaggerate their own role in precipitating its demise. As a result, many of the Western strategies tried back then, like smuggling in photocopiers and fax machines, facilitating the flow of samizdat, and supporting radio broadcasts by Radio Free Europe and the Voice of America, are given much more credit than they deserve.

Such belated Cold War triumphalism results in an egregious logical fallacy. Since the Soviet Union eventually fell, those strategies are presumed to have been extremely effective—in fact, crucial to the whole endeavor. The implications of such a view for the future of democracy promotion are tremendous, for they suggest that large doses of information and communications technology are lethal to the most repressive of regimes.

Much of the present excitement about the Internet, particularly the high hopes that are pinned on it in terms of opening up closed societies, stems from such selective and, at times, incorrect readings of history, rewritten to glorify the genius of Ronald Reagan and minimize the role of structural conditions and the inherent contradictions of the Soviet system.

It's for these chiefly historical reasons that the Internet excites so many seasoned and sophisticated decision makers who should really know better. Viewing it through the prism of the Cold War, they endow the Internet with nearly magical qualities; for them, it's the ultimate cheat sheet that could help the West finally defeat its authoritarian adversaries. Given that it's the only ray of light in an otherwise dark intellectual tunnel of democracy promotion, the Internet's prominence in future policy planning is assured.

And at first sight it seems like a brilliant idea. It's like Radio Free Europe on steroids. And it's cheap, too: no need to pay for expensive programming, broadcasting, and, if everything else fails, propaganda. After all, Internet users can discover the truth about the horrors of their regimes, about the secret charms of democracy, and about the irresistible appeal of universal human rights on their own, by turning to search engines like Google and by following their more politically savvy friends on social networking sites like Facebook. In other words, let them tweet, and they will tweet their way to freedom. By this logic, authoritarianism becomes unsustainable once the barriers to the free flow of information are removed. If the Soviet Union couldn't survive a platoon of pamphleteers, how can China survive an army of bloggers?

It's hardly surprising, then, that the only place where the West (especially the United States) is still unabashedly eager to promote democracy is in cyberspace. The Freedom Agenda is out; the Twitter Agenda is in. It's deeply symbolic that the only major speech about freedom given by a senior member of the Obama administration was Hillary Clinton's speech on Internet freedom in January 2010. It looks like a safe bet: Even if the Internet won't bring democracy to China or Iran, it can still make the Obama administration appear to have the most technologically savvy foreign policy team in history. The best and the brightest are now also the geekiest. The Google Doctrine—the enthusiastic belief in the liberating power of technology accompanied by the irresistible urge to enlist Silicon Valley start-ups in the global fight for freedom—is of growing appeal to many policymakers. In fact, many of them are as upbeat about the revolutionary potential of the Internet as their colleagues in the corporate sector were in the late 1990s. What could possibly go wrong here?

As it turns out, quite a lot. Once burst, stock bubbles have few lethal consequences; democracy bubbles, on the other hand, could easily lead to carnage. The idea that the Internet favors the oppressed rather than the oppressor is marred by what I call cyber-utopianism: a naïve belief in the emancipatory nature of online communication that rests on a stubborn refusal to acknowledge its downside. It stems from the starry-eyed digital fervor of the 1990s, when former hippies, by this time ensconced in some of the most prestigious universities in the world, went on an argumentative spree to prove that the Internet could deliver what the 1960s couldn't: boost democratic participation, trigger a renaissance of moribund communities, strengthen associational life, and serve as a bridge from bowling alone to blogging together. And if it works in Seattle, it must also work in Shanghai.

Cyber-utopians ambitiously set out to build a new and improved United Nations, only to end up with a digital Cirque du Soleil. Even if true—and that's a gigantic "if"—their theories proved difficult to adapt to non-Western and particularly nondemocratic contexts. Democratically elected governments in North America and Western Europe may, indeed, see an Internet-driven revitalization of their public spheres as a good thing; logically, they would prefer to keep out of the digital sandbox—at least as long as nothing illegal takes place. Authoritarian governments, on the other hand, have invested so much effort into suppressing any form of free expression and free assembly that they would never behave in such a civilized fashion. The early theorists of the Internet's influence on politics failed to make any space for the state, let alone a brutal authoritarian state with no tolerance for the rule of law or dissenting opinions. Whatever book lay on the cyber-utopian bedside table in the early 1990s, it was surely not Hobbes's *Leviathan*.

Failing to anticipate how authoritarian governments would respond to the Internet, cyber-utopians did not predict how useful it would prove for propaganda purposes, how masterfully dictators would learn to use it for surveillance, and how sophisticated modern systems of Internet censorship would become. Instead most cyber-utopians stuck to a populist account of how technology empowers the *people*, who, oppressed by years of authoritarian rule, will inevitably rebel, mobilizing themselves through text messages, Facebook, Twitter, and whatever new tool comes along next year. (The *people*, it must be noted, really liked to hear such theories.) Paradoxically, in their refusal to see the downside of the new digital environment, cyber-utopians ended up belittling the role of the Internet, refusing to see that it penetrates and reshapes all walks of political life, not just the ones conducive to democratization.

I myself was intoxicated with cyber-utopianism until recently. This book is an attempt to come to terms with this ideology as well as a warning against the pernicious influence that it has had and is likely to continue to have on democracy promotion. My own story is fairly typical of idealistic young people who think they are onto something that could change the world. Having watched the deterioration of democratic freedoms in my native Belarus, I was drawn to a Western NGO that sought to promote democracy and media reform in the former Soviet bloc with the help of the Internet. Blogs, social networks, wikis: We had an arsenal of weapons that seemed far more potent than police batons, surveillance cameras, and handcuffs.

Nevertheless, after I spent a few busy years circling the former Soviet region and meeting with activists and bloggers, I lost my enthusiasm. Not only were our strategies failing, but we also noticed a significant push back from the governments we sought to challenge. They were beginning to experiment with censorship, and some went so far as to start aggressively engaging with new media themselves, paying bloggers to spread propaganda and troll social networking sites looking for new information on those in the opposition. In the meantime, the Western obsession with the Internet and the monetary support it guaranteed created numerous hazards typical of such ambitious development projects. Quite predictably, many of the talented bloggers and new media entrepreneurs preferred to work for the extremely well-paid but largely ineffective Western-funded projects instead of trying to create more nimble, sustainable, and, above all, effective projects of their own. Thus, everything we did—with generous funding from Washington and Brussels—seemed to have produced the results that were the exact opposite of what my cyber-utopian self wanted.

It was tempting to throw my hands up in despair and give up on the Internet altogether. But this would have been the wrong lesson to draw from these disappointing experiences. Similarly, it would be wrong for Western policymakers to simply dismiss the Internet as a lost cause and move on to bigger, more important issues. Such digital defeatism would only play into the hands of authoritarian governments, who would be extremely happy to continue using it as both a carrot (keeping their populace entertained) and a stick (punishing those who dare to challenge the official line). Rather, the lesson to be drawn is that the Internet is here to stay, it will continue growing in importance, and those concerned with democracy promotion need not only grapple with it but also come up with mechanisms and procedures to ensure that another tragic blunder on the scale of Abu Ghraib will never happen in cyberspace. This is not a far-fetched scenario. How hard is it to imagine a site like Facebook inadvertently disclosing the private information of activists in Iran or China, tipping off governments to secret connections between the activists and their Western funders?

To be truly effective, the West needs to do more than just cleanse itself of cyber-utopian bias and adopt a more realist posture. When it comes to concrete steps to promote democracy, cyber-utopian convictions often give rise to an equally flawed approach that I dub "Internet-centrism." Unlike cyber-utopianism, Internet-centrism is not a set of beliefs; rather, it's a philosophy of action that informs how decisions, including those that deal with democracy promotion, are made and how long-term strategies are crafted. While cyber-utopianism stipulates *what* has to be done, Internet-centrism stipulates *how* it should be done. Internet-centrists like to answer every question about democratic change by first reframing it in terms of the Internet rather than the context in which that change is to occur. They are often completely oblivious to the highly political nature of technology, especially the Internet, and like to come up with strategies that assume that the logic of the Internet, which, in most cases, they are the only ones to perceive, will shape every environment than it penetrates rather than vice versa.

While most utopians are Internet-centrists, the latter are not necessarily utopians. In fact, many of them like to think of themselves as pragmatic individuals who have abandoned grand theorizing about utopia in the name of achieving tangible results. Sometimes, they are even eager to acknowledge that it takes more than bytes to foster, install, and consolidate a healthy democratic regime.

Their realistic convictions, however, rarely make up for their flawed methodology, which prioritizes the tool over the environment, and, as such, is deaf to the social, cultural, and political subtleties and indeterminacies. Internet-centrism is a highly disorienting drug; it ignores context and entraps policymakers into believing that they have a useful and powerful ally on their side. Pushed to its extreme, it leads to hubris, arrogance, and a false sense of confidence, all bolstered by the dangerous illusion of having established effective command of the Internet. All too often, its practitioners fashion themselves as possessing full mastery of their favorite tool, treating it as a stable and finalized technology, oblivious to the numerous forces that are constantly reshaping the Internet— not all of them for the better. Treating the Internet as a constant, they fail to see their own responsibility in preserving its freedom and reining in the ever-powerful intermediaries, companies like Google and Facebook.

As the Internet takes on an even greater role in the politics of both authoritarian and democratic states, the pressure to forget the context and start with what the Internet allows will only grow. All by itself, however, the Internet provides nothing certain. In fact, as has become obvious in too many contexts, it empowers the strong and disempowers the weak. It is impossible to place the Internet at the heart of the enterprise of democracy promotion without risking the success of that very enterprise.

The premise of this book is thus very simple: To salvage the Internet's promise to aid the fight against authoritarianism, those of us in the West who still care about the future of democracy will need to ditch both cyber-utopianism and Internet-centrism. Currently, we start with a flawed set of assumptions (cyber-utopianism) and act on them using a flawed, even crippled, methodology (Internet-centrism). The result is what I call the Net Delusion. Pushed to the extreme, such logic is poised to have significant global consequences that may risk undermining the very project of promoting democracy. It's a folly that the West could do without.

Instead, we'll need to opt for policies informed by a realistic assessment of the risks and dangers posed by the Internet, matched by a highly scrupulous and unbiased assessment of its promises, and a theory of action that is highly sensitive to the local context, that is cognizant of the complex connections between the Internet and the rest of foreign policymaking, and that originates not in what technology allows but in what a certain geopolitical environment requires.

In a sense, giving in to cyber-utopianism and Internet-centrism is akin to agreeing to box blindfolded. Sure, every now and then we may still strike some powerful blows against our authoritarian adversaries, but in general this is a poor strategy if we want to win. The struggle against authoritarianism is too important of a battle to fight with a voluntary intellectual handicap, even if that handicap allows us to play with the latest fancy gadgets.

chapter one

The Google Doctrine

● ● ■ ■

In June 2009 thousands of young Iranians—smartphones in their hands (and, for the more advanced, Bluetooth headsets in their ears)—poured into the stuffy streets of Tehran to protest what they believed to be a fraudulent election. Tensions ran high, and some protesters, in an unthinkable offense, called for the resignation of Ayatollah Khamenei. But many Iranians found the elections to be fair; they were willing to defend the incumbent President Mahmoud Ahmadinejad if needed. Iranian society, buffeted by the conflicting forces of populism, conservatism, and modernity, was facing its most serious political crisis since the 1979 revolution that ended the much-disliked reign of the pro-American Shah Mohammad Reza Pahlavi.

But this was not the story that most Western media chose to prioritize; instead, they preferred to muse on how the Internet was ushering in democracy into the country. "The Revolution Will Be Twittered" was the first in a series of blog posts published by the *Atlantic*'s Andrew Sullivan a few hours after the news of the protests broke. In it, Sullivan zeroed in on the resilience of the popular microblogging site Twitter, arguing that "as the regime shut down other forms of communication,

Twitter survived. With some remarkable results." In a later post, even though the "remarkable results" were still nowhere to be seen, Sullivan proclaimed Twitter to be "the critical tool for organizing the resistance in Iran" but didn't bother to quote any evidence to support his claim. Only a few hours after the protests began, his blog emerged as a major information hub that provided almost instantaneous links to Iran-related developments. Thousands of readers who didn't have the stamina to browse hundreds of news sites saw events unfolding in Iran primarily through Sullivan's eyes. (And, as it turned out, his were a rather optimistic pair.)

It didn't take long for Sullivan's version of events to gain hold elsewhere in the blogosphere—and soon enough, in the traditional media as well. Michelle Malkin, the right-wing blogging diva, suggested that "in the hands of freedom-loving dissidents, the micro-blogging social network is a revolutionary samizdat—undermining the mullah-cracy's information blockades one Tweet at a time." Marc Ambinder, Sullivan's colleague at the *Atlantic*, jumped on the bandwagon, too; for him, Twitter was so important that he had to invent a new word, "protagonal," to describe it. "When histories of the Iranian election are written, Twitter will doubtless be cast a protagonal technology that enabled the powerless to survive a brutal crackdown," wrote Ambinder on his blog. The *Wall Street Journal*'s Yochi Dreazen proclaimed that "this [revolution] would not happen without Twitter," while National Public Radio's Daniel Schorr announced that "in Iran, tyranny has run afoul of technology in the form of the Internet, turning a protest into a movement." When Nicholas Kristof of the *New York Times* asserted that in "the quintessential 21st-century conflict . . . on one side are government thugs firing bullets . . . [and] on the other side are young protesters firing 'tweets,'" he was simply registering the zeitgeist.

Soon technology pundits, excited that their favorite tool was all over the media, were on the case as well. "This is it. The big one. This is the first revolution that has been catapulted onto a global stage and transformed by social media," proclaimed New York University's Clay Shirky in an interview with TED.com. Jonathan Zittrain, a Harvard academic and the author of *The Future of the Internet and How to Stop It*, alleged that "Twitter, in particular, has proven particularly adept at organizing people and information." John Gapper, a business columnist for the *Financial Times*, opined that Twitter was "the tinderbox that fanned the spark of revolt among supporters of Mir-Hossein Moussavi." Even the usually sober *Christian Science Monitor* joined in the cyber-celebrations, noting that "the government's tight control of the Internet has spawned a generation adept at circumventing cyber road blocks, making the country ripe for a technology-driven protest movement."*

Twitter seemed omnipotent—certainly more so than the Iranian police, the United Nations, the U.S. government, and the European Union. Not only would it help to rid Iran of its despicable leader but also convince ordinary Iranians, most of whom vehemently support the government's aggressive pursuit of nuclear enrichment, that they should stop their perpetual fretting about Israel and simply go back to being their usual peaceful selves. A column in the right-wing *Human Events* declared that Twitter had accomplished "what neither the U.N. nor the European Union have [sic] been able to do," calling it "a huge threat to the Iranian regime—a pro-liberty movement being fomented and organized in short sentences." Likewise, the editorial page of the *Wall Street Journal* argued that "the Twitter-powered 'Green Revolution' in Iran . . . has used social-networking technology to do more for regime change in the Islamic Republic than years of sanctions, threats and Geneva-based haggling put together." It seemed that Twitter was improving not only democracy but diplomacy as well.

Soon enough, pundits began using the profusion of Iranian tweets as something of an excuse to draw far-reaching conclusions about the future of the world in general. To many, Iran's Twitter-inspired protests clearly indicated that authoritarianism was doomed everywhere. In a

---

* A confession is in order here: I was one of the first to fall into the Twitter Revolution trap, christening similar youth protests in Moldova, which happened a few months before Iran's, with what proved to be that sticky and extremely misleading moniker. Even though I quickly qualified it with a long and nuanced explanation, it is certainly not the proudest moment in my career, especially as all those nuances were lost on most media covering the events.

column modestly entitled "Tyranny's New Nightmare: Twitter," *Los Angeles Times* writer Tim Rutten declared that "as new media spreads its Web worldwide, authoritarians like those in Iran will have a difficult time maintaining absolute control in the face of the technology's chaotic democracy." That the Green Movement was quickly disintegrating and was unable to mount a serious challenge to Ahmadinejad didn't prevent the editorial page of the *Baltimore Sun* from concluding that the Internet was making the world safer and more democratic: "The belief that activists are blogging their lives away while governments and corporations take greater control of the world is being proven false with every tweet, every blog comment, every protest planned on Facebook."

Inspired by similar logic, Mark Pfeifle, former deputy national security advisor in the George W. Bush administration, launched a public campaign to nominate Twitter for the Nobel Peace Prize, arguing that "without Twitter, the people of Iran would not have felt empowered and confident to stand up for freedom and democracy." The Webby Awards, the Internet's equivalent of the Oscars, hailed the Iranian protests as "one of the top ten Internet moments of the decade." (The Iranian youths—or, rather, their smartphones—were in good company: The expansion of Craigslist beyond San Francisco in 2000 and the launch of Google AdWords in 2004 were among other honorees.)

But it was Gordon Brown, then the prime minister of the United Kingdom, who drew the most ridiculous conclusion from the events in Iran. "You cannot have Rwanda again because information would come out far more quickly about what is actually going on and the public opinion would grow to the point where action would need to be taken," he argued. "This week's events in Iran are a reminder of the way that people are using new technology to come together in new ways to make their views known." On Brown's logic, the millions who poured into the streets of London, New York, Rome, and other cities on February 15, 2003, to protest the impending onset of the Iraq War made one silly mistake: They didn't blog enough about it. *That* would have definitely prevented the bloodbath.

Hail the Google Doctrine

Iran's seemed like a revolution that the whole world was not just watching but also blogging, tweeting, Googling, and YouTubing. It only took a few clicks to get bombarded by links that seemed to shed more light on events in Iran—quantitatively, if not qualitatively—than anything carried by what technologists like to condescendingly call "legacy media." While the latter, at least in their rare punditry-free moments of serenity, were still trying to provide some minimal context to the Iranian protests, many Internet users preferred to simply get the raw deal on Twitter, gorging on as many videos, photos, and tweets as they could stomach. Such virtual proximity to events in Tehran, abetted by access to the highly emotional photos and videos shot by protesters themselves, led to unprecedented levels of global empathy with the cause of the Green Movement. But in doing so, such networked intimacy may have also greatly inflated popular expectations of what it could actually achieve.

As the Green Movement lost much of its momentum in the months following the election, it became clear that the Twitter Revolution so many in the West were quick to inaugurate was nothing more than a wild fantasy. And yet it still can boast of at least one unambiguous accomplishment: If anything, Iran's Twitter Revolution revealed the intense Western longing for a world where information technology is the liberator rather than the oppressor, a world where technology could be harvested to spread democracy around the globe rather than entrench existing autocracies. The irrational exuberance that marked the Western interpretation of what was happening in Iran suggests that the green-clad youngsters tweeting in the name of freedom nicely fit into some preexisting mental schema that left little room for nuanced interpretation, let alone skepticism about the actual role the Internet played at the time.

The fervent conviction that given enough gadgets, connectivity, and foreign funding, dictatorships are doomed, which so powerfully manifested itself during the Iranian protests, reveals the pervasive influence of the Google Doctrine. But while the manic surrounding Iran's Twitter Revolution helped to crystallize the main tenets of the doctrine, it did not beget those tenets. In fact, the Google Doctrine has a much finer intellectual pedigree—much of it rooted in the history of the Cold War—than many of its youthful proponents realize. The Nobel Prize–winning economist Paul Krugman was already warning about such premature triumphalism back in 1999 when he ridiculed its core beliefs in a book review. Ironically enough, the book was by Tom Friedman, his future fellow *New York Times* columnist. According to Krugman, too many Western observers, with Friedman as their cheerleader in chief, were falling under the false impression that thanks to advances in information technology "old-fashioned power politics is becoming increasingly obsolete, because it conflicts with the imperatives of global capitalism." Invariably they were reaching the excessively optimistic conclusion that "we are heading for a world that is basically democratic, because you can't keep 'em down on the farm once they have Internet access, and basically peaceful, because George Soros will pull out his money if you rattle your saber." And in a world like this, how can anything but democracy triumph in the long run?

As such, the Google Doctrine owes less to the advent of tweeting and social networking than it does to the giddy sense of superiority that many in the West felt in 1989, as the Soviet system collapsed almost overnight. As history was supposed to be ending, democracy was quickly pronounced the only game in town. Technology, with its unique ability to fuel consumerist zeal—itself seen as a threat to any authoritarian regime—as well as its prowess to awaken and mobilize the masses against their rulers, was thought to be the ultimate liberator. There is a good reason why one of the chapters in Francis Fukuyama's *The End of History and The Last Man*, the ur-text of the early 1990s that successfully bridged the worlds of positive psychology and foreign affairs, was titled "The Victory of the VCR."

The ambiguity surrounding the end of the Cold War made such arguments look far more persuasive than any close examination of their theoretical strengths would warrant. While many scholars took it to mean that the austere logic of Soviet-style communism, with its five-year plans and constant shortages of toilet paper, had simply run its course, most popular interpretations downplayed the structural deficiencies of the Soviet regime—who would want to acknowledge that the Evil Empire was only a bad joke?—preferring to emphasize the momentous achievements of the dissident movement, armed and nurtured by the West, in its struggle against a ruthless totalitarian adversary. According to this view, without the prohibited samizdat materials, photocopiers, and fax machines that were smuggled into the Soviet bloc, the Berlin Wall might have still been with us today. Once the Soviet Union's VCR movement had arrived, communism was untenable.

The two decades that followed were a mixed bag. VCR moments were soon superseded by DVD moments, and yet such impressive breakthroughs in technology failed to bring on any impressive breakthroughs in democratization. Some authoritarian regimes, like those in Slovakia and Serbia, fell. Others, like in Belarus and Kazakhstan, only got stronger. In addition, the tragedy of 9/11 seemed to suggest that history was returning from its protracted holiday in Florida and that another ubiquitous and equally reductionist thesis of the early 1990s, that of the clash of civilizations, would come to dominate the intellectual agenda of the new century. As a result, many of the once popular arguments about the liberating power of consumerism and technology faded from public view. That Al-Qaeda seemed to be as proficient in using the Internet as its Western opponents did not chime well with a view that treated technology as democracy's best friend. The dotcom crash of 2000 also reduced the fanatical enthusiasm over the revolutionary nature of new technologies: the only things falling under the pressure of the Internet were stock markets, not authoritarian regimes.

But as the Iranian events of 2009 have so clearly demonstrated, the Google Doctrine was simply put on the backburner; it did not collapse. The sighting of pro-democratic Iranians caught in a tight embrace with Twitter, a technology that many Westerners previously saw as a rather peculiar way to share one's breakfast plans, was enough to fully rehabilitate its core principles and even update them with a fancier Web 2.0 vocabulary. The almost-forgotten theory that people, once armed with a powerful technology, would triumph over the most brutal adversaries—regardless of what gas and oil prices are at the time—was suddenly enjoying an unexpected intellectual renaissance.

Had the Iranian protests succeeded, it seems fairly certain that "The Victory of Tweets" would be too good of a chapter title to go to waste. Indeed, at some point in June 2009, if only for a brief moment, it seemed as if history might be repeating itself, ridding the West of yet another archenemy—and the one with dangerous nuclear ambitions. After all, the streets of Tehran in the summer of 2009 looked much like those of Leipzig, Warsaw, or Prague in the fall of 1989. Back in '89, few in the West had the guts or the imagination to believe that such a brutal system—a system that always seemed so invulnerable and determined to live—could fall apart so peacefully. Iran, it seemed, was giving Western observers the long-awaited chance to redeem themselves over their dismal performance in 1989 and embrace the Hegelian spirit of history before it had fully manifested itself.

Whatever the political and cultural differences between the crowds that were rocking Iran in 2009 and the crowds that rocked Eastern Europe in 1989, both cases seemed to share at least one common feature: a heavy reliance on technology. Those in the streets of Eastern Europe did not yet have BlackBerries and iPhones, but their fight was, nevertheless, abetted by technologies of a different, mostly analogue variety: photocopiers and fax machines, radios tuned to Radio Free Europe and Voice of America, video cameras of Western television crews. And while in 1989 few outsiders could obtain immediate access to the most popular antigovernment leaflets or flip through clandestine photos of police brutality, in 2009 one could follow the Iranian protests pretty much the same way one could follow the Super Bowl or the Grammys: by refreshing one's Twitter page. Thus, any seasoned observers of foreign affairs—and particularly those who had a chance to compare what they saw in 1989 to what they were seeing in 2009—knew, if only intuitively, that the early signs coming from the streets of Tehran seemed to vindicate the Google Doctrine. With that in mind, conclusions about the inevitable collapse of the Iranian regime did not seem so farfetched. Only a lazy pundit would not have pronounced Iran's Twitter

Revolution a success when all the signs were suggesting the inevitability of Ahmadinejad's collapse.

The Unimaginable Consequences of an Imagined Revolution

It must have been similar reasoning—at times bordering on hubris—that led American diplomats to commit a terrible policy blunder at the height of the Iranian protests. Swayed by the monotony of media commentary, the flood of Iran-related messages on Twitter, or his own institutional and professional agendas, a senior official at the U.S. State Department sent an email to executives at Twitter, inquiring if they could reschedule the previously planned—and now extremely ill-timed—maintenance of the site, so as not to disrupt the Iranian protests. Twitter's management complied but publicly emphasized that they reached that decision independently.

The historic significance of what may have seemed like a simple email was not lost on the *New York Times*, which described it as "another new-media milestone" for the Obama administration, attesting to "the recognition by the United States government that an Internet blogging service that did not exist four years ago has the potential to change history in an ancient Islamic country." The *New York Times* may have exaggerated the amount of deliberation that the Obama administration invested in the issue (a White House spokesman immediately downplayed the significance of the "milestone" by claiming that "this wasn't a directive from Secretary of State, but rather was a low-level contact from someone who often talks to Twitter staff"), but the Gray Lady was spot on in assessing its overall importance.

Contrary to Marc Ambinder's prediction, when future historians look at what happened in those few hot weeks in June 2009, that email correspondence—which the State Department chose to widely publicize to bolster its own new media credentials—is likely to be of far greater importance that anything the Green Movement actually did on the Internet. Regardless of the immediate fate of democracy in Iran, the world is poised to feel the impact of that symbolic communication for years to come.

For the Iranian authorities, such contact between its sworn enemies in the U.S. government and a Silicon Valley firm providing online services that, at least as the Western media described it, were beloved by their citizens quickly gave rise to suspicions that the Internet is an instrument of Western power and that its ultimate end is to foster regime change in Iran. Suddenly, the Iranian authorities no longer saw the Internet as an engine of economic growth or as a way to spread the word of the prophet. All that mattered at the time was that the Web presented an unambiguous threat that many of Iran's enemies would be sure to exploit. Not surprisingly, once the protests quieted down, the Iranian authorities embarked on a digital purge of their opponents.

In just a few months, the Iranian government formed a high-level twelve-member cybercrime team and tasked it with finding any false information—or, as they put it, "insults and lies"—on Iranian websites. Those spreading false information were to be identified and arrested. The Iranian police began hunting the Internet for photos and videos that showed faces of the protesters—numerous, thanks to the ubiquity of social media—to publish them on Iranian news media websites and ask the public for help in identifying the individuals. In December 2009 the pro-Ahmadinejad *Raja News* website published a batch of thirty-eight photos with sixty-five faces circled in red and a batch of forty-seven photos with about a hundred faces circled in red. According to the Iranian police, public tip-offs helped to identify and arrest at least forty people. Ahmadinejad's supporters may have also produced a few videos of their own, including a clip—which many in the opposition believed to be a montage—that depicted a group of protesters burning a portrait of Ayatollah Khomeini. If people had believed that the footage was genuine, it could have created a major split in the opposition, alienating vast swathes of the Iranian population.

The police or someone acting on their behalf also went searching for personal details—mostly Facebook profiles and email addresses—of Iranians living abroad, sending them threatening messages and urging them not to support the Green Movement unless they wanted to hurt their relatives back in Iran. In the meantime, the authorities were equally tough on Iranians in the country, warning them to stay away from social networking sites used by the opposition. The country's police chief Gen. Ismail Ahmadi Moghaddam warned that those who incited others to protest or issued appeals "have committed a worse crime than those who come to the streets." Passport control officers at Tehran's airport asked Iranians living abroad if they had Facebook accounts; they would often double-check online, regardless of the answer, and proceed to write down any suspicious-looking online friends a traveler might have.

The authorities, however, did not dismiss technology outright. They, too, were more than happy to harvest its benefits. They turned to text messaging—on a rather massive scale—to warn Iranians to stay away from street protests in the future. One such message, sent by the intelligence ministry, was anything but friendly: "Dear citizen, according to received information, you have been influenced by the destabilizing propaganda which the media affiliated with foreign countries have been disseminating. In case of any illegal action and contact with the foreign media, you will be charged as a criminal consistent with the Islamic Punishment Act and dealt with by the Judiciary."

In the eyes of the Iranian government, the Western media was guilty of more than spreading propaganda; they accused CNN of "training hackers" after the channel reported on various cyber-attacks that Ahmadinejad's opponents were launching on websites deemed loyal to his campaign. Recognizing that the enemy was winning the battle in the virtual world, one ayatollah eventually allowed pious Iranians to use any tool, even if it contravened Shari'a law, in their online fight. "In a war, anti-Shari'a [moves] are permissible; the same applies to a cyber-war. The conditions are such that you should fight the enemy in any way you can. You don't need to be considerate of anyone. If you don't hit them, the enemy will hit you," proclaimed Ayatollah Alam Ahdi during a Friday Prayer sermon in 2010.

But the campaign against CNN was a drop in the sea compared to the accusations launched against Twitter, which the pro-Ahmadinejad Iranian media immediately took to be the real source of unrest in the

country. An editorial in *Javan*, a hard-line Iranian newspaper, accused
the U.S. State Department of trying to foment a revolution via the Internet by helping Twitter stay online, stressing its "effective role in the
continuation of riots." Given the previous history of American interference in the country's affairs—most Iranians still fret about the 1953
coup masterminded by the CIA—such accusations are likely to stick,
painting all Twitter users as a secret American revolutionary vanguard.
In contrast to the tumultuous events of 1953, the Twitter Revolution
did not seem to have its Kermit Roosevelt, Theodore Roosevelt's
grandson and the coordinator of CIA's Operation Ajax, which resulted
in the overthrow of the nationalist government of Mohammad Mosaddegh. But in the eyes of the Iranian authorities the fact that today's digital vanguards have no obvious charismatic coordinators only made
them seem more dangerous. (The Iranian propaganda officials could
not contain their glee when they discovered that Kermit Roosevelt was
a close relative of John Palfrey, the faculty codirector of Harvard's Berkman Center for Internet and Society, a think tank that the U.S. State
Department had funded to study the Iranian blogosphere.)

Other governments also took notice, perhaps out of fear that they,
too, might soon have a Twitter Revolution on their hands. Chinese authorities interpreted Washington's involvement in Iran as a warning sign
that digital revolutions facilitated by American technology companies
are not spontaneous but carefully staged affairs. "How did the unrest
after the Iranian elections come about?" pondered an editorial in the
People's Daily, the chief mouthpiece of the Communist Party. "It was
because online warfare launched by America, via YouTube video and
Twitter microblogging, spread rumors, created splits, stirred up, and
sowed discord between the followers of conservative reformist factions." Another major outlet of government propaganda, Xinhua News
Agency, took a more philosophical view, announcing that "information
technology that has brought mankind all kinds of benefits has this time
become a tool for interfering in the internal affairs of other countries."

A few months after the Iranian protests, *China National Defense*, an
official outlet of the Chinese military, ran a similar editorial, lumping
April 2010 youth protests in Moldova with those of Iran and treating

both as prime examples of Internet-enabled foreign intervention. The editorial, singling out the United States as the "keenest Western power to add the internet to its diplomatic arsenal," also linked those two protests to an ethnic uprising in China's own Xinjiang province in July 2009, concluding that more Internet control was in order, if only "to avoid the internet becoming a new poisoned arrow for hostile forces." Bizarrely, the irresponsible Iran-related punditry in Washington allowed leaders in Beijing to build a credible case for more Internet censorship in China. (The online blockade of the Xinjiang region only ended in early 2010.)

Media in the former Soviet Union took notice as well. "The Demonstrations in Iran Followed the Moldovan Scenario: The U.S. Got Burnt" proclaimed a headline on a Russian nationalist portal. A prime-time news program on the popular Russian TV channel NTV announced that the "Iranian protesters were enjoying the support of the U.S. State Department, which interfered in the internal activities of Twitter, a trendy Internet service." A newspaper in Moldova reported that the U.S. government even supplied Twitter with cutting-edge anticensorship technology.

This was globalization at its worst: A simple email based on the premise that Twitter mattered in Iran, sent by an American diplomat in Washington to an American company in San Francisco, triggered a worldwide Internet panic and politicized all online activity, painting it in bright revolutionary colors and threatening to tighten online spaces and opportunities that were previously unregulated. Instead of finding ways to establish long-term relationships with Iranian bloggers and use their work to quietly push for social, cultural, and—at some distant point in the future—maybe even political change, the American foreign policy establishment went on the record and pronounced them to be more dangerous than Lenin and Che Guevara combined. As a result, many of these "dangerous revolutionaries" were jailed, many more were put under secret surveillance, and those poor Iranian activists who happened to be attending Internet trainings funded by the U.S. State Department during the election could not return home and had to apply for asylum. (At least five such individuals got trapped in Europe.) The pundits were right: Iran's Twitter Revolution did have global repercussions. Those were, however, extremely ambiguous, and they often strengthened rather than undermined the authoritarian rule.

A Revolution in Search of Revolutionaries

Of course, American diplomats had no idea how the Iranian protests would turn out; it would be unfair to blame them for the apparent inability of the Green Movement to unseat Ahmadinejad. When the future of Iranian democracy depended on the benevolence of a Silicon Valley start-up that seemed oblivious to the geopolitical problems besetting the world, what other choice did they have but to intervene? Given what was at stake, isn't it preposterous to quibble about angry editorials in Moldovan newspapers that may have appeared even if the State Department stayed on the sidelines?

All of this is true—as long as there is evidence to assert that the situation was, indeed, dramatic. Should it prove lacking or inconclusive, American diplomats deserve more than a mere spanking. There is absolutely no excuse for giving the air of intervening into internal affairs of either private companies or foreign governments while, in reality, Western policymakers are simply standing in the corner, daydreaming about democracy and babbling their wildest fantasies into an open mic. In most cases, such "interventions" right no wrongs; instead they usually create quite a few wrongs of their own, producing unnecessary risks for those who were naïve enough to think of the U.S. government as a serious and reliable partner. American pundits go to talk shows; Iranian bloggers go to prison. The bold request sent to Twitter by the U.S. State Department could only be justified on the condition that Twitter was, indeed, playing a crucial role in the Iranian unrest and that the cause of Iranian democracy would be severely undermined had the site gone into maintenance mode for a few hours.

None of this seems to be the case. The digital witch hunts put on by the Iranian government may have been targeting imaginary enemies, created in part by the worst excesses of Western media and the hubris of Western policymakers. Two uncertainties remain to this day. First, how many people inside Iran (as opposed to those outside) were tweeting about the protests? Second, was Twitter actually used as a key tool for organizing the protests, as many pundits implied, or was its relevance limited only to sharing news and raising global awareness about what was happening?

On the first question, the evidence is at best inconclusive. There were indeed a lot of Iran-related tweets in the two weeks following the election, but it is impossible to say how many of them came from Iran as opposed to, say, its three-million-strong diaspora, sympathizers of the Green Movement elsewhere, and provocateurs loyal to the Iranian regime. Analysis by Sysomos, a social media analysis company, found only 19,235 Twitter accounts registered in Iran (0.027 percent of the population) on the eve of the 2009 elections. As many sympathizers of the Green Movement began changing their Twitter location status to Tehran to confuse the Iranian authorities, it also became nearly impossible to tell whether the people supposedly "tweeting" from Iran were in Tehran or in, say, Los Angeles. One of the most active Twitter users sharing the news about the protests, "oxfordgirl," was an Iranian journalist residing in the English county of Oxfordshire. She did an excellent job—but only as an information hub.

Speaking in early 2010, Moeed Ahmad, director of new media for Al-Jazeera, stated that fact-checking by his channel during the protests could confirm only sixty active Twitter accounts in Tehran, a number that fell to six once the Iranian authorities cracked down on online communications. This is not to understate the overall prominence of Iran-related news on Twitter in the first week of protests; research by Pew Research Center found that 98 percent of all the most popular links shared on the site during that period were Iran-related. It's just that the vast majority of them were not authored or retweeted by those in Iran.

As for the second question, whether Twitter was actually used to organize rather than simply publicize the protests, there is even less certainty. Many people who speak Farsi and who have followed the Iranian blogosphere over the years are far more doubtful than outside observers. A prominent Iranian blogger and activist known as Vahid Online, who was in Tehran during the protests, doubts the validity of the

Twitter Revolution thesis simply because few Iranians were tweeting. "Twitter never became very popular in Iran. [But] because the world was watching Iran with such [great interest] during those days, it led many to believe falsely that Iranian people were also getting their news through Twitter," says the blogger.

Twitter was used to post updates about the time and venue of the protests, but it's not clear whether this was done systematically and whether it actually brought in any new crowds onto the streets. That the Green Movement strategically chose Twitter—or, for that matter, any other Internet technology—as their favorite tool of communication is most likely just another myth. On the contrary, the Iranian opposition did not seem to be well-organized, which might explain why it eventually fizzled. "From the beginning, the Green Movement was not created and did not move forward [in an organized manner]—it wasn't like some made a decision and informed others. When you'd walk in the streets, at work, wherever you'd go, people were talking about it and they all wanted to react," says another prominent Iranian blogger, Alireza Rezaei.

The West, however, wasn't hallucinating. Tweets did get sent, and crowds did gather in the streets. This does not necessarily mean, however, that there was a causal link between the two. To put it more metaphorically: If a tree falls in the forest and everyone tweets about it, it may not be the tweets that moved it. Besides, the location and timing of protests were not exactly a secret. One didn't need to go online to notice that there was a big public protest going on in the middle of Tehran. The raging horns of cars stuck in traffic were a pretty good indicator.

In the collective euphoria that overtook the Western media during the events in Iran, dissenting voices—those challenging the dominant account that emphasized the Internet's role in fomenting the protests—received far less prominence than those who cheered the onset of the Twitter Revolution. Annabelle Sreberny, professor of global media and communications at London's School of Oriental and African Studies and an expert on the Iranian media, quickly dismissed Twitter as yet another hype—yet her voice got lost in the rest of the twitter-worshipping commentary. "Twitter was massively overrated. . . . I wouldn't argue that social media really mobilised Iranians themselves," she told the *Guardian*. Hamid Tehrani, the Persian editor of the blogging network Global Voices, was equally skeptical, speculating that the Twitter Revolution hyperbole revealed more about Western new media fantasies than about the reality in Iran. "The west was focused not on the Iranian people but on the role of western technology," says Tehrani, adding that "Twitter was important in publicising what was happening, but its role was overemphasised."

Many other members of the Iranian diaspora also felt that Twitter was getting far more attention than it deserved. Five days after the protests began, Mehdi Yahyanejad, manager of *Balatarin*, a Los Angeles–based Farsi-language news site similar to Digg.com, told the *Washington Post* that "Twitter's impact inside Iran is zero. . . . Here [in the United States], there is lots of buzz, but once you look . . . you see most of it are Americans tweeting among themselves."

That the Internet may have also had a negative impact on the protest movement was another aspect overlooked by most media commentators. An exception was Golnaz Esfandiari, an Iranian correspondent with Radio Free Europe, who, writing in *Foreign Policy* a year after the 2009 Iranian elections, deplored Twitter's "pernicious complicity in allowing rumors to spread." Esfandiari noticed that "in the early days of the post-election crackdown a rumor quickly spread on Twitter that police helicopters were pouring acid and boiling water on protesters. A year later it remains just that: a rumor."

Esfandiari also noted that the story of the Iranian activist Saeedeh Pouraghayi—who was supposedly arrested for chanting "Allah Akbar" on her rooftop, raped, disfigured, and murdered, becoming the martyr of the Green Movement—which made the rounds on Twitter, turned out to be a hoax. Pouraghayi later resurfaced in a broadcast on Iranian state television, saying that she had jumped off a balcony on the night she had been arrested and stayed low for the next few months. A reformist website later claimed that the story of her murder was planted by the Iranian government to discredit reports of other rapes. It's not obvious which side gained more from the hoax and its revelation, but this is exactly the kind of story Western journalists should have been investigating.

Sadly, in their quest to see Ahmadinejad's regime fall at the mercy of tweets, most journalists preferred to look the other way and produce upbeat copy about the emancipatory nature of the Twitter Revolution. As pundits were competing for airtime and bloggers were competing for eyeballs, few bothered to debunk the overblown claims about the power of the Internet. As a result, the myth of Iran's Twitter Revolution soon joined the gigantic pile of other urban myths about the Internet's mighty potential to topple dictators. This explains how, less than a year after the Iranian protests, a *Newsweek* writer mustered the courage to proclaim that "the revolts in Ukraine, Kyrgyzstan, Lebanon, Burma, Xinjiang, and Iran could never have happened without the web." (*Newsweek*, it must be noted, has been predicting an Internet-led revolution in Iran since 1995, when it published an article pompously titled "Chatrooms and Chadors" which posited that "if the computer geeks are right, Iran is facing the biggest revolution since the Ayatollah Khomeini.")

Unless journalists fully commit themselves to scrutinizing and, if necessary, debunking such myths, the latter risk having a corrosive effect on policymaking. As long as Twitter is presumed to have been instrumental in enabling the Iranian protests, any technologies that would allow Iranians to access Twitter by bypassing their government's censorship are also presumed to be of exceptional importance. When a newspaper like the *Washington Post* makes a case for allocating more funding to such technologies in one of its editorials, as it did in July 2010, by arguing that "investing in censorship-circumvention techniques like those that powered Tehran's 'Twitter revolution' in June 2009 could have a tremendous, measurable impact," it's a much weaker argument than appears at first glance. (The *Post*'s claim that the impact of such technologies could be "measurable" deserves close scrutiny as well.) Similarly, one should start worrying about the likely prominence of the Internet in American foreign policy on hearing Alec Ross, Hillary Clinton's senior adviser for innovation, assert that "social media played a key role in organizing the [Iranian] protests," a claim that is not very different from what Andrew Sullivan declared in June 2009. Even though Ross said this almost a year after Sullivan's hypothetical conjecture, he still cited no evidence to back up this claim. (In July 2010 Ross inadvertently revealed his own hypocrisy by also proclaiming that "there is very little information to support the claim that Facebook or Twitter or text messaging caused the rioting or can inspire an uprising.")

Where Are the Weapons of Mass Construction?

If the exalted reaction to the Iranian protests is of any indication, Western policymakers are getting lost in the mists of cyber-utopianism, a quasi-religious belief in the power of the Internet to do supernatural things, from eradicating illiteracy in Africa to organizing all of the world's information, and one of the central beliefs of the Google Doctrine. Opening up closed societies and flushing them with democracy juice until they shed off their authoritarian skin is just one of the high expectations placed on the Internet these days. It's not surprising that a 2010 op-ed in the *Guardian* even proposed to "bombard Iran with broadband"; the Internet is seen as mightier than the bomb. Cyber-utopianism seems to be everywhere these days: T-shirts urging policymakers to "drop tweets, not bombs"—a bold slogan for any modern-day antiwar movement—are already on sale online, while in 2009 one of the streets in a Palestinian refugee camp was even named after a Twitter account.

Tweets, of course, don't topple governments; people do (in a few exceptional cases, the Marines and the CIA can do just fine). Jon Stewart of *The Daily Show* has ridiculed the mythical power of the Internet to accomplish what even the most advanced military in the world has so much difficulty accomplishing in Iraq and Afghanistan: "Why did we have to send an army when we could have liberated them the same way we buy shoes?" Why, indeed? The joke is lost on Daniel Kimmage, a senior analyst with Radio Free Europe / Radio Liberty, who argues that "unfettered access to a free Internet is . . . a very practical means of countering Al Qaeda. . . . As users increasingly make themselves heard, the ensuing chaos . . . may shake the online edifice of Al Qaeda's totalitarian ideology." Jihad Jane and a whole number of other shady characters who were recruited to the terrorist cause online would be sad to learn that they did not surf the Web long enough.

By the end of 2009 cyber-utopianism reached new heights, and the Norwegian Nobel Committee did not object when *Wired Italy* (the Italian edition of the popular technology magazine) nominated the Internet for the 2010 Nobel Peace Prize, the result of a public campaign by a number of celebrities, ranging from Giorgio Armani to Shirin Ebadi, a previous winner of the Prize. (In 1991, Lennart Meri, the future president of Estonia, nominated Radio Free Europe for the same award for its role in helping to bring an end to the Soviet Union—another interesting parallel with the Cold War era.) Why did the Internet deserve the prize more than Chinese human rights activist Liu Xiaobo, who emerged as the eventual winner of the prize? Justifications given by an assortment of editors of various national editions of *Wired* magazine, the official printing organ of the Church of Cyber-Utopianism, are symptomatic of the kind of discourse that led American diplomats astray in Iran.

Riccardo Luna, the editor of the Italian edition, proposed that the Internet is a "first weapon of mass construction, which we can deploy to destroy hate and conflict and to propagate peace and democracy." Chris Anderson, the editor of the original American edition, opined that while "a Twitter account may be no match for an AK-47 . . . in the long term the keyboard is mightier than the sword." David Rowan, the editor of the British edition, argued that the Internet "gave all of us the chance to take back the power from governments and multinationals. It made the world a totally transparent place." And how can a totally transparent world fail to be a more democratic world as well?

Apparently, nothing bad ever happens on the Internet frequented by the editors of *Wired*; even spam could be viewed as the ultimate form of modern poetry. But refusing to acknowledge the Internet's darker side is like visiting Berkeley, California, cyber-utopian headquarters, and concluding that this is how the rest of America lives as well: diverse, tolerant, sun-drenched, with plenty of organic food and nice wine, and with hordes of lifelong political activists fighting for causes that don't even exist yet. But this is not how the rest of America lives, and this is certainly not how the rest of the world lives.

A further clarification might be in order at this point. The border between cyber-utopianism and cyber-naïveté is a blurry one. In fact, the reason why so many politicians and journalists believe in the power of the Internet is because they have not given this subject much thought. Their faith is not the result of a careful examination of how the Internet is being used by dictators or how it is changing the culture of resistance and dissent. On the contrary, most often it's just unthinking acceptance of conventional wisdom, which posits that since authoritarian governments are censoring the Internet, they must be really afraid of it. Thus, according to this view, the very presence of a vibrant Internet culture greatly increases the odds that such regimes will collapse.

How NASDAQ Will Save the World

Whatever one calls it, this belief in the democratizing power of the Web ruins the public's ability to assess future and existing policies, not least because it overstates the positive role that corporations play in democratizing the world without subjecting them to the scrutiny they so justly deserve. Such cyber-utopian propensity to only see the bright side was on full display in early 2010, as Google announced it was pulling out of China, fed up with the growing censorship demands of the Chinese government and mysterious cyber-attacks on its intellectual property. But what should have been treated as a purely rational business decision was lauded as a bold move to support "human rights"; that Google did not mind operating in China for more than four years prior to the pull-out was lost on most commentators.

Writing in *Newsweek*, Jacob Weisberg, a prominent American journalist and publisher, called Google's decision "heroic," while Senator John Kerry said that "Google is gutsily taking real risk in standing up for principle." The Internet guru Clay Shirky proclaimed that "what [Google is] exporting isn't a product or a service, it's a freedom." An editorial in the *New Republic* argued that Google, "an organization filled with American scientists," was heeding the advice of Andrei Sakharov, a famous Russian dissident physicist, who pleaded with his fellow Soviet scientists to "muster sufficient courage and integrity to resist the temptation and the habit of conformity." Sakharov, of course, was not selling snippet-sized advertising, nor was he on first-name terms with the National Security Agency, but the *New Republic* preferred to gloss over such inconsistencies.

Even famed journalist Bob Woodward fell under the sway of cyber-utopianism. Appearing on *Meet the Press*, one of the most popular Sunday morning TV shows in America, in May 2010 Woodward suggested that Google's engineers—"some of these people who have these great minds"—should be called in to fix the oil spill in the Gulf of Mexico. And if Google could fix the oil spill, couldn't they fix Iran as well? It seems that we are only a couple of op-eds away from having Tom Friedman pronounce that Google, with all their marvelous scanners and databases, should take over the Department of Homeland Security.

Google, of course, is not the only subject of nearly universal admiration. A headline in the *Washington Post* declares, "In Egypt, Twitter Trumps Torture," while an editorial in *Financial Times* praises social networking sites like Facebook as "a challenge to undemocratic societies," concluding that "the next great revolution may begin with a Facebook message." (Whether Facebook also presents a challenge to democratic societies is a subject that the editorial didn't broach.) Jared Cohen, the twenty-seven-year-old member of the State Department's Policy Planning staff who sent the infamous email request to Twitter during the Iranian protests, hails Facebook as "one of the most organic tools for democracy promotion the world has ever seen."

One problem that arises from such enthusiastic acceptance of Internet companies' positive role in abetting the fight against authoritarianism is that it lumps all of them together, blurring the differences in their level of commitment to defending human rights, let alone promoting democracy. Twitter, a company that received wide public admiration during the events in Iran, has refused to join the Global Network Initiative (GNI), an industry-wide pledge by other technology companies—including Google, Yahoo, and Microsoft—to behave in accordance with the laws and standards covering the right to freedom of expression and privacy embedded in internationally recognized documents like the Universal Declaration of Human Rights. Facebook, another much admired exporter of digital revolutions, refused to join GNI as well, citing lack of resources, a bizarre excuse for a company with $800 million in 2009 revenues.

While Twitter and Facebook's refusal to join GNI raised the ire of several American senators, it has not at all reflected on their public image. And their executives are right not to worry. They are, after all, friends with the U.S. State Department; they are invited to private dinners with the secretary of state and are taken on tours of exotic places like Iraq, Mexico, and Russia to boost America's image in the world.

There is more than just tech-savvy American diplomacy on full display during such visits. They also reveal that an American company does not need to make many ethical commitments to be friends with the U.S. government, at least as long as it is instrumental to Washington's foreign policy agenda. After eight years of the Bush administration, which was dominated by extremely secretive public-private partnerships like Dick Cheney's Energy Task Force, such behavior hardly provides a good blueprint for public diplomacy.

Google, despite its membership in the GNI, has much to account for as well, ranging from its increasingly carefree attitude toward privacy—hardly a cause for celebration by dissidents around the world—to its penchant for flaunting its own relationship with the U.S. government. Its much-publicized cooperation with the National Security Agency over the cyber-attacks on its servers in early 2010 was hardly an effective way to convince the Iranian authorities of the nonpolitical nature of Internet activities. There is much to admire about Google, Twitter, and Facebook, but as they begin to play an increasingly important role in mediating foreign policy, "admiration" is not a particularly helpful attitude for any policymaker.

From Milk Shakes to Molotov Cocktails

Jared Cohen's praise of Facebook's organic ability to promote democracy may be just a factual statement. Everything else being equal, a world where so many Chinese and Iranians flock to the services of American technology companies may, indeed, be a world where democracy is more likely to prevail in the long run. It's hard to disagree with this statement, especially if the other alternative is having those users opt for domestic Internet services; those tend to be much more heavily policed and censored.

That said, it's important not to lose sight of the fact that the current situation is not the result of some cunning and extremely successful American strategy to exploit Facebook. Rather, it's the result of both intellectual and market conditions at the time. Until recently, authoritarian governments simply did not give much thought to where their citizens chose to do their email and share their pasta recipes; American companies were often the first to offer their superb services, and most governments did not bother to build any barriers. They may have been piqued by the success of American platforms as opposed to local Internet start-ups, but then their domestic fast food industry was also losing ground to McDonald's; as long as no one could mistake McDonald's vanilla triple-thick shake for a Molotov cocktail, this was not something to worry about.

Nevertheless, once the likes of Jared Cohen start lauding Facebook as an organic tool for promoting democracy, it immediately stops being such. In a sense, the only reason why there was so much laxity in the regulation of Internet services operating in authoritarian states was that their leaders did not make the obvious connection between the business interests of American companies and the political interests of the American government. But as the State Department is trying to harvest the fruits of Silicon Valley's success in the global marketplace, it's inevitable that previously carefree attitudes will give way to increased suspicion. Any explicit moves by American diplomats in this space will be watched closely. Moreover, they will be interpreted according to the prevalent conspiracy theories rather than in light of the stale press releases issued by the State Department to explain its actions.

In July 2010 the Chinese Academy of Social Sciences, one of the Chinese government's finest research organizations, published a detailed report about the political implications of the Internet. It argued that social networking sites threaten state security because the United States and other Western countries "are using them to foment instability." It's hard not to see this as a direct response to the words and deeds of Jared Cohen. (The Chinese report did cite unnamed U.S. officials as saying that social networking is an "invaluable tool" for overthrowing foreign government and made good use of the U.S. government's involvement via Twitter in the Iranian unrest of 2009.) When American diplomats call Facebook a tool of democracy promotion, it's safe to assume that the rest of the world believes that America is keen to exploit this tool to its fullest potential rather than just stare at it in awe.

American diplomats have been wrong to treat the Internet, revolutionary as it might seem to them, as a space free of national prejudices. Cyberspace is far less susceptible to policy amnesia than they believe; earlier policy blunders and a long-running history of mutual animosity between the West and the rest won't be forgotten so easily. Even in the digital age, the foreign policy of a country is still constrained by the same set of rather unpleasant barriers that limited it in the analog past. As Joseph Nye and Robert Keohane, two leading scholars of international relations, pointed out more than a decade ago, "information does not flow in a vacuum but in a political space that is already occupied." Until the events in Iran, America's technology giants may have, indeed, functioned in a mostly apolitical vacuum and have been spared any bias that comes with the label "American." Such days, however, are clearly over. In the long run, refusing to recognize this new reality will only complicate the job of promoting democracy.

Why Hipsters Make Better Revolutions

In the case of Iran, Western policymakers not only misread the Internet but bragged about their own ignorance to anyone who would listen. Much to their surprise, the Iranian government believed their bluff and took aggressive countermeasures, making the job of using the Web to foster social and political change in Iran and other closed societies considerably harder. The opportunities of three years ago, when governments still thought that bloggers were mere hipsters, amusing but ultimately dismissed as a serious political movement, are no longer available. Bloggers, no longer perceived as trendy slackers, are seen as the new Solidarity activists—an overly idealistic and probably wrong characterization shared by democratic and authoritarian governments alike.

Most disturbingly, a dangerous self-negating prophecy is at work here: The more Western policymakers talk up the threat that bloggers pose to authoritarian regimes, the more likely those regimes are to limit the maneuver space where those bloggers operate. In some countries, such politicization may be for the better, as blogging would take on a more explicit political role, with bloggers enjoying the status of journalists or human rights defenders. But in many other countries such politicization may only stifle the nascent Internet movement, which could have been be far more successful if its advocacy were limited to pursuing social rather than political ends. Whether the West needs to politicize blogging and view it as a natural extension of dissident activity is certainly a complex question that merits broad public debate. But the fact that this debate is not happening at the moment does not mean that blogging is not being politicized, often to the point of no return, by the actions—as well as declarations—of Western policymakers.

Furthermore, giving in to cyber-utopianism may preclude policymakers from considering a whole range of other important questions. Should they applaud or bash technology companies who choose to operate in authoritarian regimes, bending their standard procedures as a result? Are they harbingers of democracy, as they claim to be, or just digital equivalents of Halliburton and United Fruit Company, cynically exploiting local business opportunities while also strengthening the governments that let them in? How should the West balance its sudden urge to promote democracy via the Internet with its existing commitments to other nondigital strategies for achieving the same objective, from the fostering of independent political parties to the development of civil society organizations? What are the best ways of empowering digital activists without putting them at risk? If the Internet is really a revolutionary force that could nudge all authoritarian regimes toward democracy, should the West go quiet on many of its other concerns about the Internet—remember all those fears about cyberwar, cybercrime, online child pornography, Internet piracy—and strike while the iron is still hot?

These are immensely difficult questions; they are also remarkably easy to answer incorrectly. While the Internet has helped to decrease costs for nearly everything, human folly is a commodity that still bears a relatively high price. The oft-repeated mantra of the open source movement—"fail often, fail early"—produces excellent software, but it is not applicable to situations where human lives are at stake. Western policymakers, unlike pundits and academics, simply don't have the luxury of getting it wrong and dealing with the consequences later.

From the perspective of authoritarian governments, the costs of exploiting Western follies have significantly decreased as well. Compromising the security of just one digital activist can mean compromising the security—names, faces, email addresses—of everyone that individual knows. Digitization of information has also led to its immense centralization: One stolen password now opens data doors that used not to exist (just how many different kinds of data—not to mention people—would your email password give access to, if compromised?).

Unbridled cyber-utopianism is an expensive ideology to maintain because authoritarian governments don't stand still and there are absolutely no guarantees they won't find a way to turn the Internet into a powerful tool of oppression. If, on closer examination, it turns out that the Internet has also empowered the secret police, the censors, and the propaganda offices of a modern authoritarian regime, it's quite likely that the process of democratization will become harder, not easier. Similarly, if the Internet has dampened the level of antigovernment sentiment—because people have acquired access to cheap and almost infinite digital entertainment or because they feel they need the government to protect them from the lawlessness of cyberspace—it certainly gives the regime yet another source of legitimacy. If the Internet is reshaping the very nature and culture of antigovernment resistance and dissent, shifting it away from real-world practices and toward anonymous virtual spaces, it will also have significant consequences for the scale and tempo of the protest movement, not all of them positive.

That's an insight that has been lost on most observers of the political power of the Internet. Refusing to acknowledge that the Web can actually strengthen rather than undermine authoritarian regimes is extremely irresponsible and ultimately results in bad policy, if only because it gives policymakers false confidence that the only things they need to be doing are proactive—rather than reactive—in nature. But if, on careful examination, it turns out that certain types of authoritarian regimes can benefit from the Internet in disproportionally more ways than their opponents, the focus of Western democracy promotion work should shift from empowering the activists to topple their regimes to countering the governments' own exploitation of the Web lest they become even more authoritarian. There is no point in making a revolution more effective, quick, and anonymous if the odds of the revolution's success are worsening in the meantime.

In Search of a Missing Handle

So far, most policymakers choose to be sleepwalking through this digital minefield, whistling their favorite cyber-utopian tunes and refusing to confront all the evidence. They have also been extremely lucky because the mines were far and few between. This is not an attitude they can afford anymore, if only because the mines are now almost everywhere and, thanks to the growth of the Internet, their explosive power is much greater and has implications that go far beyond the digital realm.

As Shanthi Kalathil and Taylor Boas pointed out in *Open Networks, Closed Regimes*, their pioneering 2003 study about the impact of the pre–Web 2.0 Internet on authoritarianism, "conventional wisdom . . . forms part of the gestalt in which policy is formulated, and a better understanding of the Internet's political effects should lead to better policy." The inverse is true as well: A poor understanding leads to poor policy.

If the only conclusion about the power of the Internet that Western policymakers have drawn from the Iranian events is that tweets are good for social mobilization, they are not likely to outsmart their authoritarian adversaries, who have so far shown much more sophistication in the online world. It's becoming clear that understanding the full impact of the Internet on the democratization of authoritarian states would require more than just looking at the tweets of Iranian youngsters, for they only tell one part of the story. Instead, one needs to embark on a much more thorough and complex analysis that would look at the totality of forces shaped by the Web.

Much of the current cognitive dissonance is of do-gooders' own making. What did they get wrong? Well, perhaps it was a mistake to treat the Internet as a deterministic one-directional force for either global liberation or oppression, for cosmopolitanism or xenophobia. The reality is that the Internet will enable all of these forces—as well as many others—simultaneously. But as far as laws of the Internet go, this is all we know. Which of the numerous forces unleashed by the Web will prevail in a particular social and political context is impossible to tell without first getting a thorough theoretical understanding of that context.

Likewise, it is naïve to believe that such a sophisticated and multipurpose technology as the Internet could produce identical outcomes—whether good or bad—in countries as diverse as Belarus, Burma, Kazakhstan, and Tunisia. There is so much diversity across modern authoritarian regimes that some Tolstoy paraphrasing might be in order: While all free societies are alike, each unfree society is unfree in its own way. Statistically, it's highly unlikely that such disparate entities would all react to such a powerful stimulus in the same way. To argue that the Internet would result in similar change—that is, democratization—in countries like Russia and China is akin to arguing that globalization, too, would also exert the same effect on them; more than a decade into the new century, such deterministic claims seem highly suspicious.

It is equally erroneous to assume that authoritarianism rests on brutal force alone. Religion, culture, history, and nationalism are all potent forces that, with or without the Internet, shape the nature of modern authoritarianism in ways that no one fully understands yet. In some cases, they undermine it; in many others, they enable it. Anyone who believes in the power of the Internet as I do should resist the temptation to embrace Internet-centrism and unthinkingly assume that, under the pressure of technology, all of these complex forces will evolve in just one direction, making modern authoritarian regimes more open, more participatory, more decentralized, and, all along, more conducive to democracy. The Internet does matter, but we simply don't know how it matters. This fact, paradoxically, only makes it matter even more: The costs of getting it wrong are tremendous. What's clear is that few insights would be gained by looking inward—that is, trying to crack the logic of the Internet; its logic can never be really understood outside the context in which it manifests itself.

Of course, such lack of certainty does not make the job of promoting democracy in the digital age any easier. But, at minimum, it would help if policymakers—and the public at large—free themselves of any intellectual obstacles and biases that may skew their thinking and result in utopian theorizing that has little basis in reality. The hysterical reaction to the protests in Iran has revealed that the West clearly lacks a good working theory about the impact of the Internet on authoritarianism. This is why policymakers, in a desperate attempt to draw at least some lessons about technology and democratization, subject recent events like the overthrow of communist regimes in Eastern Europe to some rather twisted interpretation. Whatever the theoretical merits of such historical parallels, policymakers should remember that all frameworks have consequences: One poorly chosen historical analogy, and the entire strategy derived from it can go to waste.

Nevertheless, while it may be impossible to produce many generalizable laws to describe the relationship between the Internet and political regimes, policymakers shouldn't simply stop thinking about these issues, commission a number of decade-long studies, and wait until the results are in. This is not a viable option. As the Internet gets more complex, so do its applications—and authoritarian regimes are usually quick to put them to good use. The longer the indecision, the greater are the odds that some of the existing opportunities for Internet-enabled action will soon no longer be available.

This is not to deny that, once mastered, the Internet could be a powerful tool in the arsenal of a policymaker; in fact, once such mastery is achieved, it would certainly be irresponsible not to deploy this tool. But as Langdon Winner, one of the shrewdest thinkers about the political implications of modern technology, once observed, "although virtually limitless in their power, our technologies are tools without handles." The Internet is, unfortunately, no exception. The handle that overconfident policymakers feel in their hands is just an optical illusion; theirs is a false mastery. They don't know how to tap into the power of the Internet, nor can they anticipate the consequences of their actions. In the meantime, all their awkward moments add up and, as was the case in Iran, have dire consequences.

Most of the Western efforts to use the Internet in the fight against authoritarianism could best be described as trying to apply a poor cure to the wrong disease. Policymakers have little control over their cure, which keeps mutating every day, so it never works the way they expect it to. (The lack of a handle does not help either.) The disease part is even more troublesome. The kind of authoritarianism they really want to fight expired in 1989. Today, however, is no 1989, and the sooner policymakers realize this, the sooner they can start crafting Internet policies that are better suited for the modern world.

The upside is that even tools without handles can be of some limited use in any household. One just needs to treat them as such and search for contexts where they are needed. At minimum, one should ensure that such tools don't hurt anyone who tries to use them with the assumption of inevitable mastery. Until policymakers come to terms with the fact that their Internet predicament is driven by such highly uncertain dynamics, they will never succeed in harvesting the Web's mighty potential.

chapter two

Texting Like It's 1989

●　●　■　■

The history of cyber-utopianism is not very eventful, but the date January 21, 2010, has a guaranteed place in its annals—probably right next to Andrew Sullivan's blog posts about Twitter's role in Tehran. For this was the day when the sitting U.S. secretary of state, Hillary Clinton, went to the Newseum, America's finest museum of news and journalism, to deliver a seminal speech about Internet freedom and thus acknowledge the Internet's prominent role in foreign affairs.

The timing of Clinton's speech could not have been better. Just a week earlier, Google announced it was considering pulling out of China—hinting that the Chinese government may have had something to do with it—so everyone was left guessing if the issue would get a mention (it did). One could feel palpable excitement all over Washington: An American commitment to promoting Internet freedom promised a new line of work for entire families in this town. All the usual suspects—policy analysts, lobbyists, consultants—were eagerly anticipating the opening salvo of this soon-to-be-lavishly-funded "war for Internet freedom." For Washington, it was the kind of universally admired quest for global justice that could allow think tanks to churn out a slew of in-depth research studies, defense contractors to design a number of cutting-edge censorship-breaking technologies, and NGOs to conduct a series of risky trainings in the most exotic locales on Earth. This is why Washington beats any other city in the world, including Iran and Beijing, in terms of how often and how many of its residents search for the term "Internet freedom" on Google. A campaign to promote Internet freedom is a genuinely Washingtonian phenomenon.

But there was also something distinctively unique about this gathering. It's not very often that the Beltway's BlackBerry mafia—the buttoned-up think-tankers and policy wonks—get to share a room with the iPhone fanboys—the unkempt and chronically underdressed entrepreneurs from Silicon Valley. Few other events could bring together Larry Diamond, a senior research fellow at the conservative Hoover Institution and a former senior adviser to the Coalition Provisional Authority in Iraq, and Chris "FactoryJoe" Messina, the twenty-nine-year-old cheerleader of Web 2.0 and Google's "Open Web Advocate" (that's his official job title!). It was a "geeks + wonks" feast.

The speech itself did not offer many surprises; its objective was to establish "Internet freedom" as a new priority for American foreign policy, and judging by the buzz that Clinton's performance generated in the media, that objective was accomplished, even if specific details were never divulged. The generalizations drawn by Clinton were rather upbeat—"information freedom supports the peace and security that provide a foundation for global progress"—and so were her prescriptions: "We need to put these tools in the hands of people around the world who will use them to advance democracy and human rights." There were too many buzzwords—"deficiencies in the current market for innovation," "harnessing the power of connection technologies," "long-term dividends from modest investments in innovation"—but such, perhaps, was the cost of trying to look cool in front of the Silicon Valley audience.

Excessive optimism and empty McKinsey-speak aside, it was Clinton's creative use of recent history that really stood out. Clinton drew a parallel between the challenges of promoting Internet freedom and the experiences of supporting dissidents during the Cold War. Speaking of her recent visit to Germany to commemorate the twentieth anniversary of the fall of the Berlin Wall, Clinton mentioned "the courageous men and women" who "made the case against oppression by circulating small pamphlets called samizdat," which "helped pierce the concrete and concertina wire of the Iron Curtain." (Newseum was a very appropriate venue to give in to Cold War nostalgia. It happens to house the largest display of sections of the Berlin Wall outside of Germany).

Something very similar is happening today, argued Clinton, adding that "as networks spread to nations around the globe, virtual walls are cropping up in place of visible walls." And as "a new information curtain is descending across much of the world . . . viral videos and blog posts are becoming the samizdat of our day." Even though Clinton did not articulate many policy objectives, they were not hard to guess from her chosen analogy. Virtual walls are to be pierced, information curtains are to be raised, digital samizdat is to be supported and disseminated, and bloggers are to be celebrated as dissidents.

As far as Washington was concerned, having Clinton utter that highly seductive phrase—"a new information curtain"—in the same breath as the Berlin Wall was tantamount to announcing a sequel to the Cold War in 3D. She tapped into the secret desires of many policymakers, who had been pining for an enemy they understood, someone unlike that bunch of bearded and cave-bound men from Waziristan who showed little appreciation for balance-of-power theorizing and seemed to occupy so much of the present agenda.

It was Ronald Reagan's lieutenants who must have felt particularly excited. Having claimed victory in the analog Cold War, they felt well-prepared to enlist—nay, triumph—in its digital equivalent. But it was certainly not the word "Internet" that made Internet freedom such an exciting issue for this group. As such, the quest for destroying the world's cyber-walls has given this aging generation of cold warriors, increasingly out of touch with a world beset by problems like climate change or the lack of financial regulation, something of a lifeline. Not that those other modern problems are unimportant—they are simply not existential enough, compared to the fight against communism. For many members of this rapidly shrinking Cold War lobby, the battle for Internet freedom is their last shot at staging a major intellectual comeback. After all, whom else would the public call on but them, the tireless and self-deprecating statesmen who helped rid the world of all those other walls and curtains?

WWW & W

It only took a few months for one such peculiar group of Washington insiders to convene a high-profile conference to discuss how a host of Cold War policies—and particularly Western support for Soviet dissidents—could be recovered from the dustbin of history and applied to the current situation. Spearheaded by George W. Bush, who, by then, had mostly retreated from the public arena, the gathering attracted a number of hawkish neoconservatives. Perhaps out of sheer disgust with the lackluster foreign policy record of the Obama administration, they had decided to wage their own fight for freedom on the Internet.

There was, of course, something surreal about George W. Bush, who was rather dismissive of the "Internets" while in office, presiding over this Internet-worship club. But then, for Bush at least, this meeting had little to do with the Web per se. Rather, its goal was to push the "freedom agenda" into new, digital territories. Seeing the internet as an ally, Bush, always keen to flaunt his credentials as the dissidents' best friend—he met more than a hundred of them while in office—agreed to host a gathering of what he called "global cyber-dissidents" in, of all places, Texas. Featuring half a dozen political bloggers from countries like Syria, Cuba, Colombia, and Iran, the conference was one of the first major public events organized by the newly inaugurated George W. Bush Institute. The pomposity of its lineup, with panels like "Freedom Stories from the Front Lines" and "Global Lessons in eFreedom," suggested that even two decades after the fall of the Berlin Wall, its veterans are still fluent in Manichean rhetoric.

But the Texas conference was not just a gathering of disgruntled and unemployed neoconservatives; respected Internet experts, like Ethan Zuckerman and Hal Roberts of Harvard's Berkman Center for Internet and Society, were in attendance as well. A senior official from the State Department—technically an Obama man—was also dispatched to Texas. "This conference highlights the work of a new generation of dissidents in the hope that it will become a beacon to others," said James Glassman, a former high-profile official in the Bush administration and the president of the George W. Bush Institute, on opening the event. According to Glassman, the conference aimed "to identify trends in effective cyber communication that spread human freedom and advance human rights." (Glassman, it must be said, is to cyber-utopianism what Thoreau is to civil disobedience; he famously coauthored a book called *Dow 36,000,* predicting that the Dow Jones was on its way toward a new height; it came out a few months before the dot-com bubble burst in 2000.)

David Keyes, a director of a project called Cyberdissidents.org, was one of the keynote speakers at the Bush event, serving as a kind of bridge to the world of the old Soviet dissidents. He used to work with Natan Sharansky, a prominent Soviet dissident whose thinking shaped much of the Bush administration's global quest for freedom. (Sharansky's *The Case for Democracy: The Power of Freedom to Overcome Tyranny and Terror* was one of the few books Bush read during his time in office; it exerted a significant influence, as Bush himself acknowledged: "If you want a glimpse of how I think about foreign policy read Natan Sharansky's book. . . . Read it. It's a great book.") According to Keyes, the mission of Cyberdissidents.org is to "make the Middle East's pro-democracy Internet activists famous and beloved in the West"— that is, to bring them to Sharansky's level of fame (the man himself sits on Cyberdissidents's board of advisers).

But one shouldn't jump to conclusions too hastily. The "cyber-cons" that attended the Texas meeting are not starry-eyed utopians, who think that the Internet will magically rid the world of dictators. On the contrary, they eagerly acknowledge—much more so than the liberals in the Obama administration—that authoritarian governments are also active on the Internet. "Democracy is not just a tweet away," writes Jeffrey Gedmin, the president of Radio Free Europe/Radio Liberty and another high-profile attendee at the event (a Bush appointee, he enjoys stellar conservative credentials, including a senior position at the American Enterprise Institute). That the cyber-cons happen to believe in the power of bloggers to topple those governments is not a sign of cyber-utopianism; rather it's the result of the general neoconservative outlook on how authoritarian societies function and on the role that dissidents—both online and offline varieties—play in transforming them. Granted, shades of utopianism are easily discernible in their vision, but this is not utopianism about technology; this is utopianism about politics in general.

The Iraqi experience may have somewhat curbed their enthusiasm, but the neoconservative belief that all societies aspire to democracy and would inevitably head in its direction—if only all the obstacles were removed—is as strong as ever. The cyber-cons may have been too slow to realize the immense potential of the Internet in accomplishing their agenda; in less than two decades it removed more such obstacles than all neocon policies combined. But now that authoritarian governments were also actively moving into this space, it was important to stop them. For most attendees at the Bush gathering, the struggle for Internet freedom was quickly emerging as the quintessential issue of the new century, the one that could help them finish the project that Ronald Reagan began in the 1980s and that Bush did his best to advance in the first decade of the new century. It seems that in the enigma of Internet freedom, neoconservatism, once widely believed to be on the wane, has found a new raison d'être—and a new lease on life to go along with it.

Few exemplify the complex intellectual connections between Cold War history, neoconservatism, and the brave new world of Internet freedom better than Mark Palmer. Cofounder of the National Endowment for Democracy, the Congress-funded leading democracy-promoting organization in the world, Palmer served as Ronald Reagan's ambassador to Hungary during the last years of communism. He is thus well-informed about the struggles of the Eastern European dissidents; he is equally knowledgeable about the ways in which the West nurtured them, for a lot of that support passed through the U.S. embassy. Today Palmer, a member of the uber-hawkish Committee on the Present Danger, has emerged as a leading advocate of Internet freedom, mostly on behalf of Falun Gong, a persecuted spiritual group from China, which is one of the most important behind-the-scenes players in the burgeoning industry of Internet freedom. Falun Gong runs several websites that were banned once the group fell out with the Chinese government in 1999. Hence its practitioners have built an impressive fleet of technologies to bypass China's numerous firewalls, making the banned sites accessible from within the country. Palmer has penned passionate pleas—including congressional testimonies—demanding that the U.S. government allocate more funding to Falun Gong's sprawling technology operation to boost their capacity and make their technology available in other repressive countries. (The U.S. State Department turned down at least one such request, but then in May 2010, under growing pressure from Falun Gong's numerous supporters, including conservative outfits like the Hudson Institute and the editorial pages of the *New York Times*, the *Washington Post*, and the *Wall Street Journal*, it relented, granting $1.5 million to the group.)

Palmer's views about the promise of the Internet epitomize the cyber-con position at its hawkish extreme. In his 2003 book *Breaking the Real Axis of Evil: How to Oust the World's Last Dictators by 2025*, his guide to overthrowing forty-five of the world's authoritarian leaders, a book that makes Dick Cheney look like a dove, Palmer lauded the emancipatory power of the Internet, calling it "a force multiplier for democracy and an expense multiplier for dictators." For him, the Internet is an excellent way to foster civil unrest that can eventually result in a revolution: "Internet skills are readily taught, and should be, by the outside democracies. Few undertakings are more cost effective than 'training the trainers' for Internet organizing." The Web is thus a powerful tool for regime change; pro-democracy activists in authoritarian states should be taught how to blog and tweet in more or less the same fashion that they are taught to practice civil disobedience and street protest— the two favorite themes of U.S.-funded trainings whose agendas are heavily influenced by the work of the American activist-academic Gene Sharp, the so-called Machiavelli of nonviolence.

With regard to Iran, for example, one of Palmer's proposed solutions is to turn diplomatic missions of "democratic states" into "freedom

houses, providing to Iranians cybercafés with access to the Internet and
other communications equipment, as well as safe rooms for meetings."
But Palmer's love for freedom houses goes well beyond Iran. He is a
board member and a former vice chair of Freedom House, another
mostly conservative outfit that specializes in tracking democratization
across the world and, when an opportune moment comes along, helping to spread it. (Because of their supposed role in fomenting Ukraine's
Orange Revolution, Freedom House and George Soros's Open Society
Foundations are two of the Kremlin's favorite Western bogeymen.) Perhaps in part thanks to pressure from Palmer, Freedom House has recently expanded its studies of democratization into the digital domain,
publishing a report on the Internet freedom situation in fifteen countries and, with some financial backing by the U.S. government, has even
set up a dedicated Internet Freedom Initiative. Whatever its emancipatory potential, the Internet will remain Washington's favorite growth
industry for years to come.

Cyber Cold War

But it would be disingenuous to suggest that it's only neoconservatives
who like delving into their former glory to grapple with the digital
world. That the intellectual legacy of the Cold War can be repurposed
to better understand the growing host of Internet-related emerging
problems is an assumption widely shared across the American political
spectrum. "To win the cyber-war, look to the Cold War," writes Mike
McConnell, America's former intelligence chief. "[The fight for Internet
freedom] is a lot like the problem we had during the Cold War," concurs
Ted Kaufman, a Democratic senator from Delaware. Freud would have
had a good laugh on seeing how the Internet, a highly resilient network
designed by the U.S. military to secure communications in case of an
attack by the Soviet Union, is at pains to get over its Cold War parent-
age. Such intellectual recycling is hardly surprising. The fight against
communism has supplied the foreign policy establishment with so
many buzzwords and metaphors—the Iron Curtain, the Evil Empire,
Star Wars, the Missile Gap—that many of them could be raised from

the dead today—simply by adding the annoying qualifiers like "cyber-," "digital," and "2.0."

By the virtue of sharing part of its name with the word "firewall," the Berlin Wall is by far the most abused term from the vocabulary of the Cold War. Senators are particularly fond of the metaphorical thinking that it inspires. Arlen Specter, a Democrat from Pennsylvania, has urged the American government to "fight fire with fire in finding ways to breach these firewalls, which dictatorships use to control their people and keep themselves in power." Why? Because "tearing down these walls can match the effect of what happened when the Berlin Wall was torn down." Speaking in October 2009 Sam Brownback, a Republican senator from Kansas, argued that "as we approach the 20th anniversary of the breaking of the Berlin Wall, we must . . . commit ourselves to finding ways to tear down . . . the cyber-walls." It feels as if Ronald Reagan's speech-writers are back in town, churning out speeches about the Internet.

European politicians are equally poetic. Carl Bildt, a former prime minister of Sweden, believes that dictatorships are fighting a losing battle because "cyber walls are as certain to fall as the walls of concrete once did." And even members of predominantly liberal NGOs cannot resist the temptation. "As in the cold war [when] you had an Iron Curtain, there is concern that authoritarian governments . . . are developing a Virtual Curtain," says Arvind Ganesan of Human Rights Watch.

Journalists, always keen to sacrifice nuance in the name of supposed clarity, are the worst abusers of Cold War history for the purpose of explaining the imperative to promote Internet freedom to their audience. Roger Cohen, a foreign affairs columnist for the *International Herald Tribune*, writes that while "Tear down this wall!" was a twentieth-century cry, the proper cry for the twenty-first century is "Tear down this firewall!" *Foreign Affairs'* David Feith argues that "just as East Germans diminished Soviet legitimacy by escaping across Checkpoint Charlie, 'hacktivists' today do the same by breaching Internet cyberwalls." And to dispel any suspicions that such linguistic promiscuity could be a mere coincidence, Eli Lake, a contributing editor for the *New Republic*, opines that "during the cold war, the dominant metaphor for describing the repression of totalitarian regimes was The Berlin Wall. To update that metaphor, we should talk about The Firewall," as if the similarity between the two cases was nothing but self-evident.

Things get worse once observers begin to develop what they think are informative and insightful parallels that go beyond the mere pairing of the Berlin Wall with the Firewall, attempting to establish a nearly functional identity between some of the activities and phenomena of the Cold War era and those of today's Internet. This is how blogging becomes samizdat (Columbia University's Lee Bollinger proclaims that "like the underground samizdat . . . the Web has allowed free speech to avoid the reach of the most authoritarian regimes"); bloggers become dissidents (Alec Ross, Hillary Clinton's senior adviser for innovation, says that "bloggers are a form of 21st century dissident"); and the Internet itself becomes a new and improved platform for Western broadcasting (New York University's Clay Shirky argues that what the Internet allows in authoritarian states "is way more threatening than Voice of America"). Since the Cold War vocabulary so profoundly affects how Western policymakers conceptualize the Internet and measure its effectiveness as a policy instrument, it's little wonder that so many of them are impressed. Blogs are, indeed, more efficient at spreading banned information than photocopiers.

The origins of the highly ambitious cyber-con agenda are thus easy to pin down; anyone who takes all these metaphors seriously, whatever the ideology, would inevitably be led to believe that the Internet is a new battleground for freedom and that, as long as Western policymakers could ensure that the old cyber-walls are destroyed and no new ones are erected in their place, authoritarianism is doomed.

Nostalgia's Lethal Metaphors

But perhaps there is no need to be so dismissive of the Cold War experience. After all, it's a relatively recent battle, still fresh in the minds of many people working on issues of Internet freedom today. Plenty of information-related aspects of the Cold War—think radio-jamming— bear at least some minor technical resemblance to today's concerns about Internet censorship. Besides, it's inevitable that decision makers in any field, not just politics, would draw on their prior experiences to understand any new problems they confront, even if they might adjust some of their previous conclusions in light of new facts. The world of foreign policy is simply too complex to be understood without borrowing concepts and ideas that originate elsewhere; it's inevitable that decision makers will use metaphors in explaining or justifying their actions. That said, it's important to ensure that the chosen metaphors actually introduce—rather than reduce—conceptual clarity. Otherwise, these are not metaphors but highly deceptive sound bites.

All metaphors come with costs, for the only way in which they can help us grasp a complex issue is by downplaying some other, seemingly less important, aspects of that issue. Thus, the theory of the "domino effect," so popular during the Cold War, predicted that once a country goes communist, other countries would soon follow—until the entire set of dominoes (countries) has fallen. While this may have helped people grasp the urgent need to respond to communism, this metaphor overemphasized interdependence between countries while paying little attention to internal causes of instability. It downplayed the possibility that democratic governments can fall on their own, without external influence. But that, of course, only became obvious in hindsight. One major problem with metaphors, no matter how creative they are, is that once they enter into wider circulation, few people pay attention to other aspects of the problem that were not captured by the original metaphor. (Ironically, it was in Eastern Europe, as communist governments began collapsing one after another, that a "domino effect" actually seemed to occur.) "The pitfall of metaphorical reasoning is that people often move from the identification of similarities to the assumption of identity— that is, they move from the realization that something is *like* something else to assuming that something is *exactly like* something else. The problem stems from using metaphors as a substitute for new thought rather than a spur to creative thought," writes Keith Shimko, a scholar of political psychology at Purdue University. Not surprisingly, metaphors often create an illusion of complete intellectual mastery of an issue, giving decision makers a false sense of similarity where there is none.

The carefree way in which Western policymakers are beginning to throw around metaphors like "virtual walls" or "information curtains" is disturbing. Not only do such metaphors play up only certain aspects of the "Internet freedom" challenge (for example, the difficulty of sending critical messages into the target country), they also downplay other aspects (the fact that the Web can be used by the very government of the target country for the purposes of surveillance or propaganda). Such metaphors also politicize anyone on the receiving end of the information coming from the other side of the "wall" or "curtain"; such recipients are almost automatically presumed to be pro-Western or, at least, to have some serious criticisms of their governments. Why would they be surreptitiously lifting the curtain otherwise?

Having previously expended so much time and effort on trying to break the Iron Curtain, Western policymakers would likely miss more effective methods to break the Information Curtain; their previous experience makes them see everything in terms of curtains that need to be lifted rather than, say, fields that need to be watered. Anyone tackling the issue unburdened by that misleading analogy would have spotted that it's a "field" not a "wall" that they are looking at. Policymakers' previous experiences with solving similar problems, however, block them from seeking more effective solutions to new problems. This is a well-known phenomenon that psychologists call the Einstellung Effect.

Many of the Cold War metaphors suggest solutions of their own. Walls need to be destroyed and curtains raised before democracy can take root. That democracy may still fail to take root even if the virtual walls are crushed is not a scenario that naturally follows from such metaphors, if only because the peaceful history of postcommunist Eastern Europe suggests otherwise. By infusing policymakers with excessive optimism, the Cold War metaphors thus result in a certain illusory sense of finality and irreversibility. Breaching a powerful firewall is in no way similar to the breaching of the Berlin Wall or the lifting of passport controls at Checkpoint Charlie, simply because patching firewalls, unlike rebuilding monumental walls, takes hours. Physical walls are cheaper to destroy than to build; their digital equivalents work the other way around. Likewise, the "cyber-wall" metaphor falsely suggests that once digital barriers are removed, new and completely different barriers won't spring up in their place—a proposition that is extremely misleading when Internet control takes on multiple forms and goes far beyond the mere blocking of websites.

Once such language creeps into policy analysis, it can result in a severe misallocation of resources. Thus, when an editorial in the *Washington Post* argues that "once there are enough holes in a firewall, it crumbles. The technology for this exists. What is needed is more capacity," it's a statement that, while technically true, is extremely deceptive. More capacity may, indeed, temporarily pierce the firewalls, but it is no guarantee that other, firewall-free approaches won't do the same job more effectively. To continue using the cyber-wall metaphor is to fall victim to extreme Internet-centrism, unable to see the sociopolitical nature of the problem of Internet control and focus only on its technological side.

Nowhere is the language problem more evident than in the popular discourse about China's draconian system of Internet control. Ever since a 1997 article in *Wired* magazine dubbed this system "the Great Firewall of China," most Western observers have relied on such mental imagery to conceptualize both the problem and the potential solutions. In the meantime, other important aspects of Internet control in the country— particularly the growing self-policing of China's own Internet companies and the rise of a sophisticated online propaganda apparatus—did not receive as much attention. According to Lokman Tsui, an Internet scholar at the University of Pennsylvania, "[the metaphor of] the 'Great Firewall' . . . limits our understanding and subsequent policy design on China's internet. . . . If we want to make a start at understanding the internet in China in all its complexity, the first step we need to take is to think beyond the Great Firewall that still has its roots in the Cold War." Tsui's advice is worth heeding, but as long as policymakers continue their collective exercise in Cold War nostalgia, it is not going to happen.

Why Photocopiers Don't Blog

Anachronistic language skewers public understanding of many other domains of Internet culture, resulting in ineffective and even counter-productive policies. The similarities between the Internet and tech-nologies used for samizdat—fax machines and photocopiers—are fewer than one might imagine. A piece of samizdat literature copied on a smuggled photocopier had only two uses: to be read and to be passed on. But the Internet is, by definition, a much more complex medium that can serve an infinite number of purposes. Yes, it can be used to pass on antigovernment information, but it can also be used to spy on citi-zens, satisfy their hunger for entertainment, subject them to subtle propaganda, and even launch cyber-attacks on the Pentagon. No deci-sions made about the regulation of faxes or photocopiers in Washing-ton had much impact on their users in Hungary or Poland; in contrast, plenty of decisions about blogs and social networking sites—made in Brussels, Washington, or Silicon Valley—have an impact on all the users in China and Iran.

Similarly, the problem with understanding blogging through the lens of samizdat is that it obfuscates many of the regime-strengthening fea-tures and entrenches the utopian myth of the Internet as a liberator. There was hardly any pro-government samizdat in the Soviet Union (even though there was plenty of samizdat accusing the government of violating the core principles of Marxism-Leninism). If someone wanted to express a position in favor of the government, they could write a let-ter to the local newspapers or raise it at the next meeting of their party cell. Blogs, on the other hand, come in all shapes and ideologies; there are plenty of pro-government blogs in Iran, China, and Russia, many of them run by people who are genuinely supportive of the regime (or at least some of its features, like foreign policy). To equate blogging with samizdat and bloggers with dissidents is to close one's eyes to what's going on in the extremely diverse world of new media across the globe. Many bloggers are actually more extreme in their positions than the government itself. Susan Shirk, an expert on Asian politics and for-mer deputy assistant secretary of state in the Clinton administration, writes that "Chinese officials . . . describe themselves as feeling under increasing pressure from nationalist public opinion. 'How do you know,' I ask, 'what public opinion actually is?' 'That's easy,' they say, 'I find out from Global Times [a nationalistic state-controlled tabloid about global affairs] and the Internet." And that public opinion may create an enabling environment for a more assertive government policy, even if the government is not particularly keen on it. "China's popular media and Internet websites sizzle with anti-Japanese vitriol. Stories related to Japan attract more hits than any other news on Internet sites and anti-Japanese petitions are a focal point for organizing on-line collective action," writes Shirk. Nor is the Iranian blogosphere any more tolerant; in late 2006 a conservative blog attacked Ahmadinejad for watching women dancers perform at a sports event abroad.

While it was possible to argue that there was some kind of linear relationship between the amount of samizdat literature in circulation or even the number of dissidents and the prospects for democratization, it's hard to make that argument about blogging and bloggers. By itself, the fact that the number of Chinese or Iranian blogs is increasing does not suggest that democratization is more likely to take root. This is where many analysts fall into the trap of equating liberalization with democratization; the latter, unlike the former, is a process with a clear end result. "Political liberalization entails a widening public sphere and a greater, but not irreversible, degree of basic freedoms. It does not imply the introduction of contestation for positions of effective governing power," write Holger Albrecht and Oliver Schlumberger, two scholars of democratization specializing in the politics of the Middle East. That there are many more voices online may be important, but what really matters is whether those voices eventually lead to any more political participation and, eventually, any more votes. (And even if they do, not all such votes are equally meaningful, for many elections are rigged before they even start.)

Which Tweet Killed the Soviet Union?

But what's most problematic about today's Cold War–inspired conceptualization of Internet freedom is that they are rooted in a shallow and triumphalist reading of the end of the Cold War, a reading that has little to do with the discipline of history as practiced by historians (as opposed to what is imagined by politicians). It's as if to understand the inner workings of our new and shiny iPads we turned to an obscure nineteenth-century manual of the telegraph written by a pseudoscientist who had never studied physics. To choose the Cold War as a source of guiding metaphors about the Internet is an invitation to conceptual stalemate, if only because the Cold War as a subject matter is so suffused with arguments, inconsistencies, and controversies—and those are growing by the year, as historians gain access to new archives—that it is completely ill-suited for any comparative inquiry, let alone the one that seeks to debate and draft effective policies for the future.

When defenders of Internet freedom fall back on Cold War rhetoric, they usually do it to show the causal connection between information and the fall of communism. The policy implications of such comparisons are easy to grasp as well: Technologies that provide for such increased information flows should be given priority and receive substantial public support.

Notice, for example, how Gordon Crovitz, a *Wall Street Journal* columnist, makes an exaggerated claim about the Cold War—"the Cold War was won by spreading information about the Free World"—before recommending a course of action—"in a world of tyrants scared of their own citizens, the new tools of the Web should be even more terrifying if the outside world makes sure that people have access to its tools." (Crovitz's was an argument in favor of giving more public money to Falun Gong-affiliated Internet groups.) Another 2009 column in the *Journal*, this time penned by former members of the Bush administration, pulls the same trick: "Just as providing photocopies and fax machines helped Solidarity dissidents in communist Poland in the 1980s"—here is the necessary qualifier without which the advice might seem less credible—"grants should be given to private groups to develop and field firewall-busting technology."

These may all be worthwhile policy recommendations, but they rest on a highly original—some historians might say suspicious—interpretation of the Cold War. Because of its unexpected and extremely fast-paced end, it begot all sorts of highly abstract theories about the power of information to transform power itself. That the end of communism in the East coincided with the beginning of a new stage in the information revolution in the West convinced many people that a monocausal relationship was at work here. The advent of the Internet was only the most obvious breakthrough, but other technologies—above all, the radio—got a lion's share of the credit for the downfall of Soviet communism. "Why did the West win the Cold War?" asks Michael Nelson, former chairman of the Reuters Foundation in his 2003 book about the history of Western broadcasting to the Soviet bloc. "Not by use of arms. Weapons did not breach the Iron Curtain. The Western invasion was by radio, which was mightier than the sword." Autobiographies of radio journalists and executives who were commanding that "invasion" in outposts like Radio Free Europe or the Voice of America are full of such rhetorical bluster; they are clearly not the ones to downplay their own roles in bringing democracy to Eastern Europe.

The person to blame for popularizing such views happens to be the same hero many conservatives widely believe to have won the Cold War itself: Ronald Reagan. Since he was the man in charge of all those Western radio broadcasts and spearheaded the undercover support to samizdat-printing dissidents, any account that links the fall of communism to the role of technology would invariably glorify Reagan's own role in the process. Reagan, however, did not have to wait for future interpretations. Proclaiming that "breezes of electronic beams blow through the Iron Curtain as if it was lace," he started the conversation that eventually degenerated into the dreamy world of "virtual curtains" and "cyber-walls." Once Reagan announced that "information is the oxygen of the modern age" and that "it seeps through the walls topped by barbed wire, it wafts across the electrified borders," pundits, politicians, and think-tankers knew they had a metaphorical treasure trove while Reagan's numerous supporters saw this narrative as finally acknowledging their hero's own gigantic contribution to ushering in democracy into Europe. (China's microchip manufacturers must have been laughing all the way to the bank when Reagan predicted that "the Goliath of totalitarianism will be brought down by the David of the microchip.")

It just took a few months to add analytical luster to Reagan's pronouncements and turn it into something of a coherent history. In 1990, the RAND Corporation, a California-based think tank that, perhaps by the sheer virtue of its propitious location, never passes up an opportunity to praise the powers of modern technology, reached a strikingly similar conclusion. "The communist bloc failed," it said in a timely published study, "not primarily or even fundamentally because of its centrally controlled economic policies or its excessive military burdens, but because its closed societies were too long denied the fruits of the information revolution." This view has proved remarkably sticky. As late as 2002, Francis Fukuyama, himself a RAND Corporation alumnus, would write that "totalitarian rule depended on a regime's ability to maintain a monopoly over information, and once modern information technology made that impossible, the regime's power was undermined."

By 1995 true believers in the power of information to crush authoritarianism were treated to a book-length treatise. *Dismantling Utopia: How Information Ended the Soviet Union*, a book by Scott Shane—who from 1988 to 1991 served as the *Baltimore Sun*'s Moscow correspondent—tried to make the best case for why information mattered, arguing that the "death of the Soviet illusion . . . [was] not by tanks and bombs but by facts and opinions, by the release of information bottled up for decades."

The crux of Shane's thesis was that as the information gates opened under glasnost, people discovered unpleasant facts about the KGB's atrocities while also being exposed to life in the West. He wasn't entirely incorrect: Increased access to previously suppressed information did expose the numerous lies advanced by the Soviet regime. (There were so many revisions to history textbooks in 1988 that a nationwide history examination had to be scrapped, as it wasn't clear if the old curriculum could actually count as "history" anymore.) It didn't take long until, to use one of Shane's memorable phrases, "ordinary information, mere facts, exploded like grenades, ripping the system and its legitimacy."

Hold On to Your Data Grenade, Comrade!

Facts exploding like grenades certainly make for a gripping journalistic narrative, but it's not the only reason why such accounts are so popular. Their wide acceptance also has to do with the fact that they always put people, rather than some abstract force of history or economics, first. Any information-centric account of the end of the Cold War is bound to prioritize the role of its users—dissidents, ordinary protesters, NGOs—and downplay the role played by structural, historical factors—the unbearable foreign debt accumulated by many Central European countries, the slowing down of the Soviet economy, the inability of the Warsaw Pact to compete with NATO.

Those who reject the structural explanation and believe that 1989 was a popular revolution from below are poised to see the crowds that gathered in the streets of Leipzig, Berlin, and Prague as exerting enormous pressure on communist institutions and eventually suffocating them. "Structuralists," on the other hand, don't make much of the crowds. For them, by October 1989 the communist regimes were already dead, politically and economically; even if the crowds would not have been as numerous, the regimes would still be as dead. And if one assumes that the Eastern European governments were already dysfunctional, unable or reluctant to fight for their existence, the heroism of protesters matters much less than most information-centric accounts suggest. Posing on the body of a dead lion that was felled by indigestion makes for a far less impressive photo op.

This debate—whether it was the dissidents or some impersonal social force that brought down communism in Eastern Europe—has taken a new shape in the growing academic dispute about whether something like "civil society" (still a favorite buzzword of many foundations and development institutions) existed under communism and whether it played any significant role in precipitating the public protests. Debates over "civil society" have immense repercussions for the future of Internet freedom policy, in part because this fuzzy concept is often endowed with revolutionary potential and bloggers are presumed to be in its vanguard. But if it turns out that the dissidents—and civil society as a whole—did not play much of a role in toppling communism, then the popular expectations about the new generation of Internet revolutions may be overblown as well. Getting it right matters because the unchecked belief in the power of civil society, just like the

unchecked power in the ability of firewall-breaching tools, would ultimately lead to bad policy and prioritize courses of action that may not
be particularly effective.

Stephen Kotkin, a noted expert of Soviet history at Princeton University, has argued that the myth of civil society as a driver of anticommunist change was mostly invented by Western academics, donors, and
journalists. "In 1989 'civil society' could not have shattered Soviet-style
socialism for the simple reason that civil society in Eastern Europe did
not then actually exist." And Kotkin has got the evidence to back it up:
In early 1989 the Czechoslovak intelligence apparatus estimated that
the country's active dissidents were no more than five hundred people,
with a core of about sixty (and even as the protests broke out in Prague,
the dissidents were calling for elections rather than a complete overthrow of the communist regime). The late Tony Judt, another gifted
historian of Eastern European history, observed that Václav Havel's
Charter 77 attracted fewer than 2,000 signatures in a Czechoslovak
population of fifteen million. Similarly, the East German dissident
movement did not play a significant role in getting people onto the
streets of Leipzig and Berlin, and such movements almost did not exist
in Romania or Bulgaria. Something like civil society did exist in Poland,
but it was also one of the few countries with virtually no significant
protests in 1989. Kotkin is thus justified in concluding that "just like
the 'bourgeoisie' were mostly an outcome of 1789, so 'civil society' was
more a consequence than a primary cause of 1989."

But even if civil society didn't exist as such, people did come out to
Prague's Wenceslas Square, choosing to spend cold November days
chanting antigovernment slogans under the ubiquitous gaze of police
forces. Whatever their role, the crowds certainly didn't hurt the cause
of democratization. If one believes that the crowds matter, then a more
effective tool of getting them into the streets would be a welcome addition; thus, the introduction of a powerful new technology—a photocopier to copy the leaflets at rates ten times faster than before—is a
genuine improvement. So are any changes in the way by which people
can reveal their incentives to each other. If you know that twenty of
your friends will join a protest, you may be more likely to join as well.

Facebook is, thus, something of a godsend to protest movements. It would be silly to deny that new means of communications can alter the likelihood and the size of a protest.

Nevertheless, if the Eastern European regimes had not already been dead, they would have mounted a defense that would have prevented any "information cascades" (the preferred scholarly term for such snowball-like public participation) from forming in the first place. On this reading, the East German regime was simply unwilling to crack down on the first wave of protests in Leipzig, well aware that it was heading for a collective suicide. Furthermore, in 1989, unlike in 1956 or 1968, the Kremlin, ruled by a new generation of leaders who still had vivid memories of the brutality of their predecessors, didn't think that bloody crackdowns were a good idea, and East Germany's senior leaders were too weak and hesitant to do it alone. As Perry Anderson, one of the most insightful students of contemporary European history, once remarked, "nothing fundamental could change in Eastern Europe so long as the Red Army remained ready to fire. Everything was possible once fundamental change started in Russia itself." To argue that it was the photocopies that triggered change in Russia and then the rest of the region is to engage in such a grotesque simplification of history that one may as well abandon practicing history altogether. This is not to deny that they played a role, but only to deny the monocausal relationship that many want to establish.

When the Radio Waves Seemed Mightier Than the Tanks

If there is a genuine lesson to be drawn from Cold War history, it is that the increased effectiveness of information technology is still severely constrained by the internal and external politics of the regime at hand, and once those politics start changing, it may well be possible to take advantage of the new technologies. A strong government that has a will to live would do its utmost to deny Internet technology its power to mobilize. As long as the Internet is tied to physical infrastructure, this is not so hard to accomplish: In virtually all authoritarian states, governments maintain control over communication networks and can turn them off at the first sign of protests. As the Chinese authorities began worrying about the growing unrest in Xinjiang in 2009, they simply turned off all Internet communications for ten months; it was a very thorough cleansing, but a few weeks would suffice in less threatening cases. Of course, they may incur significant economic losses because of such information blackouts, but when forced to choose between a blackout and a coup, many choose the former.

Even the strongest authoritarian governments are consistently challenged by protesters. It seems somewhat naïve to believe that strong authoritarian governments will balk at cracking down on protesters for fear of being accused of being too brutal, even if their every action is captured on camera; most likely, they will simply learn how to live with those accusations. The Soviet Union didn't hesitate to send tanks to Hungary in 1956 and Czechoslovakia in 1968; the Chinese didn't pause before sending tanks to Tiananmen Square, despite a sophisticated network of fax machines that was sending the information to the West; the Burmese junta didn't balk at suppressing a march by the monks, despite the presence of foreign journalists documenting their actions. The most overlooked aspect of the 2009 protests in Tehran is that even though the government was well aware that many protesters were carrying mobile phones, it still dispatched snipers on building roofs and ordered them to shoot (one such sniper supposedly shot twenty-seven-year-old Neda Agha-Soltan; her death was captured on video, and she became one of the heroes of the Green Movement, with one Iranian factory even manufacturing statues of her). There is little evidence to suggest that, at least for the kind of leaders who are least likely to receive the Nobel Peace Prize, exposure results in less violence.

Most important, governments can also take advantage of decentralized information flows and misinform their population about how popular the protest movement really is. That decentralization and multiplication of digital information would somehow make it easier for the fence-sitters to infer what is really happening in the streets seems a rather unfounded assumption. In fact, history teaches us that media could as easily send false signals; many Hungarians still remember the utterly irresponsible broadcasts by Radio Free Europe on the eve of the Soviet invasion in 1956, which suggested that American military aid would be forthcoming (it wasn't). Some of those broadcasts even offered tips on antitank warfare, urging the Hungarians to resist the Soviet occupation; they could be held at least partially responsible for the 3,000 deaths that followed the invasion. Such misinformation, whether deliberate or not, could flourish in the age of Twitter (the effort to spread fake videos purporting to show the burning of Ayatollah Khomeini's portrait in the aftermath of the Iranian protests is a case in point).

Nor is the decentralized nature of communications always good in itself, especially if the objective is to make as many people informed as fast as possible. In a 2009 interview with the *Globe and Mail*, the East German dissident Rainer Muller noted how beneficial it was that the nation's attention was not dispersed in the late 1980s: "You didn't have people looking at 200 different TV channels and 10,000 websites and e-mails from thousands of people. You could put something on a Western TV or radio station and you could be sure that half the country would know it." Few oppositional movements can boast such sizable audiences in the age of YouTube, especially when they are forced to compete with the much funnier videos of cats flushing the toilet.

While a definitive history of the Cold War remains to be written, the uniqueness of its end is not to be underestimated. Too many factors were stacked against the survival of the Soviet system: Gorbachev sent a number of cautionary signals to the communist leaders of Eastern Europe warning them against crackdowns and making it clear that the Kremlin wouldn't assist in suppressing popular uprisings; a number of Eastern European countries were running economies on the brink of bankruptcy and had a very dark future ahead of them, with or without the protests; East German police could have easily prevented the demonstrations in Leipzig, but its leaders did not exercise their authority; and a small technical change in Poland's electoral law could have prevented the Solidarity movement from forming a government that inspired other democratic movements in the region.

This is the great paradox of the Cold War's end: On the one hand, the structural conditions of countries of the Soviet bloc in late 1989 were so lethal that it seemed inevitable that communism would die. On the other hand, communist hard-liners had so much room to maneuver that absolutely nothing guaranteed that the end of the Cold War would be as bloodless as it turned out to be. Given how many things could have gone wrong in the process, it's still something of a miracle that the Soviet bloc—Romania notwithstanding—went under so peacefully. It takes a rather peculiar historical sense to look at this highly particularistic case and draw far-reaching conclusions about the role of technology in its demise and then assume that such conclusions would also hold in completely different contexts like China or Iran twenty years later. Western policymakers should rid themselves of the illusion that communism ended quickly—under the pressure of information or fax machines—or that it was guaranteed to end peacefully because the whole world was watching. The fall of communism was the result of a much longer process, and the popular protests were just its most visible, but not necessarily most important, component. Technology may have played a role, but it did so because of particular historical circumstances rather than because of technology's own qualities. Those circumstances were highly specific to Soviet communism and may no longer exist.

Western policymakers simply can't change modern Russia, China, or Iran using methods from the late 1980s. Simply opening up the information gates would not erode modern authoritarian regimes, in part because they have learned to function in an environment marked by the abundance of information. And it certainly doesn't hurt that, contrary to the expectations of many in the West, certain kinds of information could actually strengthen them.

chapter three

Orwell's Favorite Lolcat

● ● ■

"The Tits Show" sounds like a promising name for a weekly Internet show. Hosted by Russia.ru, Russia's pioneering experiment in Internet television supported by Kremlin's ideologues, the show's format is rather simple: A horny and slightly overweight young man travels around Moscow nightclubs in search of perfect breasts. Moscow nightlife being what it is, the show is never short of things to film and women to grope and interview.

"The Tits Show" is just one of more than two dozen weekly and daily video shows produced by the Russia.ru team to satisfy the quirky tastes of Russian Internet users (and "produced" they truly are: much of the site's staff are defectors from the world of professional television). Some of those shows discuss politics—there are even a few odd interviews with Russia's president, Dmitry Medvedev—but most are quite frivolous in nature. A sample episode of the "books show": an exploration of the best books about alcohol available in Moscow's bookstores.

If one reads the Western press, it's easy to get the impression that the Internet in Russia is an effective and extremely popular vehicle for attacking—if not overthrowing—the government. Nevertheless, while civic activism—raising money for sick children and campaigning to curb police corruption—is highly visible on the Russian Internet, it's still entertainment and social media that dominate (in this respect, Russia hardly differs from the United States or countries in Western Europe). The most popular Internet searches on Russian search engines are not for "what is democracy?" or "how to protect human rights" but for "what is love?" and "how to lose weight."

Russia.ru does not hide its connections to the Kremlin; senior members of the Kremlin's various youth movement even have their own talk shows. The need for such a site stems from the Kremlin's concern that the transition from the world of television, which it fully controls, to the anarchic world of the Internet might undermine the government's ability to set the agenda and shape how the public reacts to news. To that effect, the Kremlin supports, directly or indirectly, a host of sites about politics, which are usually quick to denounce the opposition and welcome every government initiative, but increasingly branches out into apolitical entertainment. From the government's perspective, it's far better to keep young Russians away from politics altogether, having them consume funny videos on Russia's own version of YouTube, RuTube (owned by Gazprom, the country's state-owned energy behemoth), or on Russia.ru, where they might be exposed to a rare ideological message as well. Many Russians are happy to comply, not least because of the high quality of such online distractions. The Russian authorities may be onto something here: The most effective system of Internet control is not the one that has the most sophisticated and draconian system of censorship, but the one that has no need for censorship whatsoever.

The Kremlin's growing online entertainment empire may explain why there is little formal censorship in Russia—the Kremlin doesn't ban access to any of its opponents' websites, with the minor exception of those created by terrorists and child molesters—and yet surprisingly little political activity. Russia.ru, with its highly skilled team and flexible budget, is just one of the many attempts to control the space; it does so by relying on entertainment rather than politics. Could it be that the vast online reservoirs of cheap entertainment are dampening the enthusiasm that the Russian youth might have for politics, thus preventing their radicalization? What if the liberating potential of the Internet also contains the seeds of depoliticization and thus dedemocratization? Could it be that just as the earlier generation of Western do-gooders mistakenly believed that Soviet office workers were secretly typing samizdat literature on their computers (rather than playing Tetris), so Westerners today harbor futile hopes that Russians are blogging about human rights and Stalin's abuses, while they are only flipping through ChatRoulette, Russia's quirky gift to the Internet?

How Cable Undermines Democracy

Here again the focus on the role of broadcasting in the Cold War keeps the West ignorant of the complex role that information plays in authoritarian societies. Two theories explain how exposure to Western media could have democratized the Soviets. One claims that Western media showed brainwashed citizens that their governments were not as innocent as they claimed to be and pushed people to think about political issues they may have previously avoided; it's what we can call "liberation by facts" theory. The second asserts that Western media spread images of prosperity and fueled consumerist angst; stories of fast cars, fancy kitchen appliances, and suburban happiness made citizens living under authoritarianism dream of change and become more active politically. This is what we can call "liberation by gadgets" theory.

While projecting the images of prosperity was easy, getting people to care about politics was more difficult—at least people who were not previously politicized. To that extent, Western broadcasting efforts included both entertainment and lifestyle programs (one of Radio Free Europe's hits was *Radio Doctor*, a program that informed listeners about recent developments in Western medicine and answered specific questions from laypeople, exposing the inefficiencies of the Soviet system in the process). Banned music was frequently broadcast as well (one survey of Belarusian youths in 1985 found that 75 percent of them listened to foreign broadcasts, mostly to catch up on music they couldn't get otherwise). In this way the West could capitalize on the cultural rigidity of communism, lure listeners with the promise of better entertainment, and secretly feed them with political messages. (Not everyone was convinced such a strategy was effective. In 1953 Walter Lippmann, one of the fathers of modern propaganda, penned a poignant op-ed, arguing that "to set up an elaborate machinery of international communication and then have it say, 'We are the Voice of America engaged in propaganda to make you like us better than you like our adversaries,' is—as propaganda—an absurdity. As a way of stimulating an appetite for the American way of life, it is like serving castor oil as a cocktail before dinner.") "Politicization" and involvement in oppositional politics were thus by-products of desire for entertainment that the West knew how to satisfy. This may not have led to the emergence of civil society, of course, but it has certainly made ideas associated with the democratic revolutions of 1989 more palatable in the end.

The media's roles in the cultivation of political knowledge in both democratic and authoritarian societies are strikingly similar. Before the rise of cable television in the West, knowledge about politics—especially of the everyday variety—was something of an accident even in democratic societies. Markus Prior, a scholar of political communications at Princeton University, argues that most Americans were exposed to political news not because they wanted to watch it but because there was nothing else to watch. This resulted in citizens who were far better politically informed, much more likely to participate in politics, and far less likely to be partisan than today. The emergence of cable television, however, gave people the choice between consuming political news and anything else—and most viewers, predictably, went for that "anything else" category, which mostly consisted of entertainment. A small cluster has continued to care about politics—and, thanks to the rise of the niche media, they have more opportunities that they could ever wish for—but the rest of the population has disengaged.

Prior's insights about the negative effects of media choice in the context of Western democracies can also shed light on why the Internet may not boost political knowledge and politicize the fence-sitters, the ones who remain undecided about whether to voice their grievances against their governments, to the degree that some of us hope. The drive for entertainment simply outweighs the drive for political knowledge—and YouTube could easily satisfy even the most demanding entertainment junkies. Watching the equivalent of "The Tits Show" in the 1970s required getting exposure to at least a five-second political commercial (even if it was the jingle of Radio Free Europe), while today one can avoid such political messages altogether.

The Denver Clan Conquers East Berlin

If policymakers stopped focusing on "virtual walls" and "information curtains," as if those are all that the Cold War could teach them, they might discover a more useful lesson on the entertainment front. The German Democratic Republic presents a fascinating case of a communist country that for virtually all of its existence could receive Western broadcasting. It would seem natural to expect that of all the other communist states GDR would have the most politically informed citizens, the most vibrant political opposition, and civil society groups as well as a burgeoning samizdat enterprise. These expectations would be in line with how the impact of information was viewed during the Cold War. It was all too easy to fall under the impression that all media consumption was political, because researchers had two limited sources for their assertions: recent émigrés and those who wrote admiring letters to the likes of Radio Free Europe. Such sources bolstered the view that consuming official narratives of events in state-run media led to apathy and disillusionment, pushing people to seek solace in foreign radio programs. Yet neither of the two groups were unbiased, and the conclusions of such studies have been repeatedly challenged. Concluding that people who wrote letters to Radio Free Europe were representative of the population at large was like walking into a bar a few blocks from Congress, interviewing a few congressional staffers deeply mesmerized by a C-SPAN broadcast on the bar's wall, and lauding the fact that most Americans are superbly informed about the nuts and bolts of national politics. (That said, not all researchers doing such quasi-detective work were dilettantes; to access the actual views of the people they interviewed and whose letters they read, they paid particular attention to Freudian slips and typos.)

Eventually there emerged a far better, more empirical way to test common Western assumptions about the role of media in authoritarian regimes. It was a stroke of luck: East Germany's geography made it difficult to block Western signals on most of its territory, and only one-sixth of the population, concentrated mostly in counties that were far from the western border, could not receive West German television (this area was widely known—and ridiculed—as *Tal Der Ahnungslosen*, "The Valley of the Clueless"). In 1961—the year the Berlin Wall went up—the country's leading youth organization, Freie Deutsche Jugend, began dispatching their youthful troops to many a rooftop to find antennae aimed at the West and either dismantle them or reorient them toward East German transmission towers. Popular anger, however, quickly drove the youngsters away, and such raids stopped. By 1973 GDR's leader, Erich Honecker, acknowledging that West German television was already widely popular, gave up and allowed all GDR's citizens—except soldiers, police, and teachers—to watch whatever they wanted, on the condition the citizens would closely scrutinize everything they saw and heard in the Western media. At the same time, Honecker urged GDR's own television to "overcome a certain type of tedium" and "to take the desire for good entertainment into account." Thus, for nearly three decades, most of GDR's citizens were in a rather peculiar situation: They could, in theory, compare how the two German regimes—one democratic and one communist—chose to portray the same events. If the conclusions of all those studies that analyzed letters sent to Radio Free Europe were right, one could expect that East Germans would be glued to news programs from the democratic West, learning of the abuses of their own regime and searching for secret antigovernment cells to join.

It's hard to say whether East Germans did practice as much media criticism as Western scholars would have subsequently wanted them to, but it seems that Western television only made them more complacent—a fact that GDR's ruling elites eventually recognized. When they insisted on removing a satellite dish that was illegally installed by the residents of the small German town of Weissenberg, the local communist officials and the mayor were quick to point out that members of their community were "'much more content' since the introduction of West German television," that their attitudes toward the East German regime had become "more positive," and that all applications for exit visas (that is, to immigrate to the West) had been withdrawn. From the early 1980s onwards, satellites were openly tolerated by the authorities.

East Germans were not all that interested in tracking the latest news from NATO. Instead, they preferred soft news and entertainment, particularly American TV series. Such shows as *Dallas*, *Miami Vice*, *Bonanza*, *Sesame Street*, and *The Streets of San Francisco* were particularly popular; even the leading Communist Party journal *Einheit* acknowledged that *Dynasty*—known in Germany as *The Denver Clan* and the most popular of the lot—was widely watched. Paul Gleye, an American Fulbright scholar who lived and taught in GDR between 1988 and 1989, remembers that whenever he brought out his map of the United States to tell East Germans about his country, "the first question often was 'Well, show me where Dallas and Denver are,'" while his students "seemed to be more interested in hearing about Montana State University when I told them it was about 850 kilometers northwest of Denver than when I described its setting in a picturesque Alpine valley in the Rocky Mountains."

Long after the Berlin Wall fell, Michael Meyen and Ute Nawratil, two German academics, conducted extensive interviews with hundreds of East Germans. They found that many of them did not even believe what they heard on the Western news. They thought that the portrayal of life in East Germany was predictably uninformed and highly ideological, while the extensive propaganda of their own government made them expect that Western news, too, was heavily shaped by the government. (Ironically, in their distrust and suspicion of the Western propaganda apparatus, they were more Chomskian than Noam Chomsky himself). When, in a separate study, East Germans were asked what changes they would like to see in their country's television programming, they voted for more entertainment and less politics. Eventually GDR's propaganda officials learned that the best way to have at least a modicum of people watch their ideological programming on GDR's own television was to schedule it when West German television was running news and current affairs programs—which East Germans found to be the least interesting.

The Opium of the Masses: Made in GDR

That the never-ending supply of Western entertainment made large parts of GDR's population useless as far as activism was concerned was not lost on German dissidents. As Christopher Hein, a prominent East German writer and dissident, stated in a 1990 interview:

> [In the GDR we had a difficult task because] the whole people could leave the country and move to the West as a man every day at 8 PM— via television. That lifted the pressure. Here is the difference between Poland, Czechoslovakia, and the Soviet Union. There the pressure continued to bear down and generated counter-pressure.... That's why I always envied the Russians and the Poles.... In general, the helpful proximity of the Federal Republic was not helpful to our own development.... Here we had no samizdat, as long as we had access to the publishing houses of West Germany.

Subsequent research based on archival data proved Hein right. East German authorities, preoccupied with their own survival, spent a lot of resources on understanding the attitudes of their young citizens. To that effect, they commissioned a number of regular studies, most of which were conducted by the ominous-sounding Central Institute for Youth Research founded in 1966. Between 1966 and 1990 it conducted several hundred surveys that studied the attitudes of high school and college students, young workers, and others; the staff of the institute could not study other demographic groups, nor could they publish their results—those were classified. The reports were declassified after German unification and have opened up a bounty of research for academics studying life in East Germany. The surveys polled respondents about regime support (e.g., asking them whether they agreed with statements like "I am convinced of the Leninist/Marxist worldview" and "I feel closely attached to East Germany").

Holger Lutz Kern and Jens Hainmueller, two German academics teaching in the United States, studied this data to understand how the relationship between life satisfaction and regime support varied according to the availability of Western broadcasting. They published their findings in a provocatively titled paper, "Opium for the Masses: How Foreign Media Can Stabilize Authoritarian Regimes." They found that those East German youth who could receive Western television were, overall, more satisfied and content with the regime; the ones who could not receive Western television—those living in the Valley of the Clueless—were much more politicized, more critical of the regime, and, most interestingly, more likely to apply for exit visas. Thus, they wrote, "in an ironic twist for Marxism, capitalist television seems to have performed the same narcotizing function in communist East Germany that Karl Marx had attributed to religious beliefs in capitalism societies when he condemned religion as 'opium of the people.'"

They described this process as "escapism": "West German Television allowed East Germans to vicariously escape life under communism at least for a couple of hours each night, making their lives more bearable and the East German regime more tolerable. . . . West German television exposure resulted in a net increase in regime support." If anything, access to excellent entertainment from the West—it took GDR authorities many years to start producing high-quality entertainment programs that could rival those from abroad—depoliticized vast swathes of the East German population, even as it nominally allowed them to learn more about the injustices of their own regime from Western news programs. Western television made life in East Germany more bearable, and by doing so it may have undermined the struggle of the dissident movement. Most interestingly, it was in the Valley of the Clueless that protests began brewing; its residents were clearly more dissatisfied with life in the country than those who found a refuge in the exciting world of *The Denver Clan*.

If we judge by the youth survey data, we might conclude that young people were particularly susceptible to escapism; moreover, we don't have much data for East German adults. The "liberation by gadgets" theory may thus have some validity. Perhaps, the adults, disappointed by the never-arriving "socialism with a human face," were much more susceptible to despair and thus easier to politicize with teasing pictures of Western capitalism. Paul Betts, a British academic who has studied consumer culture in GDR, points out that "those things that the state had supposedly overcome in the name of the great socialist experiment— subjective fancy, individual luxury, commodity fetishism, and irrational consumer desire—eventually returned as its arch nemeses. The irony is that the people apparently took these dreams of a better and more prosperous world more seriously than the state ever expected, so much so that the government was ultimately sued for false advertising." Or, as a popular joke of that period had it: "Marxism would have worked if it wasn't for cars." (It seems that the Chinese have learned the East German lesson, purchasing the entire Volvo operation from Ford in 2010.)

The East German experience shows that the media could play a much more complex and ambiguous role under authoritarianism than many in the West initially assumed. Much of the early scholarship on the subject greatly underestimated the need for entertainment and overestimated the need for information, especially of the political variety. Whatever external pressures, most people eventually find a way to accustom themselves to the most brutal political realities, whether by means of television, art, or sex.

Furthermore, the fact that the media did such a superb job at covering the fall of the Berlin Wall may have influenced many observers to believe that it played a similarly benevolent role throughout the entire history of the Cold War. But this was just a utopian dream: Whatever noble roles media take on during extraordinary crisis situations should not be generalized, for their everyday functions are strikingly different and are much more likely to be geared toward entertainment (if only because it sells better). A case in point: While many praised Twitter's role in publicizing and promoting political demonstrations in Iran, the death of Michael Jackson on June 25, 2009, quickly overtook the protests as the site's most popular topic.

Watching *Avatar* in Havana

But if the Western media made the consumerist benefits of capitalism easier to grasp than any piece of samizdat, it only gave its hopeful Eastern European viewers a rather shallow view on how democracy works and what kind of commitments and institutions it requires. As Erazim Kohák, a Czech-American philosopher whose family emigrated to the United States in 1948, memorably wrote in 1992: "The unfortunate truth is that as the former subjects of the Soviet empire dream it, the American dream has very little to do with liberty and justice for all and a great deal to do with soap operas and the Sears catalogue. . . . It is a dream made up mostly of irresponsibility, unreality, and instantly gratified greed." Kohák knew that it was affluence—"the glittering plenty we glimpsed across the border in Germany and Austria . . . freedom from care, freedom from responsibility"—rather than some abstract notion of Jeffersonian democracy that the Eastern European masses really wanted. As Kohák was quick to point out, affluence came fast in the early 1990s, without anyone giving much thought to what else democracy should mean: "When the popular Czech cartoonist Renčín draws his vision of what freedom will bring, he draws a man blissing out on a sofa, surrounded by toys and trophies—an outdoor motor, a television set with a VCR, a personal computer, a portable bar, an LP grill. There is not a trace of irony in it: this is what freedom means."

But the Russia or China of today is not the East Germany or Czechoslovakia of the late 1980s. Except for North Korea, Turkmenistan, and perhaps Burma, modern authoritarian states have embraced consumerism, and it seems to have strengthened rather than undermined their regimes. Popular culture, especially when left unchecked by appeals to some higher truth or ideal, has eroded the political commitment of even the most dissatisfied citizens. Although the jubilant Czechs installed Václav Havel, a playwright and formidable intellectual, as their leader, they couldn't resist the consumerist tornado sweeping through their lands (ironically, "Power of the Powerless" essay, Havel's most famous attack on the totalitarian system, was a fulmination against the petty-mindedness of a communist store manager). Havel should have listened to Philip Roth, who in 1990 gave him and his fellow Czech intellectuals a most precious piece of advice on the pages of the *New York Review of Books*. Roth predicted that soon the cult of the dissident intellectuals would be replaced by the cult of another, much more powerful adversary:

> I can guarantee you that no defiant crowds will ever rally in Wenceslas Square to overthrow its tyranny nor will any playwright-intellectual be elevated by the outraged masses to redeem the national soul from the fatuity into which this adversary reduces virtually all of human discourse. I am speaking about that trivializer of everything, commercial television—not a handful of channels of boring clichéd television that nobody wants to watch because it is controlled by an oafish state censor, but a dozen or two channels of boring, clichéd television that most everybody watches all the time because it is *entertaining*.

Roth could not have predicted the rise of YouTube, which has proven even more entertaining than cable. (He seems to avoid most of the pleasures of the Web; in a 2009 interview with the *Wall Street Journal*, he claimed he only uses it to buy books and groceries.)

As a writer for the *Times of London* summarized the situation, some of the former communist countries "may have escaped the grip of dictators to fall instead under the spell of Louis Vuitton."

In the absence of high ideals and stable truths, it has become nearly impossible to awaken people's political consciousness, even to fight authoritarianism. How can you, when everyone is busy buying plasma TVs (Chinese today buy TVs with the biggest screens in the world, beating Americans by four inches), shopping for stuff online (a company linked to the Iranian government launched an online supermarket the same week that the authorities decided to ban Gmail), and navigating a city with the highest number of BMWs per square meter (that would be Moscow)? Even the official media in Cuba, that stalwart of revolutionary values, now broadcast TV series like *The Sopranos*, *Friends*, and *Grey's Anatomy*. In early 2010 they reportedly broadcast a pirated version of the movie *Avatar* shortly after it opened in U.S. theaters. (The communist critics, however, remained unconvinced; "predictable . . . very simplistic . . . reiterative in its argument" was the verdict of movie buffs from *Granma*, the official daily of the Communist Party of Cuba—perhaps they didn't get the memo about the 3D glasses.) It's hardly surprising that fewer than 2 percent of Cubans tune in to the radio broadcasts funded by the U.S. government through Radio Martí, Cuba's equivalent of Radio Free Europe. Why should ordinary Cubans take any risks to listen to highly ideological and somewhat boring news about politics if they can follow the travails of Tony Soprano?

The same young people America wants to liberate with information are probably better informed about U.S. popular culture than many Americans. Teams of Chinese netizens regularly collaborate to produce Chinese-language subtitles for popular American shows like *Lost* (often they find those shows on various peer-to-peer file-sharing sites as soon as ten minutes after new episodes air in the United States). Could it also be some kind of modern-day samizdat? Maybe, but there is little indication that it poses any threat to the Chinese government. If anyone is "lost," it's the citizens, not the authorities. Even authoritarian governments have discovered that the best way to marginalize dissident books and ideas is not to ban them—this seems only to boost interest in the forbidden fruit—but to let the invisible hand flood the market with trashy popular detective stories, self-help manuals, and books on how to get your kids into Harvard (texts like *You Too Can Go to Harvard: Secrets of Getting into Famous U.S. Universities* and *Harvard Girl* are best sellers in China).

Feeling that resistance would be counterproductive, even the Burmese government has grudgingly allowed hip-hop artists to perform at state functions. The regime has also created a soccer league after years without any organized matches and increased the number of FM radio stations, allowing them to play Western-style music. There even appeared something of a local MTV channel. As a Western-educated Burmese businessman told the *New York Times* in early 2010, "The government is trying to distract people from politics. There's not enough bread, but there's a lot of circus." Once Burma is fully wired—and the junta is supportive of technology, having set up its own Silicon Valley in 2002 that goes by the very un–Silicon Valley name of Myanmar Information and Communication Technology Park—the government won't have to try hard anymore; their citizens will get distracted on their own.

Today's battle is not between David and Goliath; it's between David and David Letterman. While we thought the Internet might give us a generation of "digital renegades," it may have given us a generation of "digital captives," who know how to find comfort online, whatever the political realities of the physical world. For these captives, online entertainment seems to be a much stronger attractor than reports documenting human rights abuses by their own governments (in this, they are much like their peers in the democratic West). One 2007 survey of Chinese youth found that 80 percent of respondents believe that "digital technology is an essential part of how I live," compared with 68 percent of American respondents. What's even more interesting, 32 percent of Chinese said that the Internet broadens their sex life, compared with just 11 percent of Americans. A Fudan University poll in June 2010 of nine hundred female graduates at seventeen Shanghai universities revealed that 70 percent don't think that one-night stands are immoral, while in 2007 a Shanghai-based doctor who runs a helpline for pregnant teenagers in the city reported that 46 percent of more than 20,000 girls who called the helpline since 2005 said they had sex with boys they met on the Web. The implications of China's "hormone revolution" are not lost on the authorities, who are searching for ways to profit from it politically. The Chinese government, having cracked down on online pornography in early 2009, quickly lifted many of their bans, perhaps after realizing that censorship was a sure way to politicize millions of Chinese Internet users. Michael Anti, a Beijing-based expert on the Chinese Internet, believes this was a strategic move: "[The government must have reasoned that] if Internet users have some porn to look at, then they won't pay so much attention to political matters."

It seems highly naïve to assume that political ideals—let alone dissent—will somehow emerge from this great hodgepodge of consumerism, entertainment, and sex. As tempting as it is to think of Internet-based swinger clubs that have popped up in China in the last few years as some kind of alternative civil society, it's quite possible that, since the main ideological tenets of Chairman Mao's thought have lost much of their intellectual allure, the Chinese Communist Party would find the space to accommodate such practices. Under the pressure of globalization, authoritarianism has become extremely accommodating.

Other governments, too, are beginning to understand that online entertainment—especially spiced up with pornography—can serve as a great distraction from politics. According to reports from the official Vietnam state news agency, officials in Hanoi were flirting with the idea of setting up "an orthodox sex Website"—replete with videos—that could help couples learn more about "healthy sexual intercourse." This won't be a surprise to most Vietnamese: Much of existing Internet censorship in the country targets political resources, while leaving many pornographic sites unblocked. As Bill Hayton, a former BBC reporter in the country observes, "the Vietnamese firewall allows youngsters to consume plenty of porn but not Amnesty International reports." As online porn becomes ubiquitous, such restrictions may no longer be needed.

Unless the West stops glorifying those living in authoritarian governments, it risks falling under the false impression that if it builds enough tools to break through the barriers erected by authoritarian governments, citizens will inevitably turn into cheap clones of Andrei Sakharov and Václav Havel and rebel against repressive rule. This scenario seems highly dubious. Most likely, those very citizens would first get online to download porn, and it's not at all clear if they would return for political content. One experiment in 2007 involved Good Samaritans in the West volunteering to lend their computer bandwidth, via a tool called Psiphon, to strangers in countries that control the Internet, in the hope that, once they got their first taste of unfettered online freedom, they would use that chance to educate themselves about the horrors of their regimes. The reality was more disappointing. As *Forbes* magazine described it, once liberated, the users searched for "nude pictures of Gwen Stefani and photos of a panty-less Britney Spears." Freedom to browse whatever one wants is, of course, worth defending in its own right, but it's important to remember that, at least from a policy

perspective, such freedoms would not necessarily bring about the revolutionary democratic outcomes that many in the West expect.

Online Discontents and Their Content Intellectuals

Phillip Roth's 1990 warning to the Czechs was also a perceptive observation that their most treasured public intellectuals—those who helped
to bring democracy to the country—would soon no longer command
the power or respect they had under communism. It was inevitable that
dissident intellectuals would lose much of their appeal as the Internet
opened the gates of entertainment while globalization opened the gates
of consumerism. Another Sakharov seems inconceivable in today's Russia, and in the unlikely event that he does appear, he would probably
enjoy far less influence on Russian national discourse than Artemy
Lebedev, Russia's most popular blogger, who uses his blog to run
weekly photo competitions to find a woman with the most beautiful
breasts (the subject of breasts, one must note, is far more popular in
the Russian blogosphere than that of democracy).

But intellectuals are not blameless here either. As democracy replaced communism, many of them were bitterly disappointed by the
populist, xenophobic, and vulgar politics favored by the masses. Despite the widespread myth that Soviet dissidents were all believers in
U.S.-style democracy, many of them—including, at some point, even
Sakharov—felt extremely ambivalent about letting people rule themselves; what many of them really wanted was better-run communism.
But the triumph of liberal democracy and the consumerism that it unleashed sent many of these intellectuals into the second, perhaps somewhat less repressive, phase of their internal exile, this time combined
with despicable obscurity.

It would take a new generation of intellectuals—and unusually creative intellectuals at that—to awaken the captive minds of their fellow
citizens from their current entertainment slumber. As it turns out, there
is not much demand for intellectuals when so many social and cultural
needs can be satisfied the same way they are satisfied in the West: with
an iPad. (It helps that China knows how to manufacture them at half

the price!) The Belarusian writer Svetlana Aleksievich knows that the game is over, at least as far as serious ideas are concerned: "The point is not that we have no Havel, we do, but that they are not called for by society." And the Belarusian government, not surprisingly, doesn't seem to object to this state of affairs. On a recent trip to Belarus I discovered that some Internet Service Providers run their own servers full of illegal movies and music, available to their customers for free, while the government, which could easily put an end to such practices, prefers to look the other way and may even be encouraging them.

Consumerism is not the only reason behind the growing disengagement between the intellectuals and the masses living in authoritarian states. The Internet opened up a trove of resources for the former, allowing them to connect to their Western colleagues and follow global intellectual debates as they happen, not as their summaries are smuggled in on yellowish photocopies. But efficiency and comfort—which the Internet provides—are not necessarily the best conditions for fomenting dissent among the educated classes. The real reason why so many scientists and academics turned to dissent during Soviet times was because they were not allowed to practice the kind of science they wanted to on their own terms. Doing any kind of research in the social sciences was quite difficult even without having to follow the ideological line of the local communist cell; collaborating with foreigners was equally challenging. Lack of proper working conditions forced many academics and intellectuals either to immigrate or to stay home and become dissidents.

The Internet has solved or alleviated many of these problems, and it has proved excellent for research, but not so excellent for bringing smart and highly educated people into the dissident movement. Collaboration is now cheap and instantaneous, academics have access to more papers than they could have dreamed of, travel bans have been lifted, and research budgets have been significantly increased. Not surprisingly, by 2020 Chinese scientists are expected to produce more academic papers than American ones. Most significantly, the Internet has allowed better integration of academics and intellectuals from authoritarian states into a global intellectual sphere—they, too, can now follow debates in the *New York Review of Books*—but this has happened at the expense of severing their ties to local communities. Russian liberal intellectuals draw far larger crowds in New York, London, or Berlin than they do in Moscow, Novosibirsk, or Vladivostok, where many of them remain unknown. Not surprisingly, most of them are better informed about what's going on in Greenwich Village than in their own town hall. But their connection to politics in their native countries has also been severed; paradoxically, as they have gotten more venues to express their anger and dissent, they have chosen to retract into the nonpolitical.

It's rather depressing that none of the major Russian writers who have established a rather active presence online bothered to discuss or even mention the results of the 2008 Russian presidential elections on their blogs. Ellen Rutten, at the University of Cambridge, was the first to notice and describe the virtually nonexistent reaction to such a highly political event. She wrote that "none of the . . . [blogging] authors . . . chose to switch on the computer and react in writing to the news that must have permeated their intellectual environment." Instead, the giants of modern Russian literature decided to devote their first blog posts after the election to: (a) discussing a recent Internet conference, (b) posting a theater review, (c) describing a gigantic pie with "little cherries and peaches" spotted at a recent book fair, (d) reviewing Walt Whitman, and (e) posting a story about a man with two brains. (One could only hope that at least that last entry was an allegory meant to ridicule the Putin-Medvedev alliance.) This is definitely not what the famous Russian poet Yevgeny Yevtushenko meant when he proclaimed that "A poet in Russia is more than a poet."

This is hardly a promising environment for fighting the authoritarian chimera. All potential revolutionaries seem to be in a pleasant intellectual exile somewhere in California. The masses have been transported to Hollywood by means of pirated films they download from BitTorrent, while the elites have been shuttling between Palo Alto and Long Beach by way of TED talks. Whom exactly do we expect to lead this digital revolution? The lolcats?

If anything, the Internet makes it harder, not easier, to get people to care, if only because the alternatives to political action are so much more pleasant and risk-free. This doesn't mean that we in the West should stop promoting unfettered (read: uncensored) access to the Internet; rather, we need to find ways to supplant our promotion of a freer Internet with strategies that can engage people in political and social life. Here we should talk to both heavy consumers of cat videos and those who follow anthropology blogs. Otherwise, we may end up with an army of people who are free to connect, but all they want to connect to is potential lovers, pornography, and celebrity gossip.

The environment of information abundance is not by itself conducive to democratization, as it may disrupt a number of subtle but important relationships that help to nurture critical thinking. It's only now, as even democratic societies are navigating through this new environment of infinite content, that we realize that democracy is a much trickier, fragile, and demanding beast than we had previously assumed and that some of the conditions that enabled it may have been highly specific to an epoch when information was scarce.

The Orwell-Huxley Sandwich Has Expired

As the East German experience revealed, many Western observers like endowing those living under authoritarian conditions with magical and heroic qualities they do not have. Perhaps imagining these poor folks in a perpetual struggle against the all-seeing KGB rather than, say, relaxing in front of YouTube or playing Tetris is the only way for Western observers not to despair at their own inability to do much about the situation. Nevertheless, that this is how they choose to interpret the nature of political control under authoritarianism is not an accident. Much of Western thinking on this issue has been heavily influenced—perhaps even constrained—by two twentieth-century thinkers who spent decades thinking about the diffusion of power and control under democracy, communism, and fascism. George Orwell (1903–1950) and Aldous Huxley (1894–1963), both men of letters who managed to leave indelible marks on the world of modern political thought, have each offered us powerful and yet strikingly different visions for how modern governments would exercise control over their populations (those visions haunt millions of high school students who are to this day tasked with writing essays comparing the two). The presence of these two figures in modern public life is hard to miss: A day hardly goes by when someone in the media doesn't invoke either man to make a point about the future of democracy or the history of totalitarianism, and it's quite common to invoke both, as if one could fit any possible kind of political control in the spectrum between those two polar ends. Thus a shrewd Western politician would profess admiration of both (cue Hillary Clinton, who, when asked about books that influenced her the most, mentioned both Orwell's *1984* and Huxley's *Brave New World* in one breath).

Orwell's *1984* (1949), his most famous work and certainly one of the best novels of the twentieth century, emphasizes pervasive surveillance and mind-numbing propaganda composed in the meaningless vocabulary of "Newspeak." In Orwell's world, citizens are not entitled to any privacy; hence they treasure junk and scraps of paper, as those lie outside of the sphere controlled by the government and remind them of a much different past. Even their television sets are used to monitor their behavior. Winston Smith, the protagonist, is warned that neurologists are working to extinguish the orgasm, as full devotion to the Party requires the complete suppression of the libido.

Huxley's vision was articulated in *Brave New World* (1932) and a short later essay called *Brave New World Revisited* (1958). In Huxley's world, science and technology are put to good use to maximize pleasure, minimize the time one spends alone, and provide for a 24/7 cycle of consumption (one of the regime's slogans is "ending is better than mending!"). Not surprisingly, the citizens lose any ability to think critically and become complacent with whatever is imposed on them from above. Sexual promiscuity is encouraged from early childhood, even though sex is considered a social activity rather than the act of reproduction. The idea of a family is considered "pornographic," while social relations are organized around the maxim "everyone belongs to everyone else."

The two men knew each other and corresponded. Orwell, the younger of the two, even briefly studied French under Huxley's tutelage at Oxford. In 1940 Orwell wrote a provocative review of Huxley's book, and Huxley revisited both his own work and *1984* in his *Brave New World Revisited*. Orwell thought that while Huxley provided "a good caricature of the hedonistic Utopia," he misunderstood the nature of power in a modern totalitarian state. "[*Brave New World* was] . . . the kind of thing that seemed possible and even imminent before Hitler appeared, but it had no relation to the actual future. What we are moving towards at this moment is something more like the Spanish Inquisition, and probably far worse, thanks to the radio and the secret police," wrote Orwell in a 1940 essay.

Huxley, however, wasn't convinced. In a 1949 letter to Orwell, he expressed his doubts about the social controls described in *1984*: "The philosophy of the ruling minority in *1984* is sadism which has been carried to its logical conclusion by going beyond sex and denying it. Whether in actual fact the policy of the boot-on-the-face can go on indefinitely seems doubtful." He continued: "My own belief is that the ruling oligarchy will find less arduous and wasteful ways of governing and of satisfying its lust for power, and that these ways will resemble those which I described in *Brave New World*."

Unlike Orwell, Huxley wasn't convinced that men were rational creatures who were always acting in their best interest. As he put it in *Brave New World Revisited*, what was often missing from the social analysis of Orwell and other civil libertarians was any awareness of "man's almost infinite appetite for distractions." Huxley was being unfair to Orwell, however. Orwell did not entirely discount the power of distraction: The Proles, the lowest class in *1984*'s three-class social hierarchy, are kept at bay with the help of cheap beer, pornography, and even a national lottery. Still, it was readers' fear of the omnipotent and all-seeing figure of Big Brother that helped to make Orwell's arguments famous.

Ever since the fall of the Soviet Union, it has been something of a cliché to claim that Orwell failed to anticipate the rise of the consumer society and how closely technology would come to fulfill its desires. Huxley, too, was chided for underestimating the power of human agency to create spaces of dissent even within consumerist and hedonistic lifestyles, but it is widely assumed that he was the most prescient of the two (particularly on the subject of genetics). "Brave New World is a far shrewder guess at the likely shape of a future tyranny than Orwell's vision of Stalinist terror in Nineteen Eighty-Four. . . . Nineteen Eighty-Four has never really arrived, but Brave New World is around us everywhere," wrote the British dystopian novelist J. G. Ballard in reviewing a Huxley biography for the *Guardian* in 2002. "Orwell feared that what we hate will ruin us. Huxley feared that what we love will ruin us. This book is about the possibility that Huxley, not Orwell, was right," is how Neil Postman chose to describe the theme of his best-selling *Amusing Ourselves to Death*. "[In contrast to *Brave New World*], the political predictions of . . . *1984* were entirely wrong," writes Francis Fukuyama in *Our Posthuman Future*. Maybe, but what many critics often fail to grasp is that both texts were written as sharp social critiques of contemporary problems rather than prophecies of the future.

Orwell's work was an attack on Stalinism and the stifling practices of the British censors, while Huxley's was an attack on the then-popular philosophy of utilitarianism. In other words, those books probably tell us more about the intellectual debates that were prevalent in Britain at the time than about the authors' visions of the future. In any case, both works have earned prominent places in the pantheon of twentieth-century literature, albeit in different sections. It's in criticizing contemporary democratic societies—with their cult of entertainment, sex, and advertising—that *Brave New World* succeeded most brilliantly. Orwell's *1984*, on the other hand, is to this day seen as a guide to understanding modern authoritarianism, with its pervasive surveillance, thought control through propaganda, and brutal censorship. Both Huxley's and Orwell's books have been pigeonholed to serve a particular political purpose: one to attack the foundations of modern capitalism, the other the basis of modern authoritarianism. Huxley, offspring of a prominent British family, was concerned with the increased commercialization of life in the West (he found his eventual solace in hallucinogenic drugs, penning *The Doors of Perception*, a book that later inspired Jim Morrison to name his rock band The Doors). Orwell, a committed socialist, emerged as a favorite thinker of the Ronald Reagan right; he was "the patron saint of the Cold War," as the writer Michael Scammell dubbed him. (The

Committee for the Free World, the leading neoconservative outfit of the 1980s, even called its publishing unit "the Orwell Press.")

But two decades after the fall of the Soviet Union, the dichotomy between Orwell and Huxley's visions for the nature of political control seems outdated, if not false. It, too, is a product of the Cold War era and the propensity to engage in one-sided characterization of that ideological conflict by its participants. In reality, there was plenty of Orwellian surveillance in McCarthy-era America while there was plenty of hedonistic entertainment in Khrushchev-era USSR. The very existence of such a mental coordinate system with Orwell and Huxley at its opposite ends dictates its own extremely misleading dynamic: One can't be at both of its ends at once. To assume that all political regimes can be mapped somewhere on an Orwell-Huxley spectrum is an open invitation to simplification; to assume that a government would be choosing between reading their citizens' mail or feeding them with cheap entertainment is to lose sight of the possibility that a smart regime may be doing both.

Mash 'Em Up!

To borrow a few buzzwords from today's Internet culture, it's time to mash up and remix the two visions. To understand modern authoritarianism (and, some would argue, modern capitalism as well), we need insights from both thinkers. The rigidity of thought suggested by the Orwell-Huxley coordinate system leads many an otherwise shrewd observer to overlook the Huxleyan elements in dictatorships and, as disturbingly, the Orwellian elements in democracies. This is why it has become so easy to miss the fact that, as the writer Naomi Klein puts it, "China is becoming more like [the West] in very visible ways (Starbucks, Hooters, cellphones that are cooler than ours), and [the West is] becoming more like China in less visible ones (torture, warrantless wiretapping, indefinite detention, though not nearly on the Chinese scale)."

It seems fairly noncontroversial that most modern dictators would prefer a Huxleyan world to an Orwellian one, if only because controlling people through entertainment is cheaper and doesn't involve as much brutality. When the extremely restrictive Burmese government permits—and sometimes even funds—hip-hop performances around the country, it's not *1984* that inspires them.

With a few clearly sadistic exceptions, dictators are not in it for the blood; all they want is money and power. If they can have it simply by distracting—rather than spying on, censoring, and arresting—their people, all the better. In the long term, this strategy is far more effective than 24/7 policing, because policing, as effective as it might be in the short term, tends to politicize people and drive them toward dissent in the longer run. That Big Brother no longer has to be watching its citizens because they themselves are watching *Big Brother* on TV hardly bodes well for the democratic revolution.

Thus, as far as distraction is concerned, the Internet has boosted the power of the Huxley-inspired dictatorships. YouTube and Facebook, with their bottomless reservoirs of cheap entertainment, allow individuals to customize the experience to suit their tastes. When Philip Roth was warning the Czechs of the perils of commercial television, he was also suggesting that it could make a revolution like the one in 1989 impossible. Ironically, the Czechs had been lucky to have such hapless apparatchiks running the entertainment industry. People got bored easily and turned to politics instead. Where new media and the Internet truly excel is in suppressing boredom. Previously, boredom was one of the few truly effective ways to politicize the population denied release valves for channeling their discontent, but this is no longer the case. In a sense, the Internet has made the entertainment experiences of those living under authoritarianism and those living in a democracy much alike. Today's Czechs watch the same Hollywood movies as today's Belarusians—many probably even download them from the same illegally run servers somewhere in Serbia or Ukraine. The only difference is that the Czechs already had a democratic revolution, the results of which, luckily for them, were made irreversible when the Czech Republic joined the European Union. Meanwhile, the Belarusians were not as lucky; the prospects of their democratic revolution in the age of YouTube look very grim.

In other words, the Internet has brought the kind of creative entertainment that Roth was warning against into most closed societies without breaking down their authoritarian governance. Besides, YouTube entertainment is free of charge—unless dictators make secret donations to Hollywood—so the money saved on producing boring state entertainment can now be diverted to other budget lines.

That Internet freedom has taken on such a democracy-squashing quality does not mean that dictators were planning this all along; in most cases, it's simply the result of earlier authoritarian incompetence. Would dictators ever have allowed YouTube in their countries if anyone had asked them? Probably not. They don't always grasp the strategic value of distraction, overestimating the risks of people-led protest. By their sheer haplessness or misjudgment, they did let the Internet in, but instead of blogs ridiculing government propaganda, it's the goofy websites like lolcats that their youth are most interested in. (Rest assured: Soon enough, some think-tank report will announce that the age of "feline authoritarianism" is upon us.) Those of us concerned about the future of democracy around the globe must stop dreaming and face reality: The Internet has provided so many cheap and easily available entertainment fixes to those living under authoritarianism that it has become considerably harder to get people to care about politics at all. The Huxleyan dimension of authoritarian control has mostly been lost on policymakers and commentators, who, thanks to the influence of such critics of modern capitalism as Herbert Marcuse and Theodor Adorno, are mostly accustomed to noticing it only in their own democratic societies. Such bland glorification of those living under authoritarianism will inevitably lead to bad policies. If the ultimate Western objective is inciting a revolution or at least raising the level of political debate, the truth is that providing people with tools to circumvent censorship will be nearly as effective as giving someone with no appreciation of modern art a one-year pass to a museum. In 99 percent of cases, it's not going to work. This is not an argument against museums or anti-censorship tools; it's simply a call to use strategies that are free of utopian visions.

The Trinity of Authoritarianism

Granted, this "control by entertainment" approach is not going to work for everyone in authoritarian societies; some people already have so many grudges against their governments that flooding them with entertainment would not change their minds. In addition, Western governments and foundations will always find ways to politicize the locals from the outside, even if it involves fueling ethnic or religious tensions, a foolproof way to spark hatred in the age of YouTube. Thus, if only to maintain power over those who have preserved the ability to think for themselves, some Orwellian elements of political control will need to be present. Despite the reductionist models that have made many in the West believe that information can destroy authoritarianism, information also plays an instrumental role in enabling propaganda, censorship, and surveillance, the three main pillars of Orwell-style authoritarian control.

The Internet hasn't changed the composition of this "trinity of authoritarianism," but it has brought significant changes to how each of these three activities is practiced. The decentralized nature of the Internet may have made comprehensive censorship much harder, but it may have also made propaganda more effective, as government messages can now be spread through undercover government-run blogs. The opportunity to cheaply encrypt their online communications may have made "professional" activists more secure, but the proliferation of Web 2.0 services—and especially social networking—has turned "amateur" activists into easier targets for surveillance.

While there is nothing we in the West can do about the growing appeal of YouTube and lolcats—online entertainment is poised to remain an important, if indirect, weapon in the authoritarian arsenal—it's possible to do something about each of those three authoritarian pillars. The danger here, of course, is that Western leaders might, once again, frame the solutions in intimately familiar Cold War terms, where the default response to the censorship practices of the Soviet Union was to blast even more information through the likes of Radio Free Europe.

This is an urge that needs to be resisted. The strategy behind the existence of Radio Free Europe and other similar broadcasting services during the Cold War was relatively straightforward. By funding more radio broadcasts, Western policymakers wanted to ensure that authoritarian propaganda would be countered, if not weakened; the draconian system of censorship would be undermined; and more listeners would doubt the central premises of communism as a result.

With technologies like the radio, it was relatively easy to grasp how certain inputs produced certain outputs. Thus, when the Soviet authorities were jamming its radio stations, the West reacted by turning up the volume—in part because, being in charge of all the programming, it was confident that exposure to its broadcasts would have the desired effect of politicizing the masses. The Soviets couldn't just take the Western radio signal and use it to fight back, nor could listeners avoid political programming and opt for entertainment only (as already stated, not all Western radio programs were political, but even lifestyle shows were usually aimed at revealing the moral bankruptcy of the Soviet system).

There is no such certainty about the Internet; the West has far less command over it as an instrument than it did in the case of radio. The Internet is a much more capricious technology, producing side effects that can weaken the propaganda system but enhance the power of the surveillance apparatus or, alternatively, that can help to evade censorship but only at the expense of making the public more susceptible to propaganda. The Internet has made the three information pillars of authoritarianism much more interconnected, so Western efforts to undermine one pillar might ruin its efforts to do something about the other two.

Take one example: While it is tempting to encourage everyone to flock to social networking sites and blogs to avoid the control of the censors, it would also play into the hands of those in charge of surveillance and propaganda. The more connections between activists it can identify, the better for the government, while the more trust users have in blogs and social networks, the easier it is to use those networks to

promote carefully disguised government messages and boost the propaganda apparatus. The only way to avoid making painful mistakes and
strengthening authoritarianism in the process of promoting Internet
freedom is to carefully examine how surveillance, censorship, and propaganda strategies have changed in the Internet era.

chapter four

Censors and Sensibilities

• • ■ ■

Western propaganda produced during the Cold War may not have been very convincing, but it was effective on at least one level: It cultivated a certain myth of authoritarianism that is hard to dispel a full decade into the twenty-first century. Many Western observers still imagine authoritarian states to be populated by hyperactive doubles of Arthur Koestler—smart, uncompromising in the face of terror, eager to take on existential risks in the name of freedom—while being governed by an intractable array of ridiculous Disney characters—stupid, distracted, utterly uninterested in their own survival, and constantly on the verge of group suicide. Struggle and opposition are the default conditions of the former; passivity and incompetence are the default condition of the latter. All it takes to change the world, then, is to link the rebels with each other, expose them to a stream of shocking statistics they have never seen, and hand out a few shiny gadgets. Bingo! A revolution is already on its way, for perpetual revolt, according to this view, is the natural condition of authoritarianism.

This highly stylized account of modern authoritarianism tells us more about Western biases than it does about modern authoritarian regimes. The persistence of modern authoritarianism can be explained by a whole variety of factors—energy endowment, little or no previous experience with democratic forms of rule, covert support from immoral Western democracies, bad neighbors—but an uninformed citizenry that cries out to be liberated by an electronic bombing of factoids and punchy tweets is typically not one of them. Most citizens of modern-day Russia or China do not go to bed reading *Darkness at Noon* only to wake up to the jingle of Voice of America or Radio Free Europe; chances are that much like their Western counterparts, they, too, wake up to the same annoying Lady Gaga song blasting from their iPhones. While they might have a strong preference for democracy, many of them take it to mean orderly justice rather than the presence of free elections and other institutions that are commonly associated with the Western model of liberal democracy. For many of them, being able to vote is not as valuable as being able to receive education or medical care without having to bribe a dozen greedy officials. Furthermore, citizens of authoritarian states do not necessarily perceive their undemocratically installed governments to be illegitimate, for legitimacy can be derived from things other than elections; jingoist nationalism (China), fear of a foreign invasion (Iran), fast rates of economic development (Russia), low corruption (Belarus), and efficiency of government services (Singapore) have all been successfully co-opted for these purposes.

Thus, to understand the impact of the Internet on authoritarianism, one needs to look beyond the Web's obvious uses by opponents of the government and study how it has affected legitimacy-boosting aspects of modern authoritarian rule as well. Take a closer look at the blogospheres in almost any authoritarian regime, and you are likely to discover that they are teeming with nationalism and xenophobia, sometimes so poisonous that official government policy looks cosmopolitan in comparison. What impact such radicalization of nationalist opinion would have on the governments' legitimacy is hard to predict, but things don't look particularly bright for the kind of flawless democratization that some expect from the Internet's arrival. Likewise, bloggers uncovering and publicizing corruption in local governments could be—and are—easily co-opted by higher-ranking politicians and made part of the anti-corruption campaign. The overall impact on the strength of the regime in this case is hard to determine; the bloggers may be diminishing the power of local authorities but boosting the power of the federal government. Without first developing a clear understanding of how power is distributed between the center and the periphery and how changes in this distribution affect the process of democratization, it is hard to predict what role the Internet might play.

Or look at how Wikis and social networking sites, not to mention various government online initiatives, are improving the performance of both governments and businesses they patronize. Today's authoritarian leaders, obsessed with plans to modernize their economies, spew out more buzzwords per sentence than an average editorial in *Harvard Business Review* (Vladislav Surkov, one of the Kremlin's leading ideologues and the godfather of Russia's Silicon Valley, has recently confessed that he is fascinated by "crowdsourcing"). Authoritarian regimes in Central Asia, for example, have been actively promoting a host of e-government initiatives. But the reason why they pursue such modernization is not because they want to shorten the distance between the citizen and the bureaucrat but because they see it as a way to attract funds from foreign donors (the likes of IMF and the World Bank) while also removing the unnecessary red-tape barriers to economic growth.

Dress Your Own Windows

Authoritarian survival increasingly involves power sharing and institution building, two processes that many political scientists have traditionally neglected. Even such shrewd observers of modern politics as Zbigniew Brzezinski and Carl Friedrich told readers of their 1965 classic, *Totalitarian Dictatorship and Autocracy*, to forget institutions altogether: "The reader may wonder why we do not discuss the 'structure of government,' or perhaps 'the constitution' of these totalitarian systems. The reason is that these structures are of very little importance."

Such rigid conceptual frameworks may have helped in understanding Stalinism, but this is too simplistic of a perspective to explain much of what is going on inside today's authoritarian states, which are busy

organizing elections, setting up parliaments, and propping up their ju-
diciaries. If authoritarian regimes are bold enough to allow elections,
for reasons of their own, what makes us think they wouldn't also allow
blogs for reasons that Western analysts may not be able to understand
yet?

"Institutions, students of authoritarianism often claim, are but 'window-
dressing,'" writes Adam Przeworski, a professor of political science at
New York University. "But why would some autocrats care to dress their
own windows?" Why, indeed? In the last thirty years, political scientists
have unearthed plenty of possible motivations: Some rulers want to
identify the most talented bureaucrats by having them compete in sham
elections; some want to co-opt their potential enemies by offering them
a stake in the survival of the regime and placing them in impotent parliaments and other feeble, quasi-representative institutions; some simply want to talk the democracy talk that helps them raise funds from
the West, and institutions—especially if those are easily recognizable
institutions commonly associated with liberal democracy—are all the
West usually asks for.

But it seems that the most innovative dictators not only organize
sham elections but also manage to surround themselves with the gloss
of modern technology. How else to explain a 2009 parliamentary election in the former Soviet republic of Azerbaijan, where the government
decided to install five hundred Web cams at election stations? It made
for good PR, but it didn't make the elections any more democratic, for
most manipulations had occurred before the election campaign even
started. And such a move may have had more sinister implications. As
Bashir Suleymanly, executive director of Azerbaijan's Elections Monitoring and Democracy Teaching Center, told reporters on the eve of
the election, "local executive bodies and organizations that are financed
from the state budget instruct their employees for whom they should
give their vote and frighten them by the webcams that record their participation and whom they vote for." Russian authorities, too, believe
that the kind of transparency fostered by the Web cams may bolster
their democratic credentials. After devastating summer fires destroyed
many villages in the heat wave of 2010, the Kremlin installed Web cams

at the building sites for new houses, so that the process could be observed in real time (that didn't stop complaints from the future owners of the houses; living in the provinces, they didn't have computers, nor did they know how to use the Internet).

Institutions matter, and dictators love building them, if only to prolong their stay in power. The relative usefulness of the Internet—especially the blogosphere—has to be analyzed through a similar institutional lens. Some bloggers are simply too useful to get rid of. Many of them are talented, creative, and extremely well-educated individuals—and only short-sighted dictators would choose to fight them, when they can be used more strategically instead. For example, it is much more useful to build an environment where these bloggers can serve as symbolic tokens of "liberalization," packaged for either domestic or foreign consumption, or at least where they can be counted on to help generate new ideas and ideologies for otherwise intellectually starving governments.

Not surprisingly, efforts to institutionalize blogging have already begun in many authoritarian states. Officials in some Gulf states are calling for the creation of blogging associations, while one of Russia's top bureaucrats recently proposed to set up a "Bloggers' Chamber" that can set standards of acceptable behavior in the blogosphere, so that the Kremlin does not have to resort to formal censorship. In reality, of course, such blogging chambers are likely to be staffed by pro-Kremlin bloggers, which is yet another way to hide the fact that the Russian government manages to practice Internet control without formally banning all that many websites. Such efforts may fail—and the West can only hope that they do—but they suggest that authoritarian governments have an operational view of blogging that is light-years ahead of the idea of bloggers as twenty-first-century dissidents.

If we view all Internet activity in authoritarian states as being primarily political and oppositional, we are likely to miss much of what makes it so rich and diverse. While the Western media pay a lot of attention to how China's "human flesh search engines"—people who name and shame misbehaving public officials and other Internet users by publicizing their personal details—are making the Chinese government more accountable, they rarely report that the Chinese government, too, has found ways to co-opt these same "search engines" to score propaganda points. When in March 2010 an Internet user from the Chinese city of Changzhou complained about pollution in Beitang River and accused the chief of the local environmental protection bureau of failing to do his job, demanding his resignation, the local administration mobilized the local "human flesh search engines" to track down the complainer, so that he could be rewarded with 2,000 yuan.

One of the temptations that Western observers should avoid is to interpret the fact that authoritarian governments are adjusting their operating methods as a sign of democratization. This is a common fallacy among those who do not yet understand that it is perpetual change, not stagnation, that has enabled authoritarianism to survive for so long. A modern authoritarian state is much like the Ship of Theseus in Greek mythology: It's been rebuilt so many times that even those navigating it are no longer sure if any of the original wood remains.

Although prominent Western blogger-academics like *Instapundit*'s Glenn Reynolds laud the power of mobile phones and argue that "converting an unresponsive and murderous Stalinist/Maoist tyranny into something that responds to cellphone calls is not an achievement to be sneezed at," we should not just pat ourselves on the back, clap hands, and praise the inexorable march of Internet freedom. A tyranny that responds to cellphone calls is still a tyranny, and its leaders may even enjoy fiddling with their iPhone apps. Nor should we automatically assume that tyrannies do not want to respond to cellphone calls. The supposed gains of "democratization" may look considerably less impressive if they are seen as indirectly facilitating the survival of dictatorships, even if in slightly modified form.

The Kremlin Likes Blogs and So Should You

Contrary to the usual Western stereotypes, modern dictators are not just a loose bunch of utterly confused loonies lounging around in their information-resistant bunkers, counting their riches, Scrooge McDuck-style, and waiting to get deposed, oblivious to what is happening outside. Quite the opposite: They are usually active consumers and producers of information. In fact, finding ways to understand and gather information—especially about threats to the regime—is one invariable feature of authoritarian survival. But dictators can't just go and interview random people in the streets; they almost always have to go to intermediaries, usually the secret police.

This, however, rarely gives an accurate view of what's happening, if only because nobody wants to take responsibility for the inevitable malfunctions of the authoritarian system. That's why throughout history rulers always tried to diversify their news sources. In fact, Mahmoud Ahmadinejad's Internet strategy has a rich intellectual tradition to draw on. Back in the nineteenth century, Iran's monarch Nasi al-Din Shah was enthusiastically installing telegraph lines throughout the country, requiring *daily* reports even from the most minor bureaucrats in the tiniest of villages, primarily as a means of cross-checking reports received from their higher-ups. This was in line with the advice offered by Iran's eleventh-century vizier Nizam al-Mulk in his celebrated *Book of Government*: Each king should have dual sources of information.

The noted social scientist Ithiel de Sola Pool, one of the leading thinkers about technology and democracy of the last century, played an important role in shaping how the West understands the role that information plays in authoritarian states. "The authoritarian state is inherently fragile and will quickly collapse if information flows freely," wrote Pool, giving rise to a view that has become widely shared—and, undoubtedly, made Pool and his numerous followers overestimate the liberating power of information. (Pool, a disillusioned ex-Trotskyite, also famously overestimated the power of Western broadcasting, using letters that Eastern Europeans sent to Radio Free Europe as one of his main sources.) Such technological utopianism stems from a rather shallow reading of the politics and regime dynamics of authoritarian states. For if one presumes, like Pool, that authoritarian structures rest on little else than the suppression of information, as soon as the West finds a way to poke holes in those structures, it follows that democracy promotion boils down to finding ways to unleash the information flood on the oppressed.

On closer examination, views like Pool's appear counterintuitive and for good reason. Surely there are benefits to having access to more sources of information, if only because a regime can flag emerging threats. (On this point, Iranian rulers of the past were a bit more sophisticated than many contemporary Western academics.) That diverse and independent information can help heighten—or at least preserve—their power has not been lost on those presiding over authoritarian states. One insightful observer of the final years of the Soviet era remarked in 1987: "There surely must be days—maybe the morning after Chernobyl—when Gorbachev wishes he could buy a Kremlin equivalent of the *Washington Post* and find out what is going on in his socialist wonderland." (Gorbachev did acknowledge that Western radio broadcasts were instrumental in helping him follow the short-lived putsch in August 1991, when he was locked up in his Sochi dacha.)

Well, there is no need to hunt for the Russian equivalents of the *Washington Post* anymore. Even in the absence of a truly free press, Dmitry Medvedev can learn almost everything he needs from the diverse world of Russian blogs. As he himself has confessed, this is how he starts many of his mornings. (Medvedev is also a big fan of ebooks and the iPad.) And he doesn't have to spend much time searching for complaints. Anyone with a grudge against a local bureaucrat can leave a complaint as a comment on Medvedev's blog, a popular practice in Russia. And to score some bonus propaganda points, Medvedev's subordinates like to take highly publicized action in response to such complaints, replacing the crumbling infrastructure and firing the corrupt bureaucrats. This, however, is done selectively, more for the propaganda value it creates than for the purpose of fixing the system. No one knows what happens to the complaints that are too critical or border on whistle-blowing, but quite a few angry messages are removed from the blog very quickly. (Vladimir Putin, Medvedev's predecessor as president and currently Russia's prime minister, also likes to collect complaints by having people call in to his yearly TV address; when in 2007 a police officer told the switchboard operator he wanted to complain about corruption in his unit, his call was traced, and he was reprimanded.) Similarly, while the Chinese authorities are blocking openly antigovernment content, they appear quite tolerant of blog posts that expose local corruption.

Authorities in Singapore regularly monitor blogs that provide policy criticism and claim to incorporate suggestions from netizens into their policymaking. Thus, while the themes of many blogs in modern-day authoritarian regimes are clearly not to the governments' taste, there are plenty of others they approve of or even try to promote.

Dictators and Their Dilemmas

While it's becoming clear that few authoritarian regimes are interested in completely shutting down all communications, if only because they want to stay abreast of emerging threats, censorship of at least some content is inevitable. For the last three decades, conventional wisdom suggested that the need to censor put authoritarian regimes into a corner: They either censored and thus suffered the economic consequences, for censorship is incompatible with globalization, or they didn't censor and thus risked a revolution. Hillary Clinton said as much in her Internet freedom speech: "Countries that censor news and information must recognize that from an economic standpoint, there is no distinction between censoring political speech and commercial speech. If businesses in your nations are denied access to either type of information, it will inevitably impact on growth." Reporting on the role of technology in powering Iran's Twitter Revolution, the *New York Times* expressed a similar opinion: "Because digital technologies are so critical today to modern economies, repressive governments would pay a high price for shutting them out completely, if that were still possible."

This binary view—that dictators cannot globalize unless they open up their networks to hordes of international consultants and investment bankers scouring their lands in search of the next acquisition target—has become known as "dictator's dilemma" and found many supporters among policymakers, especially when the latter discuss the benevolent role of the Internet. But the existence of a direct connection between economic growth and modern-day Internet censorship is not self-evident. Could it be yet another poorly examined and rather harmful assumption that stems from the Cold War?

In 1985 George Schultz, the then U.S. secretary of state, was one of the first to articulate the popular view when he said that "totalitarian societies face a dilemma: either they try to stifle these technologies and thereby fall further behind in the new industrial revolution, or else they permit these technologies and see their totalitarian control inevitably eroded." And those governments were doomed, according to Schultz: "They do not have a choice, because they will never be able entirely to block the tide of technological advance." Schultz's view, expressed in a high-profile article in *Foreign Affairs*, gained a lot of supporters. A 1989 editorial in the *New Republic* just a week after the Chinese government cleared the protestors out of Tiananmen Square argued that the choice facing the dictators was either to "let the people think for themselves and speak their minds . . . —or smell your economy rot."

This was music to the ears of many Eastern Europeans at the time, and the ensuing collapse of the Soviet system seemed to vindicate the *New Republic*'s determinism. In fact, such predictions were the intellectual product of the optimism of that era. Anyone following the zeitgeist of the late 1980s and early 1990s couldn't have missed the connection between two popular theories at the time, one pertaining to technology and one to politics, that, in a rather mysterious twist, bore virtually the same name. One theory, developed by the futurist Alvin Toffler, posited that the rapid technological change of the period would give rise to the "Third Wave Society," marked by democratized access to knowledge and the dawn of the information age. For Toffler, information technology followed two other revolutionary waves, agriculture and industrialization, ushering in a completely new period in human history.

The second theory, developed by the Harvard political scientist Samuel Huntington, posited that the period was marked by the emergence of "the third wave" of worldwide democratization, with more and more countries choosing democratic forms of governance. (It was "third" because, according to Huntington, it followed the first wave, which lasted from the early nineteenth century until the rise of fascism in Italy, and the second, which lasted from the end of the Second World War until the mid-1960s.)

It was too tempting not to see those two third waves as coinciding at some point in recent history, and 1989 looked like the best candidate. Such views often implied the existence of a strong causality between the march of democracy around the globe and the onset of the information revolution, a relationship that was often inferred but only rarely demonstrated. "Dictator's dilemma" has become a useful moniker, a way to capture the inevitability of authoritarian collapse when faced with fax machines, photocopiers, and so forth. Following George Schultz's lead, between 1990 and 2010 plenty of senior U.S. government officials, including James Baker, Madeleine Albright, and Robert Gates, spoke of "dictator's dilemma" as if it were common sense. But it was Columbia University's outspoken economist Jagdish Bhagwati who captured the essence of "dictator's dilemma" most eloquently: "The PC [personal computer] is incompatible with the C.P. [Communist Party]." As a free-spirited intellectual Bhagwati can, of course, believe whatever he wants without having to pay attention to the developments in the real world, but political leaders don't have that luxury, if only because the effectiveness of future policies is at stake. The danger of succumbing to the logic of "dictator's dilemma," as well as other similar beliefs about the inevitable triumph of capitalism or the end of history, is that it suffuses political leaders with a dangerous sense of historic inevitability and encourages a lazy approach to policymaking. If authoritarian states are facing such a serious, even lethal dilemma, why risk tipping the scales with some thoughtless interventions? Such unwarranted optimism inevitably leads to inaction and paralysis.

Thomas Friedman, the *New York Times* foreign affairs columnist, in his typical fashion, trivialized—and did much to popularize— the "dictator's dilemma" fallacy by coining a new buzzword: "Microchip Immune Deficiency Syndrome" (MIDS). MIDS is a "disease that can afflict any bloated, overweight, sclerotic system in the post–Cold War era. MIDS is usually contracted by countries and companies that fail to inoculate themselves against changes brought about by the microchip and the democratization of technology, finance and information." Thanks to the Internet, authoritarian governments are doomed: "Within a few

years, every citizen of the world will be able to comparison shop between his own ... government and the one next door." (For some reason, however, Americans, with all their unfettered access to the
Internet, don't hail Friedman's advice, failing to do much government-
shopping on their own and see that other governments have far more
reasonable approaches to, for example, imprisoning their citizens.)
Nicholas Kristof, Friedman's more sober colleague at the *New York
Times*, is also a strong believer in the inevitability of the information-
driven authoritarian collapse, writing that "by giving the Chinese
people broadband," the Chinese leaders are "digging the Communist
Party's grave."

Thus, it's still common to assume that the Internet would eventually
tear authoritarianism apart by dealing it a thousand lethal information
blows. Tough leaders can't survive without information technology,
but they will crumble even if they let it in, for their citizens, desperate
for Disneyland, Big Macs, and MTV, will rush to the streets demanding
fair elections. The problem with this view is that when it comes to assessing the empirical evidence and considering the case of the Internet,
it's hard to think of a state that actually didn't survive the challenges
posed by the dilemma. Save for North Korea, all authoritarian states
have accepted the Internet, with China having more Internet users than
there are people in the United States. Where the pundits and the policymakers have failed is in understanding the sophistication and flexibility of the censorship apparatus built on top of the Internet. One
crucial assumption behind "dictator's dilemma" was that it would be
impossible to design precise censorship mechanisms that could block
openly political Internet activity and yet allow any Internet activity—
perhaps even make it faster—that helped to foster economic growth.
This assumption has proved to be false: Governments have mastered
the art of keyword-based filtering, thus gaining the ability to block websites based on the URLs and even the text of their pages. The next logical stage would be for governments to develop ways in which to restrict
access to content based on concrete demographics and specific user behavior, figuring who exactly is trying to access what, for what possible

reason, what else they have accessed in the previous two weeks, and so forth before making the decision whether to block or allow access to a given page.

In the not so distant future, a banker perusing nothing but *Reuters* and *Financial Times* and with other bankers as her online friends, would be left alone to do anything she wants, even browse Wikipedia pages about human rights violations. In contrast, a person of unknown occupation, who is occasionally reading *Financial Times* but is also connected to five well-known political activists through Facebook, and who has written blog comments that included words like "democracy" and "freedom," would only be allowed to visit government-run websites (or, if she is an important intelligence target, she'll be allowed to visit other sites, with her online activities closely monitored).

When Censors Understand You Better Than Your Mom Does

Is such customization of censorship actually possible? Would censors know so much about us that they might eventually be able to make automated decisions about not just each individual but each individual acting in a particular context?

If online advertising is anything to judge by, such behavioral precision is not far away. Google already bases the ads it shows us on our searches and the text of our emails; Facebook aspires to make its ads much more fine-grained, taking into account what kind of content we have previously "liked" on other sites and what our friends are "liking" and buying online. Imagine building censorship systems that are as detailed and fine-tuned to the information needs of their users as the behavioral advertising we encounter every day. The only difference between the two is that one system learns everything about us to show us more relevant advertisements, while the other one learns everything about us to ban us from accessing relevant pages. Dictators have been somewhat slow to realize that the customization mechanisms underpinning so much of Web 2.0 can be easily turned to purposes that are much more nefarious than behavioral advertising, but they are fast learners.

By paying so much attention to the most conventional and certainly blandest way of Internet control—blocking access to particular URLs— we may have missed more fundamental shifts in the field. Internet censorship is poised to grow in both depth, looking deeper and deeper into the kinds of things we do online and even offline, and breadth, incorporating more and more information indicators before a decision to censor something is made.

When in the summer of 2009 the Chinese government announced that it would require all computers sold in the country to have one special piece of software called GreenDam installed on them, most media accounts focused on how monumental the plan seemed to be or how poorly the authorities handled GreenDam's rollout. As a result of heavy domestic and international criticism, the plan was scrapped, but millions of computers in Chinese schools and Internet cafés still continue to use the software to this day.

Internal politics aside, GreenDam stood out for its innovative embrace of predictive censorship, a precursor of highly customized censorship that awaits us in the near future. It went beyond mechanically blocking access to a given list of banned resources to actually analyzing what the user was doing, guessing at whether such behavior was allowed or not. It was definitely not the smartest software on the Internet; some users even reported that it blocked their access to any websites starting with the letter f in their URL.

It's not the implementation but the underlying principle that should have stood out. GreenDam is extremely invasive, taking a thorough look at the nature of the activities users engage in. It is programmed to study users' computer behavior—from browsing websites to composing text files to viewing pictures—and try to prevent them from engaging in activities it doesn't like (mostly by shutting down the corresponding applications, e.g., the Internet browser or word processor). For example, the color pink is GreenDam's shorthand for pornography; if it detects too much pink in the photos being viewed, it shuts down the photo-viewing application (while photos of nude dark-skinned people, perversely, pass the civility test).

Most disturbingly, GreenDam also features an Internet back door through which software can communicate with its "headquarters" and share behavioral insights about the user under surveillance. This could teach other GreenDam computers on the network about new ways to identify unwanted content. GreenDam is a censorship system with immense potential for distributed self-learning: The moment it discovers that someone types "demokracy" instead of "democracy" to avoid detection, no other users will be able to take advantage of that loophole.

Think of this as the Global Brain of Censorship. Every second it can imbibe the insights that come from millions of users who are trying to subvert the system and put them to work almost immediately to make such subversions technically impossible. GreenDam is a poor implementation of an extremely powerful—and dangerous—concept.

Time to Unfriend

But governments do not need to wait until breakthroughs in artificial intelligence to make more accurate decisions about what it is they need to censor. One remarkable difference between the Internet and other media is that online information is hyperlinked. To a large extent, all those links act as nano-endorsements. If someone links to a particular page, that page is granted some importance. Google has managed to aggregate all these nano-endorsements—making the number of incoming links the key predictor of relevance for search results—and build a mighty business around it.

Hyperlinks also make it possible to infer the context in which particular bits of information appear online without having to know the meaning of those bits. If a dozen antigovernment blogs link to a PDF published on a blog that was previously unknown to the Internet police, the latter may assume that the document is worth blocking without ever reading it. The links—the "nano-endorsements" from antigovernment bloggers—speak for themselves. The PDF is simply guilty by association.

Thanks to Twitter, Facebook, and other social media, such associations are getting much easier for the secret police to trace.

If authoritarian governments master the art of aggregating the most popular links that their opponents share on Twitter, Facebook, and other social media sites, they can create a very elegant, sophisticated, and, most disturbingly, accurate solution to their censorship needs. Even though the absolute amount of information—or the number of links, for that matter—may be growing, it does not follow that there will be less "censorship" in the world. It would simply become more fine-tuned. If anything, there might be less one-size-fits-all "wasteful" censorship, but this is hardly a cause for celebration.

The belief that the Internet is too big to censor is dangerously naïve. As the Web becomes even more social, nothing prevents governments— or any other interested players—from building censorship engines powered by recommendation technology similar to that of Amazon and Netflix. The only difference, however, would be that instead of being prompted to check out the "recommended" pages, we'd be denied access to them. The "social graph"—a collection of all our connections across different sites (think of a graph that shows everyone you are connected to on different sites across the Web, from Facebook to Twitter to YouTube)—a concept so much beloved by the "digerati," could encircle all of us.

The main reason why censorship methods have not yet become more social is because much of our Internet browsing is still done anonymously. When we visit different sites, the people who administer them cannot easily tell who we are. There is absolutely no guarantee that this will still be the case five years from now; two powerful forces may destroy online anonymity. From the commercial end, we see stronger integration between social networks and different websites— you can now spot Facebook's "Like" button on many sites—so there are growing incentives to tell sites who you are. Many of us would eagerly trade our privacy for a discount coupon to be used at the Apple store. From the government end, growing concerns over child pornography, copyright violations, cybercrime, and cyberwarfare also make it more likely that there will be more ways in which we will need to prove our identity online.

The future of Internet control is thus a function of numerous (and rather complex) business and social forces; sadly, many of them originating in free and democratic societies. Western governments and foundations can't solve the censorship problem by just building more tools; they need to identify, publicly debate, and, if necessary, legislate against each of those numerous forces. The West excels at building and supporting effective tools to pierce through the firewalls of authoritarian governments, but it is also skilled at letting many of its corporations disregard the privacy of their users, often with disastrous implications for those who live in oppressive societies. Very little about the currently fashionable imperative to promote Internet freedom suggests that Western policymakers are committed to resolving the problems that they themselves have helped to create.

We Don't Censor; We Outsource!

Another reason why so much of today's Internet censorship is invisible is because it's not the governments who practice it. While in most cases it's enough to block access to a particular critical blog post, it's even better to remove that blog post from the Internet in its entirety. While governments do not have such mighty power, companies that enable users to publish such blog posts on their sites can do it in a blink. Being able to force companies to police the Web according to a set of some broad guidelines is a dream come true for any government. It's the companies who incur all the costs, it's the companies who do the dirty work, and it's the companies who eventually get blamed by the users. Companies also are more likely to catch unruly content, as they know their online communities better than government censors. Finally, no individual can tell companies how to run those communities, so most appeals to freedom of expression are pointless.

Not surprisingly, this is the direction in which Chinese censorship is evolving. According to research done by Rebecca MacKinnon, who studies the Chinese Internet at New America Foundation and is a former CNN bureau chief in Beijing, censorship of Chinese user-generated content is "highly decentralized," while its "implementation is left to the Web companies themselves."

To prove this, in mid-2008 she set up anonymous accounts on a dozen Chinese blog platforms and published more than a hundred posts on controversial subjects, from corruption to AIDS to Tibet, to each of them. MacKinnon's objective was to test if and how soon they would be deleted. Responses differed widely across companies: The most vigilant ones deleted roughly half of all posts, while the least vigilant company censored only one. There was little coherence to the companies' behavior, but then this is what happens when governments say "censor" but don't spell out what it is that needs to be censored, leaving it for the scared executives to figure out. The more leeway companies have in interpreting the rules, the more uncertainty there is as to whether a certain blog post will be removed or allowed to stay. This Kafkaesque uncertainty can eventually cause more harm than censorship itself, for it's hard to plan an activist campaign if you cannot be sure that your content will remain available.

This also suggests that, as bad as Google and Facebook may look to us, they still probably undercensor compared to most companies operating in authoritarian countries. Global companies are usually unhappy to take on a censorship role, for it might cost them dearly. Nor are they happy to face a barrage of accusations of censorship in their own home countries. (Local companies, on the other hand, couldn't care less: Social networking sites in Azerbaijan probably have no business in the United States or Western Europe, nor are their names likely to be mispronounced at congressional hearings.)

But this is one battle that the West is already losing. Users usually prefer local rather than global services; those are usually faster, more relevant, easier to use, and in line with local cultural norms. Look at the Internet market in most authoritarian states, and you'll probably find at least five local alternatives to each prominent Web 2.0 start-up from Silicon Valley. For a total online population of more than 300 million,

Facebook's 14,000 Chinese users, by one 2009 count, are just a drop in the sea (or, to be exact, 0.00046 percent).

Companies, however, are not the only intermediaries that could be pressured into deleting unwanted content. RuNet (the colloquial name for the Russian-speaking Internet), for example, heavily relies on "communities," which are somewhat akin to Facebook groups, and those are run by dedicated moderators. Most of the socially relevant online activism in Russia happens on just one platform, LiveJournal. When in 2008 the online community of automobile lovers on LiveJournal became the place to share photos and reports from a wave of unexpected protests organized by unhappy drivers in the far eastern Russian city of Vladivostok, its administrators immediately received requests from FSB, KGB's successor, urging them to delete the reports. They complied, although they complained about the matter in a subsequent report that they posted to the community's webpage (within just a few hours that post disappeared as well). Formally, though, nothing has been blocked; this is the kind of invisible censorship that is most difficult to fight.

The more intermediaries—whether human or corporate—are involved in publishing and disseminating a particular piece of information, the more points of control exist for quietly removing or altering that information. The early believers in "dictator's dilemma" have grossly underestimated the need for online intermediaries. Someone still has to provide access to the Internet, host a blog or a website, moderate an online community, or even make that community visible in search engines. As long as all those entities have to be tied to a nation state, there will be ways to pressure them into accepting and facilitating highly customized censorship that will have no impact on economic growth.

Wise Crowds, Unwise Causes

Thailand's extremely strict lèse-majesté laws make it illegal to publish—including in blog and tweet form—anything that may offend the country's royal family. But effectively policing the country's rapidly expanding blogosphere has proved very challenging for the Thai police.

In early 2009 a Thai MP loyal to the king proposed a new solution to this intractable problem. A new site, called ProtectTheKing.net, was set up so that Thai users could report links to any websites they believed to be offensive to the monarchy. According to the BBC, the government blocked 5,000 submitted links in the first twenty-four hours. Not surprisingly, the site's creators "forgot" to provide a way in which to complain about sites that were blocked in error.

Similarly, Saudi Arabia allows its citizens to report any links they find offensive; 1,200 of those are submitted to the country's Communications & Information Technology Commission on a daily basis. This allows the Saudi government to achieve a certain efficiency in the censorship process. According to *Business Week*, in 2008 the Commission's censorship wing employed only twenty-five people, although many of them came from top Western universities, including Harvard and Carnegie Mellon.

The most interesting part about the Saudi censorship scheme is that it at least informs the user why a website has been blocked; many other countries simply show a bland message like "the page cannot be displayed," making it impossible to discern whether the site is blocked or simply unavailable because of some technical glitches. In the Saudi case, banned porn websites carry a message that explains in detail the reasons for the ban, referencing a *Duke Law Journal* article on pornography written by the American legal scholar Cass Sunstein and a 1,960-page study conducted by the U.S. attorney general's Commission on Pornography in 1986. (At least for most nonlawyers, those are probably far less satisfying than the porn pages they were seeking to visit.)

The practice of "crowdsourcing" censorship is becoming popular in democracies as well. Both the British and the French authorities have similar schemes for their citizens to report child pornography and several other kinds of illegal content. As there are more and more websites and blogs to check for illegal material, it's quite likely that such crowdsourcing schemes will become more common.

The Thai, Saudi, and British authorities rely on citizens' goodwill, but a new scheme in China actually offers monetary awards to anyone submitting links to online pornography. Found a porn site? Report it to the authorities, and get paid. The scheme may have backfired, however. When it was first introduced in early 2010, there was also a considerable spike in those searching for pornography. Who knows how many of the reported videos have first been downloaded and saved to local hard drives? More important, how many pages containing nonsexual content could be found and dealt with in such a manner?

In some cases, the state does not need to become directly involved at all. Tech-savvy groups of individuals loyal to a particular cause or national government will harness their networks to censor their opponents, usually by dismantling their groups on social networking sites. The most famous of such networks is a mysterious online organization that calls itself Jewish Internet Defense Force (JIDF). This pro-Israel advocacy group made headlines by compiling lists of anti-Israeli Facebook groups, infiltrating them to become their administrators, and ultimately disabling them. One of its most remarkable accomplishments was deleting nearly 110,000 members from a 118,000-strong Arabic-language group sympathetic to Hezbollah. In some such cases, Facebook administrators are quick enough to intervene before the group is completely destroyed, but often they aren't. The online social capital that took months to develop goes to waste in a matter of hours. It's important to understand that increasingly it is communities—not just individual bloggers—that produce value on today's Internet. Thus modern censorship will increasingly go beyond just blocking access to particular content and aim to erode and destroy entire online communities instead.

Denial-of-Philosophy

If philosophy is your passion, Saudi Arabia would not top your list of places to spend a year abroad. Perhaps because the discipline encourages independent thinking and questioning of authority (or simply aggravates the problem of unemployment), the subject is banned at universities, and so are philosophy books. Explaining his resistance to the introduction of philosophy as a subject in the Saudi high school curriculum, the director of planning at the Jeddah Educational Administration noted in

December 2005 that "philosophy is a subject derived from the Greeks and the Romans. . . . We do not need this kind of philosophy because the Holy Quran is rich in Islamic philosophy."

The modern elements within Saudi Arabia's civil society were hopeful of finally getting some autonomy in cyberspace. And their hopes were not in vain: The Internet quickly filled the void, with nearly free and easy access to philosophy books, video lectures, and scholarly magazines. But there was no centralized repository of links to such content, so several U.S.-educated Saudis started an Internet forum, Tomaar, to discuss all things related to philosophy and share links to interesting content. The site enjoyed tremendous success; in just a few months, the site branched out beyond philosophy, with its users discussing Middle East politics and controversial social issues (since the site was in Arabic, non-Saudi users frequented it as well). At its peak, the site had more than 12,000 active members, who contributed an average of 1,000 posts a day.

But it was a short-lived triumph. Before long, the Saudi government noticed the phenomenal success of Tomaar and quickly banned all Saudi users from accessing the site. This, however, was an easy problem to solve. In the last decade or so, plenty of tools had emerged to circumvent such government bans; their creation was fueled mostly by the excessive censorship of the Chinese authorities. In essence, governments cannot erase the content they do not like, especially if it's hosted on a foreign server; what they can do is to ban their own nationals from accessing that content by requiring ISPs to simply stop serving requests for a particular URL. But it's possible to trick the ISPs by connecting to a third-party computer and using that computer's Internet connection to access the content you need; all that the government would see is that you are connected to some random computer on the Net, but they won't know that you are accessing content they don't like. Tomaar's fans made good use of such censorship-circumvention tools and were able to use the site despite the ban. (Of course, once too many users connect to one computer or its address is publicized, the authorities may understand what is taking place and ban access to it as well.)

But their jubilation did not last long. Shortly thereafter the website became inaccessible to even those users who relied on censorship-circumvention tools. It appeared that the site was enjoying such popularity that it was simply overloaded with Internet traffic. The American company that hosted Tomaar wrote the site administrators to inform them that it was terminating their contract, making the site a "digital refugee." Something eerie was happening, and Tomaar's administrators could not figure out what it was (none of them was a techie—one worked as a salesman in a high-end consumer electronic store, and another was a financial consultant in a bank).

It took some time before it became clear that Tomaar was a target of a protracted cyber-attack that aimed to make the website unavailable. The type of attack in question—the so-called Distributed-Denial-of-Service (DDoS) attack—is an increasingly popular way of silencing one's opponents. Much like pubs and salons, all websites have certain occupancy limits. Popular sites like CNN.com can handle millions of simultaneous sessions, while most amateur sites can barely handle a hundred or two hundred simultaneous visits. A DDoS attack seeks to take advantage of such resource constraints by sending fake visitors to targeted websites. Where do such fake visitors come from? They are generated by computers that have been infected with malware and viruses, thus allowing a third party to establish full command over them and use their resources however it sees fit. Nowadays, the capacity to launch such attacks is often bought and sold on eBay for a few hundred dollars.

Since the attack originates from thousands of computers, it's almost always impossible to identify its mastermind. This was true in Tomaar's case. While it seemed logical that the Saudi government would be interested in silencing the site, there is no concrete evidence to assert that connection. But Tomaar's hosting company did not drop them for nothing: DDoS attacks eat a lot of traffic, it takes quite some time to clean up afterward, and it's the hosting companies that have to pay the bills. This is how online dissent can easily turn into a preexisting condition. If you have something sensitive to say and it can attract DDoS attacks, most hosting companies would think twice before signing you up as their client. Since businesses are also frequent targets of DDoS attacks, there exists a commercial market in protection services (for example, banning computers from certain parts of the world from being able to visit your site), but they sell at rates that are not affordable to most volunteer-funded sites. Eventually, Tomaar did find a new home, but cyber-attacks continued. The site was regularly down for one week out of four, with DDoS attacks eroding its community's spirit and draining the pockets of its founders, who were naïve enough to believe that online dissent is as cheap as their monthly hosting fee.

Cases like Tomaar's are increasingly common, especially among activist and human rights organizations. Burma's exiled media—Irrawaddy, Mizzima, and the Democratic Voice of Burma—all experienced major cyber-attacks (the heaviest wave occurred on the first anniversary of the Saffron Revolution in 2008); ditto the Belarusian oppositional site Charter97, the Russian independent newspaper *Novaya Gazeta* (the one that employed the slain Russian journalist Anna Politkovskaya), the Kazakh oppositional newspaper *Respublika*, and even various local branches of Radio Free Europe / Radio Liberty.

Individual bloggers fall victims to such attacks as well. In August 2009, on the first anniversary of the Russian-Georgian war, Cyxymu, one of the most popular Georgian bloggers, found himself under such an intensive DDoS attack that it even took down powerful websites like Twitter and Facebook, where he had duplicate accounts. Here was a case of a dissenting voice who could not say what he wanted because all the platforms where he established online identities came under severe DDoS attacks and put immense pressure on the administrators running those platforms; they, of course, found it quite tempting to simply delete his account to enable all other users to continue with their business.

DDoS attacks present a serious and poorly understood threat to freedom of expression on the Internet, as they are increasingly used not only against the websites of institutions and companies but also against individual bloggers. In the past, conventional wisdom dictated that all it took to give voice to marginalized communities was to get them online and maybe pay their Internet bill. Not anymore. Being heard online—at least beyond the first few tweets and blog posts—increasingly involves strategizing about server administration, creating back-up plans in case of a DDoS emergency, and even budgeting for extremely expensive anti-DDoS protection services.

The worst part about DDoS-type restrictions on freedom of expression is that they lead to significant undercounting of the total amount of Internet censorship around the world. Our traditional notion of censorship is still strongly influenced by the binary logic of "blocked/unblocked," which in cases like those of Cyxymu or *Novaya Gazeta* simply do not make much sense. The sites may be technically unblocked, but their users still cannot access them one week out of the month.

To solve this kind of problem, not only do Western governments and international institutions need to create new metrics for tracking Internet censorship, they also need to go beyond the usual panacea offered against Internet censorship, like circumvention tools that allow access to banned content. The problem with DDoS is that even users in countries that do not block the Internet would not be able to access sites that are under attack; circumvention tools don't work in those situations. It's no longer the case of brutal Soviet agents jamming Radio Free Europe; it's the case of mostly unknown individuals—perhaps on the Kremlin's payroll, perhaps not—erecting roadblocks around the building from which the new Radio Free Europe is supposed to broadcast. Antijamming equipment is not going to help if nobody can actually get in and produce the broadcasts.

Tearing Down the Wrong Walls

Those of us in the West who care about defending online freedom of expression can no longer afford to think about censorship based on obsolete models developed during the Cold War. The old model assumed that censorship was expensive and could only be carried out by one party—the government. Today, however, while many kinds of censorship are still expensive (e.g., software like GreenDam), others are cheap and getting cheaper (e.g., DDoS attacks). This allows the governments to deflect the blame—they're not doing the censorship, after all—and thus also significantly undercounts total censorship in the world. In many cases, governments don't have to do anything at all; plenty of their loyal supporters will be launching DDoS attacks on their own. The democratization of access to launching cyber-attacks has resulted in the democratization of censorship; this is poised to have chilling effects on freedom of expression. As more and more censorship is done by intermediaries (like social networking sites) rather than governments, the way to defend against censorship is to find ways to exert commercial—not just political—pressure on the main actors involved.

It's also becoming clear that authoritarian governments can and will develop sophisticated information strategies that will allow them to sustain economic growth without loosening their grip on the Internet activities of their opponents. We certainly don't want to spend all our energy tearing down some imaginary walls—making sure that all information is accessible—only to discover that censorship is now being outsourced to corporations or those who know how to launch DDoS attacks. This is yet another reason why "virtual walls" and "information curtains" are the wrong metaphors to assist us in conceptualizing the threat to Internet freedom. They invariably lead policymakers to opt for solutions for breaking through the information blockade, which is fine and useful, but only as long as there is still something on the other end of the blockade. Breaking the firewalls to discover that the content one seeks has been deleted by a zealous intermediary or taken down through a cyber-attack is going to be disappointing.

There are plenty of things to be done to protect against this new, more aggressive kind of censorship. One is to search for ways to provide mirrors of websites that are under DDoS attacks or to train their administrators, many of whom are self-taught and may not be managing the crisis properly, to do so. Another is to find ways to disrupt, mute, or even intentionally pollute our "social graph," rendering it useless to those who would like to restrict access to information based on user demographics. We may even want to figure out how everyone online can pretend to be an investment banker seeking to read *Financial Times!* One could also make it harder to hijack and delete various groups from

Facebook and other social networking sites. Or one could design a way to profit from methods like "crowdsourcing" in fighting, not just facilitating, Internet censorship; surely if a group of government royalists troll the Web to find new censorship targets, another group could also be searching for websites in need of extra protection?

Western policymakers have a long list of options to choose from, and all of them should be carefully considered not just on their own terms, but also in terms of the negative unintended consequences—often, outside of the geographic region where they are applied—that each of them would inevitably generate. Of course, it's essential to continue funding various tools to access banned websites, since blocking users from visiting certain URLs is still the dominant method of Internet control. But policymakers should not lose sight of new and potentially more dangerous threats to freedom of expression on the Internet. It's important to stay vigilant and be constantly on the lookout for new, yet invisible barriers; fighting the older ones, especially those that are already crumbling anyway, is a rather poor foundation for effective policy. Otherwise, cases like Russia, which has little formal Internet filtering but plenty of other methods of flexing the government's muscles online, will continue puzzling Western observers.

The main thing to keep in mind, though, is that different contexts give rise to different problems and are thus in need of custom-made solutions and strategies. Clinging to Internet-centrism—that pernicious tendency to place Internet technologies before the environment in which they operate—gives policymakers a false sense of comfort, a false hope that by designing a one-size-fits-all technology that destroys whatever firewall it sees, they will also solve the problem of Internet control. The last decade, characterized, if anything, by a massive increase both in the amount and in the sophistication of control, suggests that authoritarian regimes have proved highly creative at suppressing dissent through means that are not necessarily technological. As such, most of the firewalls to be destroyed are social and political rather than technological in nature.

The problem is that technologists who have been designing tools to break technological rather than political firewalls—and often have been doing it with the financial support of Western governments and foundations—are the ones who control the public conversation. It's in their direct interest to overstate the effectiveness of their own tools and downplay the presence of other nontechnological threats to the freedom of expression. In doing this, they mislead policymakers, who then make poor decisions about the allocation of resources to fight Internet control. Shiyu Zhou, the founder of a Falung Gong technology group that designs and distributes software to access sites banned by the Chinese government, says that "the entire battle over the Internet has boiled down to a battle over resources" and that "for every dollar we [America] spend, China has to spend a hundred, maybe hundreds of dollars" in an interview to the *New York Times* as part of an argument that more funding should be allocated to promote such tools in Iran. This is at best misleading and at worst disingenuous, a throwback to the Cold War debates about closing the missile gap, but this time by overspending the enemy on digital weapons.

This kind of argument only perpetuates myths like "dictator's dilemma" and suggests that authoritarian governments are more vulnerable to the threat of technology than they really are. But even if such sly manipulation of public opinion can be overcome, one still has to remember that no solutions to the censorship problem can be designed in isolation from the other two problems—surveillance and propaganda. The decentralized nature of the Internet makes it relatively easy to set up an infinite number of copies for every byte of information shared over the Web. This ability does not come free, however, even if the financial costs are marginal, for it also allows the creation of new, faster, and often more legitimate publishing outlets that can make government propaganda more believable. Moreover, it opens up opportunities for tracking how information spreads online, enabling the authorities to learn more about those who spread it. Information wants to be free, but so do those exchanging it.

chapter five

Hugo Chavez Would Like to Welcome You to the Spinternet

• • • ■

For years Venezuela's President Hugo Chavez was the world's least likely person to join Twitter. Brevity is not exactly one of his virtues: In the last ten years, Chavez spent more than 1,500 hours denouncing capitalism on *Alo Presidente*, his own TV show. In a broadcast that aired in March 2010, the self-proclaimed leader of the Bolivarian Revolution even attacked the Internet as "a battle trench" that was bringing "a current of conspiracy"; anyone who used "Twitter, the Internet [and] text messaging" to criticize his regime was engaging in "terrorism." Chavez had plenty of reasons to worry about the Internet. A jailed judge had started using Twitter to keep in touch with her followers from prison, while the director of an opposition TV station used it to denounce a conspiracy to oust him. Chavez's outburst was more than a rhetorical flourish. It seems that, much like his American detractors, he was also under the impression that Twitter was the driving force behind the protests in Iran.

As the Venezuelan opposition started using Twitter to mobilize its supporters, Chavez changed his mind. In late April 2010, Diosdado Cabello, the head of Venezuela's communications watchdog and an aide close to Chavez, announced on his own Twitter account that his boss was about to join the site. "Comrades, @chavezcandanga has been reserved, soon we will have messages there from our Comandante," tweeted Cabello. (In Spanish, *candanga* means "the devil," but Venezuelans also use the term to describe someone naughty and wild.) Within twenty-four hours of signing up for Twitter, Chavez had 50,000 followers, and within just one month, he gained more than 500,000, making him one of the most popular foreign politicians on the predominantly English-speaking site. His love affair with technology quickly expanded to other devices and platforms; in July 2010 he even widely praised "a little apparatus" (an iPod) that his daughter gave him. "I have like 5,000 songs," boasted Chavez. "It's tiny, I remember having to go around with a bunch of cassette tapes before." The Bolivarian Revolution was turning high-tech.

So far, the tweeting Chavez, unlike his real combative self, has been charming and polite. Responding to criticism from a sixteen-year-old Mexican girl who accused him of being a dictator, Chavez responded quite politely: "Hello Mariana, the truth is I'm an anti-dictator, and I love my beautiful Mexico." When a Venezuelan Twitter user named Desiree tweeted out her admiration for Chavez, the latter responded, "My dear Desiree, I send you a kiss." Chavez also promised to convince his buddy Evo Morales, Bolivia's president, to start using Twitter as well. And he is putting Twitter to some creative uses as well; only three months after he set up an account on the service, he boasted that he had already received nearly 288,000 requests for help from citizens. In July 2010 Chavez made international headlines by tweeting about his quixotic quest to exhume the remains of his hero Simón Bolívar, the nineteenth-century aristocrat who liberated much of Latin America from Spanish rule. "What impressive moments we've lived tonight! Rise up, Simón, as it's not time to die!" tweeted the Venezuelan leader.

The secret of Chavez's popularity on Twitter lies not only in his charisma but also in his leverage of government resources to bolster his crusade. Just a few days after joining the service, Chavez dispelled any illusions that Twitter was a minor and temporary distraction for him. "I've created my own *Chavezcandanga* mission to answer the messages, and we're even going to create a fund for the mission to provide many things that are now missing and that are urgent," said Chavez in a televised cabinet meeting. To that effect, he promised to allocate two hundred staffers—funded with public monies—to help him win the Twitter war. Holding up a BlackBerry in front of the camera, Chavez told his TV audience that Twitter was his "secret weapon" and dismissed the idea he was using a capitalist tool. "The Internet can't be just for the bourgeoisie; it's for the ideological battle as well," declared Chavez, boasting that he was gaining two hundred new followers per minute.

But I Saw It on the Spinternet!

The evolution of Chavez's reaction to Twitter—strong ideological opposition followed by a wide embrace—is typical of how authoritarian regimes respond to the Internet. First, they believe that the Web is something frivolous that will go away as fast at it has appeared. To their great disappointment, it never does; worse, sooner or later it is embraced by the opposition, if there is one, who use it chiefly to avoid the government's tight control of the media. This is when many authoritarian governments begin to experiment with censorship. Much here depends on the political situation in their countries. For some, Internet censorship would be acceptable, because they already censor other media; for others, direct censorship would not be an option, as they prefer to crack down on free media through more indirect means, ranging from tax inspections to intimidation of individual journalists. When Internet censorship is impractical, politically indefensible, or prohibitively expensive, governments begin to experiment with propaganda and, in a few extreme cases, pervasive surveillance.

Unlike many authoritarian governments, Chavez's regime has always preferred softer means of intervention and control, trying to avoid the harsher methods embraced by the Chinese and Iranian governments. Thus, in 2007 Chavez refused to renew the concession given to Globovision, a popular—and extremely critical—TV channel, essentially forcing it to move to cable, while in 2009 his communications minister shut down more than sixty radio stations, stating that they lacked the necessary licenses and promising to use their frequencies for community media. When it comes to Twitter, from which the government cannot really withdraw a license, Chavez's choice is not between censorship and free speech but between staying out of the Twitter space altogether—and thus, risking losing control of online conversations—and trying to infuse such conversations with his own ideology.

This is not how it was supposed to be. Many early predictions about the Internet posited that it would rid the world of government propaganda. Frances Cairncross in her 1997 best seller, *The Death of Distance*, a defining text in the cyber-utopian canon, predicted that "free to explore different points of view, on the Internet or on the thousands of television and radio channels that will eventually be available, people will become less susceptible to propaganda from politicians." Her prediction proved to be wrong: Governments have learned that they can still manipulate online conversations by slightly adjusting how they manufacture and package their propaganda, with some of their older and otherwise stale messages finding new life and appealing to new audiences. It was hard to predict that xenophobic and anti-American messages would sound more persuasive when delivered by edgy and supposedly independent bloggers.

But this raises an even broader question: Why does government propaganda—and especially propaganda based on lies and intentional misrepresentation of facts—still work in an age when one could find plenty of credible evidence online to disprove it? It works for the same unfortunate reasons that myths about Barack Obama's missing birth certificate or the myths about 9/11 being an inside job work for so many audiences in America. The easy availability of evidence to the contrary is not enough to dispel such myths, for they are not always based on rational examination of evidence. In addition, certain structural conditions of public life under an authoritarian regime might make such government-induced myths harder to dispel. Barbara Geddes, a noted political scientist at the University of California at Los Angeles, who studied sources of popular support for authoritarian states around the world, discovered that a particular sector in the population is most susceptible to government propaganda. Usually it is what we can best describe as the middle class: people with some basic education who earn a good living and are neither poor and ignorant nor rich and sophisticated. (These two latter groups, Geddes found, were least susceptible to government propaganda: the former because they could not even understand what the government wanted and the latter because they could easily see through it.)

Thus, high rates of exposure to government propaganda may not necessarily make people aware of the fact that they are being brainwashed, let alone allow them to read between the lines. Conventional wisdom about government propaganda in authoritarian states has been best summed up by none other than Ithiel de Sola Pool, who said, "When regimes impose daily propaganda in large doses, people stop listening." Geddes disagreed, "We must suppose that it must take an uncommonly educated populace before governmental control of information flows begins to boomerang in any serious way." Mere exposure to information does not by itself decrease support for authoritarian governments; it does not guarantee an increase in media literacy or sophistication. Simply getting a country's population online is not going to trigger a revolution in critical thinking; judging by the recent global hysteria over how the Internet might be dumbing us down, some people clearly believe that the opposite is more likely.

It's hardly surprising, then, that authoritarian governments from Russia to Iran and from China to Azerbaijan are busy turning the Internet into the Spinternet—a Web with little censorship but lots of spin and propaganda—which reinforces their ideological supremacy. The age of new media, with its characteristic fragmentation of public discourse and decentralization of control, has made the lives of propaganda officials toiling in stuffy offices of authoritarian governments considerably easier.

Elude the Cat, Empower the Masses

When in January 2009 Li Qiaoming, a twenty-four-year old peasant living near Yuxi City, a major hub of Yunnan Province in China's southwest, was arrested for illegal logging, little did he know that he had only two weeks to live. Locked into Cell Number 9 of the Puning County jail, he accidentally hit his head against the door while playing "elude the cat"—the Chinese equivalent of "hide-and-seek"—with fellow inmates. Or so went the local police department's explanation to Li's parents when they were told to go and pick up his corpse.

Within hours Li's death became a cause célèbre in the Chinese blogosphere. Netizens were quick to accuse the Yunnan police of a nasty and poorly veiled cover-up. QQ.com—one of the most popular sites in China—attracted more than 70,000 comments on the issue, and the accusations spread like wildfire. The Chinese authorities had a major cyber-riot on their hands.

The way they chose to deal with the incident shows the growing evolution of China's Internet controls and is poised to enter future textbooks of Internet propaganda. Instead of dispatching censors to delete hundreds of thousands of angry comments, they publicly reached out to Chinese Internet users and invited them to become "netizen investigators," asking them to help inspect Cell Number 9 and write a report that could eliminate any doubts about what had really happened.

It seemed like a reasonable solution and helped to quell the tensions— at least temporarily. More than one thousand candidates applied for the job; a committee of fifteen was duly formed and dispatched to the detention facility to produce a conclusive report about Qiaoming's death.

"Past experience has shown that the doubts of the netizens will not shift or recede on their own over time," said Wu Hao, a senior official with Yunnan's propaganda department who steered the campaign, adding that "a matter of Internet public opinion must be solved by Internet methods." Those methods pretended to prioritize open and decentralized decision making; there was no better way to showcase the democratization of China's governance than to form a "netizen" commission.

The commission itself was just a formality. The investigators were not even allowed to view the tapes from the cell's surveillance camera. Predictably, the report they produced was inconclusive; all they could say was that they lacked the evidence to rule either way. But the police, who must have known the real cause of Li's death well in advance, used the Internet buzz generated by the release of the report to publicly apologize to his parents, revealing that he had been beaten by other inmates. Score another propaganda point for showing the humility of the Chinese police.

This was digital public relations at its best. What could have become a major liability for the government became an opportunity to flaunt their commitment to democratization. China's propaganda visionaries could give most Western colleagues a run for their money. Given that censorship was an easy way out, they chose wisely. The delivery was not perfect: After China's "human flesh search engines" did a background check on the fifteen citizen inspectors that composed the commission, it turned out that nearly all of them were current or former employees of the state-owned media. This was probably a gaffe that the Chinese authorities are likely to rectify in the future. (That said, China's state-owned media are in the habit of hiring little-known actors to pose as passers-by when they shoot street interviews about important matters with the ordinary folk. The actors can lie with a straight face at least.)

Most interestingly, the South Korean government, perhaps following China's lead, has recently used a similar scheme to counter Internet rumors and calm its own netizens on an issue that was far more explosive. When many in the blogosphere and even some in the traditional media began doubting the official explanation that the South Korean warship *Cheonan* was drowned by a North Korean torpedo, suspecting that North Korea was just used as a bogeyman to either cover up incompetence or some more sinister domestic operation, the government chose not to further antagonize the bloggers. Instead, it announced that it would select twenty Twitter users, ten Internet bloggers, thirty student journalists, five representatives from portal Internet sites, and five government officials to tour the wreckage by randomly picking them from among the applicants. But the South Korean government went one step further than the Chinese, allowing all participants to take photos and videos—a state-engineered triumph of citizen journalism. Their plans, however, may have been spoiled by the North Korean authorities, who were unbelievably quick to colonize the capitalist cyberspace as well. In August 2010 they took their anti-South propaganda campaign to Twitter, setting up an account—supposedly through their foreign supporters— meant to challenge the South Korean version of events a hundred forty characters at a time.

What Barbara Streisand Could Teach Nicolae Ceauşescu

In the last few years of the Soviet Union's existence, its most progressive leaders were fond of touting—half-jokingly, of course—their commitment to the "Sinatra doctrine": the notion that Central and Eastern European states were free to go their own way, very much along the lines of Sinatra's song "My Way." On the Internet, though, Sinatra is out of luck; the buzzword du jour is the "Streisand effect"—the notion that the more you try to get something off the Internet, the more you fuel everyone's interest in it, thus defeating the purpose of your original intervention.

The concept was coined by Mike Masnick, a popular American technology blogger, to describe Barbara Streisand's desperate attempts to remove photos of her Malibu house from the Internet. Shot by a professional photographer to document coastal erosion under the auspices of the California Coastal Records Project, the pictures did not receive much public attention until Streisand filed a $50 million lawsuit.

To the feisty world of technology blogs, it seemed as if Barbara Streisand was declaring a war on both the Internet and common sense. It didn't take long for a massive online network of solidarity to be formed. Hundreds of other bloggers began posting the Malibu photos on their own blogs. Sure enough, Streisand's Malibu property became a considerably more discussed subject than ever before.

In hindsight, Streisand would have been better off doing nothing at all and letting the initial set of photos lie online in almost guaranteed obscurity (after all, they were part of a set of 12,000 photos). Instead, she chose the censorship route—and paid for it dearly. Furthermore, she paved the way for countless other celebrities and VIPs. The Church of Scientology, the Russian oligarch Alisher Usmanov, Sony Corporation, Britain's Internet Watch Foundation, and the popular social news site Digg.com are just some of the high-profile victims of the Streisand Effect. Confronting online content they didn't want to be public, they instinctively fought back—without realizing how such aggression might backfire. Even when some of them scored a victory in temporarily removing the harmful content from the Internet, they usually helped fuel more interest in what it was they were trying to hide.

In fact, an entire organization called WikiLeaks has been built to ensure that all controversial documents that someone wants to get off the Web have a dedicated and well-protected place to stay online. Even the omnipotent American military are finding it hard to take sensitive content off the Internet, as they discovered when WikiLeaks released the video of a 2007 Baghdad air strike that killed several Reuters news staffers as well as a trove of documents related to the war in Afghanistan.

The logic behind the Streisand Effect, however, does not have much to do with the Internet. Throughout history there has hardly been a more effective way to ensure that people talk about something than to ban discussions about it. Herostratus, a young Greek man who in 356 BCE set fire to the Temple of Artemis at Ephesus, may be the world's first documented case of the Streisand Effect. Herostratus's ultimate punishment—that is, in addition to being executed—was for his act to be forgotten, on strict orders from the Ephesean authorities, who banned anyone from ever mentioning his name. And here we are, discussing the story of this narcissistic pyromaniac thousands of years later (the Ephesean authorities could not foresee that Herostratus would be immortalized on his own Wikipedia page).

Even though it's publicity-conscious Hollywood stars, corporations, and activist organizations that are most troubled by the Streisand Effect, authoritarian governments are worried as well. For most of their existence, authoritarian governments believed—and not without good reason—they could control the spread of information by limiting access to publishing tools and tracking how they were being used by those with access. Nicolae Ceaușescu, the cruel dictator who ruled Romania from 1965 to 1989, made it a crime to own typewriters without registering them. At one point he even considered getting every Romanian to submit a handwriting sample. This was his way of preventing mass correspondence between his citizens and Western media like Radio Free Europe (nevertheless, even such drastic measures couldn't stop all the letters).

When the costs of trafficking in information—whether financial or reputational—are too high, there are fewer opportunities to spread it around; the Streisand Effect is not likely to undermine official information flows. But when almost everyone has access to cheap means of self-publishing as well as concealing their identity, the Streisand Effect becomes a real threat. It often renders traditional forms of censorship counterproductive. Most political bloggers start salivating at the thought of starving the government's censorship beast by reproducing what the government wants to ban.

But it would be wrong to conclude that the Streisand Effect means the demise of information control. Censors can simply switch to other, less obvious, and less intrusive means of minimizing the negative impact of the information being spread online. Instead of censoring it—which, in many cases, could only confer additional credibility on whatever criticism was expressed in the original blog post or article—the government may choose to go for the Spinternet option: Counter the blog post with effective propaganda rather than a blanket ban.

In countries where even ardent supporters of democratization are often paranoid about foreign intervention, all it takes to discredit a blogger is to accuse her of being funded by the CIA, MI6, or Mossad (or, even better, all three simultaneously!). If that accusation is repeated by a hundred other bloggers—even if some of them look rather dubious—most sane critics of the government think twice before reposting that blogger's critical message. The best way to create such a culture of mistrust is for governments to cultivate extremely agile rapid-response blogging teams that fight fire with fire.

The benefits of such an approach have not been lost on Western policy wonks. In 2008 Cass Sunstein, a prominent American legal scholar who now heads the White House Office of Information and Regulatory Affairs, coauthored a punchy policy paper that recommended the U.S. government practice "cognitive infiltration" of Internet groups that are spreading conspiracy theories, suggesting that "government agents or their allies (acting either virtually or in real space, and either openly or anonymously) will undermine the crippled epistemology of believers by planting doubts about the theories and stylized facts that circulate within such groups." As far as Hugo Chavez or Mahmoud Ahmadinejad are concerned, those pushing for greater U.S.-style democratization of their countries are also part of some kind of conspiracy theory. Not surprisingly, they are the ones who put Sunstein's recommendations into practice by hiring their own contingents of government-funded but technically independent online commentators. (Also not surprisingly, Sunstein's suggestions are hardly politically accepted in the Western world. It's hard to imagine anyone supporting White House–funded covert bloggers engaging in deliberations with the conservative bloggers they disagree with.)

Russia's First Pornographer Meets Russia's Sarah Palin

One of the most persistent myths in public discussions about the Internet is that authoritarian governments are just weak and ineffective bureaucracies that fear or don't understand technology and are thus bound to misuse it, even if, in defiance of "dictator's dilemma," they let it in. This is, of course, a misguided perspective: after all, it is Hugo Chavez, not Hillary Clinton, who is tweeting from his BlackBerry. (Asked about Twitter a few days after the protests began in Iran, Clinton couldn't muster anything but "I wouldn't know a Twitter from a tweeter. But apparently it is very important.")

The reason why so many authoritarian governments have proved adept at getting their online propaganda act right is because they have been surrounded by some of the brightest Internet visionaries in their countries. It's a mistake to think that such governments rely on poor advice from the wrong people; they are listening to some of the best advice they can get. In part, this is happening because working for the government entails a lot of undisputed advantages. Despite the seeming irrelevance of the ideals of communism in today's highly globalized world, some of the brightest young people in China are still very much driven to join the ranks of the Communist Party, if only because it provides a significant boost to their career. (A popular slogan among Chinese youth goes like this: "Before you enter the Party you're sweating all over; after you enter the Party you can relax. Before you enter the Party you're full of revolution; after you enter the Party you meld into the crowd.") One Chinese university surveyed students and intellectuals who were Party members and found out that almost half of them joined the Party in the hope that it would help them find a good job. Since modern authoritarian regimes don't really project a set of coherent values or ideologies—China and Russia haven't been able to successfully communicate what they are really about post-Lenin and post-Mao—many talented young people are not as disturbed working for their governments as they might have been had their leaders actually believed what they were saying in public.

The Kremlin has been particularly successful in cultivating strong connections with Russia's vibrant Internet culture, leveraging it to the government's own ideological advantage. Take Yevgeny Kaspersky, the founder of the popular antivirus software Kaspersky Lab and one of the brightest minds on matters of Internet security, who was persuaded to join Russia's Public Chamber, a quasi-state institution composed of pro-Kremlin celebrities, businessmen, and intellectuals that puts the stamp of civil-society approval on whatever initiative the Kremlin demands of them. On joining the Chamber, Kaspersky began advocating that online anonymity may need to go to keep the Internet working.

But no one embodies the immense sophistication of the Russian Internet propaganda machine better than Konstantin Rykov. Born in 1979, Rykov is an undisputed godfather of the Russian Internet and has had a major part to play in some of its most bizarre countercultural projects since the mid-1990s. In 1998 he was one of the founders of an e-zine with the suggestive URL of fuck.ru, mixing reports about Moscow's nightlife with jokes, light entertainment, and articles about art. He has even fashioned himself as Russia's first pornographer in numerous media interviews.

Few of Rykov's earlier collaborators would have predicted that just a decade later he would become a respected deputy of the Russian Duma—actually, one of its youngest members—also moonlighting as the Kremlin's unofficial ambassador to "all things Internet." Gone was the radicalism of the 1990s; Rykov the deputy established himself as a strong defender of family values, arguing for a crackdown on violence and pornography in the media, and proposing to ban all children younger than nine from surfing the Internet on their own.

What made Rykov so appealing to the Kremlin was that in the intervening years he had built a successful propaganda empire, spanning both traditional and new media. On the traditional side, Rykov founded "Poplit," a publishing house that specializes in popular fiction and aggressively uses social media to push its works to the broader public. Rykov's Internet presence, clustered mostly around his company New Media Stars, consists of sites like zaputina.ru ("'For Putin!'"), a short-lived Internet campaign that urged Putin to seek a third term as president; dni.ru, a popular Internet tabloid; dozory.ru, a science-fiction Internet game; and vz.ru, something of a Russian *Slate* with a strong conservative bent.

Rykov, one of the founders of Russia.ru, the hedonistic Internet TV station that produces "The Tits Show," is thus responsible for churning out heaps of highly propagandist video content. One of the highlights of Rykov's propaganda career came when his company produced a documentary called *War 08.08.08: The War of Treason*, which dealt with the highly sensitive subject of 2008's war between Russia and Georgia. Cut from hours of video footage that Georgian soldiers themselves had supposedly shot on their mobile phones—later confiscated by Russians—the film analyzed the war from a highly ideological position, portraying Georgians in the worst possible light imaginable.

The film quickly became a viral sensation, with 2.5 million people watching it online. The success of the online campaign owes much to the zeal with which producers of the film embraced all forms of digital distribution, putting it on all important peer-to-peer networks and encouraging piracy. Putting the film on Russia.ru helped to make it visible.

The film made a smooth transition to TV screen, too: On the first anniversary of the war, the film was shown on one of Russia's national patriotic channels. And to ensure absolute media dominance, it was followed up by a book, which was well received by Russian bloggers and journalists. This is a level of media convergence that most Western content producers can only dream of.

As one of the most knowledgeable people about the Internet in today's Russia, Rykov understands the Streisand Effect and is thus trying to steer the government away from pure and highly visible Internet censorship and toward softer, propaganda-driven methods of control. Speaking in 2009, he left little doubt about his views on the issue: "Censorship is not compatible with the Internet. So long as the Internet exists, censorship is impossible." Russian leaders follow Rykov's lead and eagerly beat the drum of "there is no Internet censorship in Russia," if only to position themselves as more progressive than their Chinese neighbors. This is Rykov's propaganda genius at its best: The Kremlin manages to score propaganda points simply by *not* censoring the Internet.

With people like Rykov shaping the Kremlin's online propaganda strategy, getting Russians to see through what they read on the Internet—even if there is no Internet censorship—won't be easy. What's worse, Rykov's efforts are now being complemented by a growing cohort of younger Internet gurus, who appeared in part thanks to the hard work of various Kremlin-affiliated youth movements. (The creation of state-supported youth movements was the Kremlin's way to minimize the threat of a potential color revolution, as pro-Western youth movements played an influential role in toppling the governments in Serbia, Georgia, and Ukraine.) Of those gurus, the most remarkable character is Maria Sergeyeva, a twenty-five-year-old member of Young Guard, one of the Kremlin's pocket youth organizations.

Sergeyeva, a stunning blonde and a student of philosophy, writes a very popular personal blog in which she ruminates about the need to support the dying Russian auto industry, extols Catherine the Great, tells all immigrants "to "go home," and occasionally posts photos from the coolest parties around town. "I was brought up to be a patriot from day one, " she said in an interview with the *Times of London*. "My love

for Russia came with my mother's milk. I loved listening to my grand-parents' heroic tales from the war. [Vladimir] Putin has given us sta-bility and economic growth. It's good that he's hardline and tough."
Imagine a blogging Pat Buchanan trapped in the body of Paris Hilton:
That's Sergeyeva in a nutshell. Thanks to a series of pro-Kremlin short video clips and blog posts that went viral, Sergeyeva even gained noto-riety in the international press, with the *Daily Mail* calling her "Putin's Poster Girl" and the *New York Daily News* calling her "Russia's Sarah Palin."

The Kremlin badly needs people like Sergeyeva to reach younger au-diences that are unreachable via the platforms the government already controls—radio, television, and newspapers. Bringing the young people back into the Kremlin's sway—in part, by making the Kremlin look "cool"—is such a high priority that in 2009 Vladimir Putin, in his ad-dress to a national hip-hop convention, proclaimed that "break danc-ing, hip-hop, and graffiti" are more entertaining than "vodka, caviar and nesting dolls." (The Kremlin's attempts to co-opt the hip-hop move-ment are quite similar to their attempts to co-opt the blogosphere; un-like Russian rock and pop singers, rappers eagerly take on political subjects and sing of corruption, police brutality, and the Kremlin's dis-regard for the poor.)

When needed, the Kremlin is also prepared to make alliances between the Internet, the youth movements, and religion. In 2009 Boris Yake-menko, the founder of the Kremlin's Nashi youth movement, penned an op-ed advising the Russian Orthodox Church what to do about the Internet. He called for the creation of "Internet-missionaries . . . who could argue and convince . . . who could snatch out those who have been sucked into the World Wide Web and direct them to people who could direct them to the church. . . . The victory of the church online is essential in the struggle for the youth." It took less than a year before the head of the Russian Orthodox Church heeded Yakemenko's call and urged his followers to become active online.

Further amplifying the influence that the likes of Rykov and Sergeyeva have on the national discourse, the Kremlin is happy to turn them into national celebrities by broadcasting fragments of their speeches or highlighting their numerous public activities in the state-controlled traditional media. Maksim Kononenko, a prominent conservative blogger close to Rykov, even got to co-host his own prime-time TV show on a national channel; this only added to his popularity in the blogosphere. Thus new and old media reinforce each other; the more national prominence the Kremlin's new propagandists get on television or in print, the more people pay attention to their work online. Bloggers affiliated with the Kremlin start with an unfair advantage and nearly unlimited resources; it's little wonder they become more visible than their liberal opponents. Ironically, the Kremlin is aggressively exploiting its existing powers to profit from the decentralized nature of the Web.

They are putting money into training as well. In 2009 a Kremlin-affiliated think tank launched The Kremlin's School of Bloggers, a series of public talks and workshops given by the leading ideologues and propagandists of the current regime. The origins of this project illustrate how seemingly benign actions by the West could trigger destructive counter-actions by the very governments the West seeks to undermine. The founding of this project was a direct response to another "school," pompously called The School of Bloggers, that was organized by the Glasnost Foundation, an organization funded in part by America's own National Endowment for Democracy (of which the hawkish Ambassador Palmer—the one seeking to overthrow dictators with the help of the Internet—is a cofounder).

Once conservative Russian bloggers, led by none other than Maksim Kononenko, discovered that a foundation so closely associated with the U.S. government was somehow involved in funding schools of bloggers (there was more than one) across Russia, the blogosphere was brimming with all sorts of conspiracy theories and suggestions on how to counter the "virtual threat" to the Russian sovereignty, which eventually coalesced into The Kremlin's School of Bloggers. (It certainly didn't help that the person the Glasnost Foundation chose to lead this project was a prominent Russian journalist once accused of spying for the Japanese.) Of course, much of the controversy could have been easily avoided if the project carried a less catchy name. Few pro-Kremlin supporters would have ever even noticed a seminar called "Basic Electronic Technologies for Professionals in the Non-Profit Sector: An Introduction." But given the growing politicization of blogging, any training course or conference that has the word "blog" in its title is increasingly perceived as a guerrilla camp for launching the next color revolution. If anything, the publicity associated with the initial School of Bloggers made the Kremlin painfully aware that it needs to be present in the new media space or risk losing it to the West.

The mastermind behind The Kremlin's School of Bloggers is Alexey Chadayev, age thirty-two, now the top ideological functionary in the ruling United Russia Party and one of the most sophisticated intellectual apologists of Putin's regime. In 2006 Chadayev even published a serious work on the subject with the revealing title of "Putin and His Ideology," which United Russia later proposed to consider "the official interpretation of the government's ideology." Chadayev, who wrote his master's thesis on how subcultures form on the Web, entered politics by means of technology: He designed a website for a prominent liberal Russian politician. A few years later, he even ran an anti-Putin online campaign, before switching sides and going to work for the Kremlin.

Chadayev is the opposite of the populist and anti-intellectual Sergeyeva and is not afraid to flaunt his erudition. He is particularly fond of discussing the relevance of thinkers like Slavoj Zizek, Jacques Lacan, and Gilles Deleuze to the Kremlin's propaganda strategy on both his blog and, more recently, his Twitter account. In July 2010 a series of his angry tweets even forced the head of the Kremlin's human rights commission, one of the few Russian liberals in a position of power, to resign. Chadayev, who once confessed that "everything I have now, I owe to the Internet," is extremely well-versed in both the latest Internet trends and cutting-edge methods of modern propaganda. He also keeps an eye on Russia's diverse youth culture, and particularly the growing cluster of people who eschew the state-controlled media, form small online communities, and consume all their news from the Web instead. "The current task in front of United Russia is to find a common language with those people, find a point of entry to those online communities" was how he described the professional agenda for him and his party.

Like Rykov and Sergeyeva, Chadayev knows how to make messages resonate on the Internet—mainly because the Internet is where he matured professionally and intellectually. His is the kind of sophisticated behind-the-scenes tinkering that is easy to miss, especially were one to judge the Russian blogosphere only based on the number of new voices participating in the online conversation. With advisors like this by their side, it's little wonder that the Kremlin feels no strong need to control the Web. For them, it's not just a space to be controlled but an exciting new playground for propaganda experiments.

Fifty Cents Gets You a Long Way on the Spinternet

While Rykov and Sergeyeva do not have to conceal their relationship with the Kremlin, since they are comfortable producing branded propaganda, some governments are exploring more anonymous and creative models. China's Spinternet, for example, is much more decentralized, with local and regional authorities playing a crucial role in shaping public discourse in their own areas of the blogosphere. Collectively, China's pro-government Internet commentators are known as the Fifty-Cent Party, with "Fifty-Cent" referring to what they supposedly earn for each pro-government comment.

David Bandurski, a China analyst at Hong Kong University, who keeps a close eye on the evolution of the Fifty-Cent Party, says that their mission is "to safeguard the interests of the Communist Party by infiltrating and policing a rapidly growing Chinese Internet." Part of a giant propaganda machine, they deliberately engage in online discussions, steering them in ideologically appropriate directions and, according to Bandurski, "neutralizing undesirable public opinion by pushing pro-Party views through chat rooms and Web forums." Bandurski estimates that there may be as many as 280,000 Fifty-Centers. Not only are some of them regularly paid for their online contributions, but various government bodies also organize routine training sessions to hone their argumentative skills.

The government doesn't hide its role in manipulating online conversations. Wu Hao, the official in charge of damage control in the "elude the cat" episode, acknowledges that "when there is the situation that [online] opinion leans totally to one side, then we will indeed put some different voices out there to allow the public to make their own judgment independently." In other words, the Chinese officials don't mind the public reaching their own conclusions independently but they'll do their best to manipulate the evidence. Li Xiaolin, the head of the propaganda department in the Chinese city of Shaoguan, suggests that many of the activities of the Fifty-Centers only aim at countering rumors rather than spreading propaganda: "Sometimes a rumor is like a snowball. It will become bigger and bigger, especially on the internet. If there is a lack of communication, it will create a market for rumors. If communication goes well, there is no space for rumors."

The very idea of a Fifty-Cent Party—a hybrid between old government-run propaganda models and new and agile forms of persuasion that can be used outside of the government apparatus—fits the Chinese leaders' fascination with "public-opinion guidance," where the government and the citizens mutually reinforce what each other does, with the government, of course, playing the leading role. Chinese communist intellectuals are also acutely aware that the propaganda model must adapt to the Internet age, with some of them aggressively touting the benefits of a more proactive ideological approach to the Internet. As Huang Tianhan and Hui Shugang, two young Chinese academics, argue: "We must . . . realize that there is a huge gap between the traditional forms of propaganda and education and the ways of modern mass media. This forces us to implement creative changes in traditional forms of propaganda, and to use modern high technology to fine-tune, enrich, and perfect the content and forms of our culture—in order to make it easier for young people to accept and be influenced by education." (The language of propaganda is subject to such "creative changes" as well. At a 2010 gathering of nine hundred officials and students at the Central Committee's Party

School, China's vice president, Xi Jingping, urged Chinese officials to purge their speeches of any "unhealthy" writing that may undermine efficiency by ridding them of "empty words" and political jargon.)

The emergence of Fifty-Cent Party commentators on the Chinese Internet is an important stage in the country's constantly evolving propaganda strategy; in its latest reiteration, it is characterized by greater decentralization, increased reliance on the private sector, and a radical internationalization. In her 2009 book *Marketing Dictatorship: Propaganda and Thought Work in Contemporary China*, Anne-Marie Brady, one of the world's foremost experts on the evolution of China's propaganda, notes a shift toward more scientific ways of producing and thinking about propaganda among Chinese officials. Since the Tiananmen crackdown, they have been paying more attention to public relations, mass communications, and social psychology. According to Brady, after the Tiananmen tragedy, which was arguably the direct result of temporarily slowing down the propaganda machine and allowing freer discussions to take hold in the 1980s, the Party turned to the old slogan "seize with both hands; both hands must be strong," which meant that both economic development and propaganda should serve as sources of political legitimacy.

Luckily for the CCP, plenty of Western intellectuals, especially in the first half of the twentieth century, also saw propaganda as essential to the functioning of a modern capitalist state. Not surprisingly, the works of American propaganda theorists like Harold Lasswell (signature quote: "we must not succumb to democratic dogmatisms about men being the best judges of their own interests") and Walter Lippmann ("the public must be put in its place . . . so that each of us may live free of the trampling and the roar of a bewildered herd") have been translated into Chinese and, according to Brady, are quite popular with China's own propaganda officers.

The Chinese propaganda leaders, in other words, look to the West and imbibe its vast intellectual resources, using them for their own antidemocratic purposes. (Something similar is happening in Russia, where Kremlin-affiliated young intellectuals use their blogs to share download links to pirated editions of key Western academic texts in economics, psychology, philosophy, and political science.) Brady notes that "the re-invention of the British Labour party under Tony Blair became a model for the CCP's own repackaging in the 1990s." Peter Mandelson, who played a key role in reshaping the Labour Party, was invited to give a talk at the Central Party School in 2001 and share his insights. Brady believes that Chinese propaganda officials used Blair's spin doctors as a model for managing the media during political crises in the wake of the SARS crisis of 2002 and 2003. Chinese officials also paid visits to left-wing parties in Germany to study their transformation in the last few decades. Given that most propaganda tricks used by the Chinese regime today come from Western textbooks, it won't be shocking if one day we'll discover that the Fifty-Cent Party was inspired by the widely spread corporate practice of "astroturfing," or faking support from the grass roots to seek political or corporate benefits. It's as if the Mad Men have set up an office in Beijing.

The Chinese experience has inspired other governments, both authoritarian and democratic, to create their own cyber-brigades of loyal Internet commentators. In 2009 the Nigerian government sought to enlist more than seven hundred Nigerians abroad and at home and create a so-called Anti-Bloggers Fund intended to raise a new generation of pro-government bloggers to engage in online battles with antigovernment opponents. Their compensation was cybercafé vouchers and blogging allowances. In late 2009, editorials in official Cuban newspapers began calling for pro-government Cuban journalists to "man the cyber-trenches" and defend the revolution online by setting up their own blogs, leaving critical comments on antigovernment blogs, and reprinting the best pro-government blog posts in the official media. Before they decided to pursue China's "elude the cat" tactic and co-opt their critics, the South Korean authorities accused officials in the North Korean government of using fake identities to precipitate a war (the North Koreans were allegedly spreading rumors that the sinking of the warship was not caused by a North Korean torpedo, as the South Korean government was claiming, and that the evidence presented so far had been fabricated). In May 2010 the ruling party of Azerbaijan, concerned with the fact that antigovernment activists had been aggressively using Facebook and YouTube to disseminate oppositional materials, hosted a meeting with pro-government youth groups in which it was decided that the nascent Azeri Spinternet movement would be given a dedicated office, from which the staff could engage in online battles with opponents of the regime.

Before he jumped on the Twitter bandwagon, Hugo Chavez announced the formation of Communicational Guerrilla, a network of seventy-five students between ages thirteen and seventeen. Dressed in khaki jackets and wearing red bandanas tied around their necks, they were supposedly trained to "fight against imperialist messages," either on social networks online, on walls, in pamphlets, or "through direct intervention."

Egypt is not far behind. Noticing that Facebook had been used to publicize antigovernment protests in 2008, Egyptian authoritarians decided to embrace the site as well—it was too popular to be banned. As Gamal Mubarak, the son of Hosni Mubarak and his likely successor, began giving online interviews, more than fifty Facebook groups, all of them supposedly of the grassroots variety, sprang up online to nominate him for the presidency.

For all the anti-Internet sentiment exhibited by the Iranian authorities following the 2009 protests, they seem to have grasped the message that they need to become active players in cyberspace. In 2010 Iran's hard-liners launched their own social networking site, Valayatmadaran (the name is a reference to "followers of the *velayat*," or Iran's Supreme Leader Ayatollah Ali Khamenei). It offers the standard package expected of such a site: Its members (by mid-2010 around 3,000) can become friends, post pictures (cartoons ridiculing the Green Movement seem to be particularly popular), videos, and links to interesting articles. The only twist is that the site's members seem to be united by little else than the highly ambitious goal of fighting "evil," although there is also space to discuss more prosaic issues like "the rule of the supreme jurist" and "women and family."

In a sense, the appearance of such a site is just the next step in the country's long-running strategy of co-opting new media. Iran has been training a new generation of religious bloggers since 2006, when the Bureau for the Development of Religious Web Logs was set up at Qom, the center of religious scholarship in the country. Most of the bureau's activities target women. The clerics may have grudgingly accepted the fact that blogging women are here to stay, but they are still trying to shape the topics of their conversations. In 2006 Iran proudly hosted Quranic Bloggers Festival, an event that also featured a blogging contest that purported to "help expand presence of the holy Quran on the internet."

The Revolutionary Guards, too, have been aggressive in cyberspace. In late 2008, they even pledged to launch 10,000 blogs under the supervision of the paramilitary Basij forces to counter the secular bloggers. All of this could come in quite handy during emergencies. The most remarkable—but also the most overlooked—fact about Iran's Twitter Revolution was that two weeks after the protests began, the number of pro-government messages on Twitter increased two-hundred-fold compared with the period immediately after the election. And most likely this was not because Iran's Twitterati suddenly fell in love with Ahmadinejad.

Small Doses of Propaganda Are Still Bad for You

Whether they steer public opinion by training an army of Fifty-Centers—masking themselves as the "real" voices of the people, on a crusade to expose the biased and Western-funded opinions of those opposing the government—or by empowering a number of charismatic Internet personalities like Rykov or Sergeyeva, authoritarian governments have proved remarkably adept at shaping the direction, if not always the outcome, of most sensitive online conversations.

Obviously, not all of these schemes work. Some of the online propaganda efforts are still clumsy, as the "elude the cat" episode demonstrates; some cannot entirely diffuse social discontent because spinning efforts come too late or the issue is so big that no propaganda can suppress it. Yet it's high time that we disabuse ourselves of the naïve belief that the Internet makes it easier to see the truth and avoid government shaping of news agenda. The fact that public discourse in the Internet era has become decentralized—allowing everyone to produce and disseminate their own views and opinions at almost no cost—does not by itself herald an era of transparency and honesty.

The existing imbalance of power between state structures and their opponents means that from the beginning the more powerful side—in virtually all cases, the state—is better positioned to take advantage of this new decentralized environment. Decentralization, if anything, creates more points of leverage over the public discourse, which, under certain conditions, can make it easier and cheaper to implant desired ideas into the national conversation.

Free and democratic societies do not have much to boast of here either. It's the Internet culture that we have to thank for the persistence of many recent urban myths, from the idea of "death panels" to the belief that climate change is a hoax. And these otherwise crazy ideas persist even in the absence of a well-funded propaganda office; the dynamics of collective belief under authoritarian conditions could make truth even harder to establish (not to mention preserve).

It's not the *New York Times* that those living under authoritarian conditions hold up as a benchmark against what they read online; it's that original bastion of fair and balanced reporting, the newspaper *Pravda*. And compared to *Pravda* (which means "truth" in Russian) or *Izvestiya* (another highly propagandist paper of the Communist regime; its name means "news"), almost anything ever published online—no matter how anonymous or profane—looks more believable. The old Soviet joke said it best: "In the *Truth* (*Pravda*) there is no news, and in the *News* (*Izvestiya*) there is no truth." Today, most people in authoritarian countries are operating in a media environment where there is some truth and there is some news, but the exact balance is unclear, and misjudgments are inevitable.

It's not surprising, then, that surveys continuously reveal that Russians trust what they read online more than what they hear on television or read in newspapers (and not just Russians—plenty of Americans still seem to believe that Barack Obama was born in Kenya). History has made them intimately familiar with the methods of *Pravda*-based propaganda—and it takes some imagination and experience with Internet culture to understand how such methods can be applied to the online environment. The myth that the Internet is somehow unsuitable to government propaganda is as strong among its direct recipients as it is among their Western sympathizers.

It's not so hard to discern what it is that governments are trying to achieve by flooding blogs and social networks with artificially engineered content. In most cases, the goal is to create an impression that moderate, pro-democracy, pro-Western positions are less popular with "netizens" than they are in reality, while also trying to convert more "undecided" citizens to their causes. At a certain point, economies of scale begin to kick in: The presence of paid commentators may significantly boost the number of genuine supporters of the regime, and the new converts can now do some proselytizing on their own, without ever asking for their fifty cents.

Thus, all the government has to do is to "seed" a pro-government movement at some early stage—inject it with the right ideology and talking points and perhaps some money—and quietly withdraw into the background. All the heavy lifting can then be done by actual ideologically committed proponents of a given political system (of which, unfortunately, there are plenty, even in the most brutal regimes).

Darning Mao's Socks, One SMS at a Time

Modern propaganda does not discriminate between platforms, easily penetrating text messages, computer games, blog posts, and, most recently, ringback tones. In 2009 millions of customers of the state-controlled China Mobile, who perhaps were not feeling patriotic enough on the country's National Day, woke up to discover that the company replaced their usual ringback tone with a patriotic tune sang by the popular actor Jackie Chan and a female actress. Just like their Russian counterparts, Chinese propagandists have embraced media diversity and love to package their messages in more than one platform, as perhaps the only way to secure exposure to the younger audiences that are hard to reach with traditional media. China's is the kind of communism that doesn't mind imbibing the worst advertising excesses of its capitalist adversaries; the excesses, in fact, allow it to keep on going. These days even the website of China's Defense Ministry has a section with music downloads; one can enjoy jingoistic music all one wants.

It's through innovative platforms like computer games that even old and seemingly obsolete messages can find a new life. Two Chinese games, *Learn from Lei Feng* and *Incorruptible Warrior*, show that creativity, hedonism, and gaming do not always have to part ways with authoritarianism. (The provenance of such games is not always clear; some are funded by the government, some are merely subsidized, and some are produced by the private sector in hopes to curry favor with the government.)

The protagonist of *Learn from Lei Feng*, a brave but plain soldier in the People's Liberation Army who died on duty at the tender age of twenty-two, is a real and much venerated character in the history of communist China. Chairman Mao seized on his life story for its immense propaganda value, commemorating Lei Feng on book covers, posters, and stamps. In keeping with the spirit of the original story, the protagonist of *Learn from Lei Feng* has simple but Party-strengthening tasks to accomplish: darn socks, volunteer on building sites, and fight enemy agents. And in case his health is running low, the Party secretary is always there to help. The sock-darning quest is clearly worth it: Lei Feng's ultimate reward is a pile of Chairman Mao's collected works.

Those playing *Incorruptible Warrior* have a more ambitious mission to accomplish. They need to coordinate an assortment of prominent characters from Chinese history in their quest to fight and kill corrupt officials, usually accompanied by bikini-clad mistresses and muscular henchmen. Who doesn't like to punch a corrupt local mayor in the face? Released in the summer of 2007, the game almost immediately attracted over 10,000 players; its site even had to temporarily shut down so that more players could be accommodated.

One of the most interesting and overlooked features of today's globalized world is how much and how quickly authoritarian governments
seem to learn from each other; any new innovations in Internet control
by the most advanced are likely to trickle down to others. China's experiments with game propaganda, for example, seem to have inspired
Russian lawmakers, who in early 2010 passed a law to provide tax
breaks to Russian gaming companies that would produce games with
a patriotic bent. A few months later Vietnam's Ministry of Information
and Communication proposed a similar bill that aimed at encouraging
local firms to develop more online games while at the same time restricting foreign game imports. ("Incubators of new media suppression"
is how Chris Walker, director of studies at Freedom House, dubbed the
likes of Russia and China.)

But it's not just games. The use of text messaging for propaganda
purposes—known as "red-texting"—reveals another creative streak
among China's propaganda virtuosos. The practice may have grown
out of a competition organized by one of China's mobile phone operators to compose the most eloquent Party-admiring text message. Fast
forward a few years, and senior telecom officials in Beijing are already
busily attending "red-texting" symposia.

"I really like these words of Chairman Mao: 'The world is ours, we
should unite for achievements. Responsibility and seriousness can con-
quer the world and the Chinese Communist Party members represent
these qualities.' These words are incisive and inspirational." This is a
text message that thirteen million mobile phone users in the Chinese
city of Chongqing received one day in April 2009. Sent by Bo Xilai, the
aggressive secretary of the city's Communist Party who is speculated
to have strong ambitions for a future in national politics, the messages
were then forwarded another sixteen million times. Not so bad for an
odd quote from a long-dead Communist dictator.

It appears, then, that the Internet is not going to undermine the propaganda foundations of modern authoritarianism. The emergence of fast,
decentralized, and anonymous communications channels will certainly

reshape the ways in which government messages are packaged and disseminated, but it will not necessarily make those messages less effective. But they don't have to be decentralized; the emergence of cool, edgy, and government-loyal Internet personalities—like in the Russian case—may also boost governments' attempts at controlling the Internet conversation.

Can supporters of democracy in the West stop or at least thwart the growth of the Spinternet? Maybe. Should they be trying? This is a much harder question to answer. We're not completely powerless in the face of the Spinternet. Western governments could fight online propaganda in a variety of ways. For example, they might create some kind of a ratings site for all Russian or Chinese commentators, where their reputation can be ranked. Alternatively, all comments coming from one IP address might be aggregated under a unique online profile, thus exposing the propagandists working from the offices of the government's propaganda department or even their PR consultants. But just because one could fight spin does not mean that one should. In most cases, such Western interventions would also erode online anonymity and put dissidents' lives on the line. This may not be such a high price to pay in democracies—many still remember the news storm that erupted once it became known that the CIA is fond of editing Wikipedia pages—but in authoritarian states it may also unintentionally expose enemies of the regime.

Spin-fighting tools are, perversely, often the best instruments of monitoring and identifying dissent, and they should thus be used very carefully. The inherent tension between fighting propaganda and keeping the Internet anonymous could leave our hands tied. If we ignore this tension, we are only playing into the hands of repressive regimes.

Another reason why fighting spin is difficult is because the authoritarian hydra has too many heads. Just look at the way authoritarian governments are now busy hiring lobbying firms in Washington and Brussels to advance their political agenda, increased transparency may simply prompt them to shift their online propaganda activities to Western PR firms, making their spinning even more sophisticated.

The best that Western governments can do is to educate, whether in person or remotely, those running websites that oppose the government about how to build communities, make their content visible, and find ways to resist being overloaded with pro-government commentators. Although the proliferation of spin is a natural feature of the modern Internet, it may still be possible to outspin the spinners.

chapter six

Why the KGB Wants
You to Join Facebook

● ● ■

Imagine that you are a target of some deeply mysterious spying operation. While you happily poke your online friends, tweet your breakfast plans, and shop for Christmas presents, all your online activity is being secretly reported to an unknown party. Imagine that someone has also broken into your computer and is using it to launch DDoS attacks. They could be targeting Saudi websites about philosophy or dissident Georgian bloggers. You have no idea that your computer is part of this mysterious cyber-army, let alone who is being attacked or why. It's as if a stranger has been secretly reading your diary and also using it to clobber a passerby.

This is precisely what happened to a number of brave activists from Vietnam who in 2009 were protesting the building of a new bauxite mine in their country. (The project is a joint venture between Chalco, a subsidiary of China's state-run aluminum company Chinalco, and the Vietnamese government.) Their computers were compromised, allowing an unknown third party not only to monitor their online activity but also to attack other online targets in Vietnam and elsewhere. But theirs was not a case of basic computer illiteracy, where pressing the wrong button or visiting a weird porn site could surrender months of hard work to a nasty virus. It's quite likely that the Vietnamese dissidents did no such thing, avoiding any suspicious-looking sites and attachments. What could have gone wrong?

Vietnam, nominally still ruled by a Communist Party, boasts a burgeoning Internet culture, with antigovernment bloggers mounting frequent campaigns about social issues, especially the poorly regulated sprawling urban development. The government, concerned that its tight hold on public life is beginning to loosen, has been trying to reassert control, preferably without drawing much ire from Vietnam's trading partners in the West. The authorities, all too keen on harvesting the information benefits of globalization, do not shy away from computers or the Internet outright. In April 2010 they embarked on an ambitious crusade to supply farmers in more than one thousand communes with free computers, so that they could, as one official put it, "contact and consult with . . . scientists . . . [about] the current epidemics on their breeds and seeds." The government was even kind enough to organize computer training courses for the farmers.

Those opposing the government's paradigm of "modernization at all costs" are unlikely to be invited to attend such courses. In 2009 two of the most vocal blogs that challenged the government, *Bauxite Vietnam* and *Blogosin*, became targets of powerful DDoS attacks similar to those launched against Tomaar and Cyxymu. Soon, *Bauxite Vietnam* was forced down the "digital refugee" route, eventually emerging on a Google-owned blogging service, while the blogger behind *Blogosin* told his readers that he was quitting blogging altogether to "focus on personal matters." Those attacks made it quite clear that the Vietnamese government was up-to-date on the rapidly evolving nature of Internet control and wouldn't stop at just blocking access to particular websites.

Most likely, the antimine activists, careful as they were, inadvertently hit a government trap that allowed the secret police to establish remote control over their computers. And what a trap it was: Someone broke into the server that hosted the website of the Vietnamese Professionals Society (VPS), a trusted diaspora organization, and replaced one of the most popular downloads, a simple computer program that facilitated typing in the Vietnamese language, to an almost identical file—"almost" because it also contained a virus. Anyone who downloaded and installed the software risked turning their computer into a powerful spy and attack hub. Such breaches of security are generally hard to detect, for everything seems to be working normally, and no suspicious activity is taking place.

The Vietnamese activists might have never actually discovered that their every online move was being followed were it not for the buzz generated by the high-profile cyber-attacks that hit Google in December 2009. While investigating the mysterious origins of those sly attacks, researchers from McAffee, a computer security firm, accidentally unearthed the clandestine spying operation in Vietnam and initially believed the two to be related (they were not). The media buzz generated by McAffee's unexpected discovery—also heavily publicized by Google through their own channels—probably generated enough coverage in the Western press to protect the Vietnamese activists from immediate persecution, even if a lot of their private data might have been compromised nevertheless. It's impossible to say how many similar spying operations go undetected, putting authoritarian governments ahead of their opponents.

Never Trust Anyone with a Website

But many such surveillance campaigns—especially when heavily publicized in the media—have effects that extend far beyond the mere gathering of information. Knowing that they might be watched by government agents but not knowing how exactly such surveillance happens, many activists might lean toward self-censorship or even stop engaging in risky online behavior altogether. Thus, even if authoritarian governments cannot actually accomplish what the activists fear, the pervasive climate of uncertainty, anxiety, and fear only further entrenches their power.

Such schemes have much in common with the design of the perfect prison, the panopticon, described by the nineteenth-century British utilitarian philosopher Jeremy Bentham. The point of such systems is to exert control over prisoners' behavior, even when nobody is watching them, by never letting the prisoners know if they are being watched. Governments, of course, are quite happy to overstate their actual capabilities, for such boasting works to their advantage. Thus, in January 2010, when Ahmadi Moghaddam, Iran's police chief, boasted that "the new technologies allow us to identify conspirators and those who are violating the law, without having to control all people individually," he must have known that his words would have an effect even if he had greatly exaggerated his capability. When no one quite knows just how extensive government surveillance really is, every new arrest of a blogger—whether it's based on genuine surveillance practices, tips-off from the public, intuition, or flipping through a phone book—will help deter subversive action, especially from those who are not full-time dissidents.

Security has never been among the Internet's strong sides, and the proliferation of social media in the last decade has only made things worse. Even the most protected email service won't protect your password if there is a keylogger—software that can record and transmit your every keystroke—installed on your computer (or that slow and funky computer in a random Internet café that you once had to use). Nor does one need to break into your email to read some of it. Mounting and hiding a tiny, almost invisible digital camera behind your back is enough. Similarly, even secure, encrypted services like Skype will be of little consolation if a secret police operator occupies an apartment next to yours and sticks a parabolic microphone out the window. As long as most virtual activities are tied to physical infrastructure—keyboards, microphones, screens—no advances in encryption technology could eliminate all the risks and vulnerabilities.

But as security professionals attest, while it's possible to minimize the risks created by the infrastructure, it's much harder to discipline the users of a technology. Many sophisticated attacks originate by manipulating our trust networks, like sending us an email from a person we know or having us download files from trusted websites, as happened in the case of the Vietnamese activists. When we visit a website of an organization we trust, we do not expect to be hit with malware any more than we expect to be poisoned at a dinner party; we trust that the links we click on won't lead to sites that will turn our computers into mini-panopticons. Such trust has undoubtedly made the Internet an appealing place to do business or just waste so many hours of our lives. Few of us spend much time pondering the security settings on our favorite sites, especially if no sensitive data is divulged. But a low level of awareness is precisely what makes compromising the security of such sites so tempting, especially if these are niche sites catering to particular audiences. An attack can infect computers of all independent journalists, brave human rights defenders, or revisionist historians without triggering any suspicions from more computer-savvy user groups.

Poorly secured sites of specific communities thus enable the kind of attacks—many of which invariably result in more surveillance—that may not succeed were members of such communities targeted individually. This is what happened to Reporters Without Borders (RSF), a prominent international NGO defending freedom of expression, in July 2009, when someone inserted a malicious link into an email that RSF sent to its supporters. The link was placed next to the text of a 13,000-strong petition demanding the release of the documentary filmmaker Dhondup Wangchen from prison. Once clicked, it did lead to what looked like a genuine petition—so one would not suspect anything inappropriate—but the website also contained a security trap, infecting the computers of anyone who clicked on the malicious link. Alerted to the problem, RSF promptly removed the link, but it is difficult to estimate how many computers were compromised.

Even popular and much better-staffed organizations are not immune to embarrassing vulnerabilities that could cause damage to everyone in their social and professional circle. In early 2009 the website of the *New York Times*, which relies on banner ads provided by third parties, inadvertently served malware to some of its visitors. Such gaffes are poised to become even more widespread, as more and more websites incorporate a bevy of third-party services (e.g., Facebook's "like" button), surrendering full control over what kind of data flows through their site. When even the website of the *New York Times* feeds you viruses, there is little on the Internet you can safely surf on autopilot.

The Internet runs on trust, but its dependence on trust also opens up numerous vulnerabilities. Its effectiveness as a tool of carving out spaces of dissent and, in exceptional cases, even campaigning against authoritarian governments has to be judged on a much wider set of criteria than just the cost and ease of communications. It's quite obvious that in a world where there are no other uses for the Internet, email is a cheaper, more effective, and more secure alternative to the handwritten letter. But in a world like ours, where the Internet has many other functions, it would be a mistake to evaluate the practice of email in isolation from other online activities: browsing, chatting, typing, gaming, file sharing, and downloading and viewing porn. Each of these activities creates multiple vulnerabilities that alter the risk calculus.

It's important to avoid falling victim to Internet-centrism and focusing only on the intrinsic qualities of online tools at the expense of studying how those qualities are mitigated by the contexts in which the tools are used. Sending and receiving email on an Internet café's computer where the previous customer was downloading porn from illegal websites may not be a tremendous improvement over hand-delivering a typewritten letter. Yet this is the environment in which many activists in the developing world, short on money and equipment or simply hiding from the all-seeing eye of the secret police, are forced to work. Understanding the full gamut of risks and vulnerabilities that activists expose themselves to requires a bit more investigative work than simply comparing the terms of service that come with all newly created email accounts.

Why Databases Are Better Than Stasi Officers

Information may, indeed, be the oxygen of the modern age, as Ronald Reagan famously alleged, but it could be that peculiar type of oxygen that helps to keep dictators on life support. What reasonable dictator passes up an opportunity to learn more about his current or future enemies? Finding effective strategies to gather such information has always been a priority for authoritarian governments. Often such strategies were intrusive, such as placing bugs in dissidents' apartments and wiretapping their phone conversations, as happened in many countries of the Soviet bloc. But sometimes governments found more creative ways to do it, especially if they were simply trying to gauge public sentiment rather than peep inside the minds of particular dissidents.

The Greek military regime, for example, tried to keep track of everyone's reading habits by monitoring their choice of newspapers, thus quickly learning about their political leanings. The Greek generals would have loved the Internet. Today one could simply data-mine Amazon.com's wish lists—collections of books, films, and other items—that customers freely self-disclose. In 2006 the technology consultant Tom Owad conducted a quirky experiment: In less than a day he downloaded the wish lists of 260,000 Americans, used the publicly disclosed names and some limited contact information of Amazon's customers to find their full addresses, and then placed those with interesting book requests—like Orwell's *1984* or the Quran—on a map of the United States.

How do other old-school surveillance tactics score in the digital age? At first glance, it may seem they don't do so well. As a vast chunk of political communication has migrated online, there is little to be gained from bugging dissidents' apartments. Much of the digital information is swapped in silence, punctuated, perhaps, only by keystroke sounds; even the most advanced recording equipment cannot yet decipher those. Not surprisingly, analog bugs have long been replaced by their digital equivalent, making surveillance easier and less prone to error and misinterpretation; instead of recording the sounds of keyboard strokes, the secret police can now record the keyboard strokes themselves.

The Lives of Others, a 2006 Oscar-winning German drama, with its sharp portrayal of pervasive surveillance activities of the Stasi, GDR's secret police, helps to put things in perspective. Focusing on the meticulous work of a dedicated Stasi officer who has been assigned to snoop on the bugged apartment of a brave East German dissident, the film reveals just how costly surveillance used to be. Recording tape had to be bought, stored, and processed; bugs had to be installed one by one; Stasi officers had to spend days and nights on end glued to their headphones, waiting for their subjects to launch into an antigovernment tirade or inadvertently disclose other members of their network. And this line of work also took a heavy psychological toll on its practitioners: the Stasi anti-hero of the film, living alone and given to bouts of depression, patronizes prostitutes—apparently at the expense of his understanding employer.

As the Soviet Union began crumbling, a high-ranking KGB officer came forward with a detailed description of how much effort it took to bug an apartment:

> Three teams are usually required for that purpose: One team monitors the place where that citizen works; a second team monitors the place where the spouse works. Meanwhile, a third team enters the apartment and establishes observation posts one floor above and one floor below the apartment. About six people enter the apartment wearing soft shoes; they move aside a bookcase, for example, cut a square opening in the wallpaper, drill a hole in the wall, place the bug inside, and glue the wallpaper back. The artist on the team airbrushes the spot so carefully that one cannot notice any tampering. The furniture is replaced, the door is closed, and the wiretappers leave.

Given such elaborate preparations, the secret police had to discriminate and go only for well-known high-priority targets. The KGB may have been the most important institution of the Soviet regime, but its resources were still finite; they simply could not afford to bug everyone who looked suspicious. Despite such tremendous efforts, surveillance did not always work as planned. Even the toughest security officers— like the protagonist of the German film—had their soft spots and often developed feelings of empathy for those under surveillance, sometimes going so far as to tip them off about upcoming searches and arrests. The human factor could thus ruin months of diligent surveillance work.

The shift of communications into the digital realm solves many of the problems that plagued surveillance in the analog age. Digital surveillance is much cheaper: Storage space is infinite, equipment retails for next to nothing, and digital technology allows doing more with less. Moreover, there is no need to read every single word in an email to identify its most interesting parts; one can simply search for certain keywords—"democracy," "opposition," "human rights," or simply the names of the country's opposition leaders—and focus only on particular segments of the conversation. Digital bugs are also easier to conceal. While seasoned dissidents knew they constantly had to search their own apartments looking for the bug or, failing that, at least tighten their lips, knowing that the secret police was listening, this is rarely an option with digital surveillance. How do you know that someone else is reading your email?

To its credit, a few weeks after Google discovered that someone was trying to break into the email accounts of Chinese human rights dissidents, it began alerting users if someone else was also accessing their account from a different computer at that time. Few other email providers followed Google's lead—it would be seen as yet another unjustified expense—so this incident hardly put an end to the practice of secret police reading dissidents' email.

More important, the Internet has helped to tame the human factor, as partial exposure, based on snippets and keywords of highlighted text, makes it less likely for police officers to develop strong emotional bonding with their subjects. The larger-than-life personalities of fearless dissidents that melted the icy heart of the Stasi officer in *The Lives of Others* are barely visible to the Internet police, who see the subjects of surveillance reduced to one-dimensional, boring database entries. The old means of doing surveillance usually began with a target and only then searched for the crimes one could ascribe to it. Today, the situation is the reverse: Crimes—antigovernment slogans or suspicious connections to the West—are detected first, and their perpetrators are located later. It's hard to imagine Iranian Internet police developing sympathy for the people they investigate based on snippets of texts detected by the system, for they already know of their guilt and can always dig up more textual evidence if needed.

That technology helps to eliminate the indecision and frailty (and, more often than not, common sense and humanity) associated with human decision makers was not lost on the Nazis. Testifying at the

Nuremberg trials in 1946, Albert Speer, who served as Hitler's chief architect and later as the minister of armaments and war production, said that "earlier dictators during their work of leadership needed highly qualified assistants, even at the lowest level, men who could think and act independently. The totalitarian system in the period of modern technical development can dispense with them; the means of communication alone make it possible to mechanize the subordinate leadership." It's undoubtedly barbaric to be blaming Nazi atrocities on the evils of technology alone, but Speer had a point: The world is yet to meet a database that cried over its contents.

Tremendous cost savings introduced by digital surveillance technologies have also made it possible to shift surveillance personnel to more burning tasks. In a 2009 interview with *Financial Times*, a marketing manager for TRS Solutions, a Chinese data-mining firm that offers an Internet-monitoring service to the Chinese authorities, boasted that China's Internet police—thanks in part to the innovations developed by TRS Solutions—now only need one person where ten were required previously. But it's too early to celebrate; it's unlikely that the other nine were laid off. Most probably they were shifted to perform more analytical tasks, connecting the dots between hundreds of digital snippets gathered by automated computer systems. As the TRS manager pointed out, business is booming: "[The Chinese authorities have] many different demands—early warning, policy support, competitive spying between government departments. In the end, this will create a whole industry." Perhaps, this is not the kind of Internet-friendly industry celebrated by the proponents of wikinomics, who rarely acknowledge that, while the Internet has indeed helped to cut the unnecessary slack from many an institution, it has also inadvertently boosted the productivity of the secret police and their contractors in the private sector. A book on "wikiethics" is long overdue.

Say Hi. You're on Camera!

It's not just text that has become easier to search, organize, and act on; video footage is moving in that direction as well, thus paving the way for even more video surveillance. This explains why the Chinese government keeps installing video cameras in its most troubling cities. Not only do such cameras remind passersby about the panopticon they inhabit, they also supply the secret police with useful clues (in 2010 47,000 cameras were already scanning Urumqi, the capital of China's restive Xinjiang Province, and that number was projected to rise to 60,000 by the end of the year). Such revolution in video surveillance did not happen without some involvement from Western partners.

Researchers at the University of California at Los Angeles, funded in part by the Chinese government, have managed to build surveillance software that can automatically annotate and comment on what it sees, generating text files that can later be searched by humans, obviating the need to watch hours of video footage in search of one particular frame. (To make that possible, the researchers had to recruit twenty graduates of local art colleges in China to annotate and classify a library of more than two million images.) Such automation systems help surveillance to achieve the much needed scale, for as long as the content produced by surveillance cameras can be indexed and searched, one can continue installing new surveillance cameras.

But as the maddening pace of innovation in data analysis expands the range of what is possible, surveillance is poised to become more sophisticated as well, taking on many new features that only seemed like science fiction in the not-so-distant past. Digital surveillance is poised to get a significant boost as techniques of face-recognition improve and enter the consumer market. The face-recognition industry is so lucrative that even giants like Google can't resist getting into the game, feeling the growing pressure from smaller players like Face.com, a popular tool that allows users to find and automatically annotate unique faces that appear throughout their photo collections. In 2009 Face.com launched a Facebook application that first asks users to identify a Facebook friend of theirs in a photo and then proceeds to search the social networking site for other pictures in which that friend appears. By early 2010, the company boasted of scanning 9 billion pictures and identifying 52 million individuals. This is the kind of productivity that would make the KGB envious.

One obvious use of face-recognition technology would be to allow Iranian authorities to quickly learn the identity of the people photographed during street protests in Tehran. For why should the Iranian government embark on expensive investigations if they can get their computers to match the photos taken during the protests—many of them by the very activists appearing on them—with more casual photos uploaded on social networking profiles by the same activists? That said, governments and law-enforcement agencies had been using face-recognition technologies for a while before they became a commercially viable business. What is most likely to happen in the case of Iran is that widely accessible face-recognition technologies will empower various solo agents, socially and politically conservative cyber-vigilantes who do not work for the government but would like to help its cause. Just as hordes of loyal Thais surf the Web in search of websites criticizing the monarchy or hordes of pro-government Chinese are on the lookout for highly sensitive blog posts, hordes of hard-line Iranians will be checking photos from the antigovernment protests against those in massive commercial photo banks, populated by photos and names harvested from social networking sites, that are sure to pop up, not always legally, once face-recognition technology goes fully mainstream. The cyber-vigilantes may then continue stalking the dissidents, launch DDoS attacks against their blogs, or simply report them to authorities.

Search engines capable of finding photos that contain a given face anywhere on the Internet are not far on the horizon either. For example, SAPIR, an ambitious project funded by the European Union, seeks to create an audiovisual search engine that would first automatically analyze a photo, video, or sound recording; then extract certain features to identify it; and finally use these unique identifiers to search for similar content on the Web. An antigovernment chant recorded from the streets of Tehran may soon be broken down into individual voices, which in turn can then be compared to a universe of all possible voices that exist on amateur videos posted on YouTube.

Or consider Recognizr, the cutting-edge smartphone application developed by two Swedish software firms that allows anyone to point their mobile phone at a stranger and immediately query the Internet about what is known about this person (or, to be more exact, about this person's face). Its developers are the first to point to the tremendous privacy implications of their invention, promising that strict controls would eventually be built into the system. Nevertheless, it takes a leap of faith to believe that once the innovation genie is out of the bottle, no similar rogue applications would be available for purchase and download elsewhere.

How to Lose Face on Facebook

One gloomy day in 2009, the young Belarusian activist Pavel Lyashkovich learned the dangers of excessive social networking the hard way. A freshman at a public university in Minsk, he was unexpectedly called to the dean's office, where he was met by two suspicious-looking men who told him they worked for the KGB, one public organization that the Belarusian authorities decided not to rename even after the fall of communism (they're a brand-conscious bunch).

The KGB officers asked Pavel all sorts of detailed questions about his trips to Poland and Ukraine as well as his membership in various antigovernment movements.

Their extensive knowledge of the internal affairs of the Belarusian opposition—and particularly of Pavel's own involvement in them, something he didn't believe to be common knowledge—greatly surprised him. But then it all became clear, when the KGB duo loaded his page on vkontakte.ru, a popular Russian social networking site, pointing out that he was listed as a "friend" by a number of well-known oppositional activists. Shortly thereafter, the visitors offered Lyashkovich to sign an informal "cooperation agreement" with their organization. He declined—which may eventually cost him dearly, as many students sympathetic to the opposition and unwilling to cooperate with authorities have been expelled from universities in the past. We will never know how many other new suspects the KGB added to its list by browsing Lyashkovich's profile.

Belarus is not an isolated case, and other governments are quickly beginning to understand the immense intelligence value of information posted to social networking sites. Some even want to run their own sites, perhaps to save on surveillance costs. In May 2010, having banned Facebook and sensing the unmet and growing demands for social networking services among their population, Vietnam's Ministry of Information and Communications moved in to open their own social networking site, staffed with three hundred computer programmers, graphic designers, technicians, and editors. It is hard to say if it will become popular—with a name like GoOnline, it seems like a long shot—but from a government's perspective, it is even easier to spy on members of a social network once it knows all their passwords.

Democratic governments have also succumbed to such practices. The Indian police in the disputed territory of Kashmir, for example, are paying close attention to anything Kashmir-related that is posted on Facebook. On finding something suspicious, they call the users, ask about their activities, and order them to report to police stations. (This has prompted many activist users in Kashmir to start registering under false names, a practice that Facebook, keen not to dilute the quality of its superb user base with false entries, strongly discourages.)

Not all social networking is harmful, of course. Being part of a network carries many advantages. For example, it's much easier and cheaper to reach other members when such a need arises (e.g., before an upcoming protest). But membership in a network is something of a double-edged sword: Its usefulness can easily backfire if some segments get compromised and their relationships with other members become common knowledge. Before the advent of social media, it took a lot of effort for repressive governments to learn about the people dissidents are associated with. The secret police may have tracked one or two key contacts, but creating a comprehensive list—with names, photos, and contact information—was extremely expensive. In the past, the KGB resorted to torture to learn of connections between activists; today, they simply need to get on Facebook.

Unfortunately, there is still a widespread belief that authoritarian governments and their security services are too dumb and technophobic to go on social networking sites in search of such data. In his 2007 book *Children of Jihad* the U.S. State Department's Jared Cohen writes that "the Internet is a place where Iranian youth can operate freely, express themselves, and obtain information on their own terms. [They] can be anyone and say anything they want as they operate free from the grips of the police-state apparatus. . . . It is true that the government tries to monitor their online discussions and interactions, but this is a virtually impossible enterprise." This is simply factually wrong, as proven by the aftermath of the 2009 protests; for someone charged with developing effective Internet policy on Iran, Cohen is given to dangerously excessive cyber-utopianism. (One could only hope that it was not Cohen's Panglossian optimism that Condoleezza Rice, who hired him to work for the State Department's policy planning unit, was praising when she said that "Jared had insights into Iran that we [in the U.S. government] didn't have.") As it turns out, the Iranian authorities did spend a lot of time analyzing social networking sites in the aftermath of the elections and even used some of the information they gleaned to send warnings to Iranians in the diaspora. During the 2009 witch hunt trials in Iran, authorities used a dissident's membership in an academic mailing list run by Columbia University as proof that he was spying for Western powers.

Thus, even if an online social network is of minimum intelligence value, being friends with the wrong people provides evidence that can be used in court. Previously such information was hard to discover; often dissidents took extra efforts to conceal it. Belinda Cooper, an American activist who spent the late 1980s in GDR and was a member of several dissident environmental groups, writes that one of the rules practiced by the dissidents entering and leaving East Germany was to "never bring address books when going to the east (as border guards could and would photocopy them)." Today the situation has changed dramatically, as the lists of our friends on Facebook are available for anyone to see. Unfortunately, staying out of Facebook is not a reasonable option for most dissidents. They need to be present in these spaces to counter government propaganda, to raise awareness about their work in the West, to mobilize support for their causes among domestic audiences, and so forth. They may do so anonymously, of course, but anonymity also makes their involvement far less effective. Sakharov's advocacy would have been far less successful if he hadn't practiced it openly.

Numerous academic studies confirm that every time we share personal data on a social networking site, we make it more likely that someone might use it to predict what we are like, and knowing what we are like is a good first step toward controlling our behavior. A 2009 study by researchers at MIT has shown that it is possible to predict—with a striking degree of accuracy—the sexual orientation of Facebook users by analyzing their online friends. This is hardly good news for those in regions like the Middle East, where homosexuality still carries a heavy social stigma.

Another 2009 study conducted by researchers at the University of Cambridge, whose report is titled "Eight Friends Are Enough," found that based on the limited information that Facebook discloses to search engines like Google, it is possible to make accurate inferences about information that is not being disclosed.

Many of the functions that make social networking sites so easy to use—for example, to find one's friends who are already members of the site—also make it easy to trace identities behind emails or even trace users' activities across various other sites. Most of us know how easy it is to check whether our friends have already signed up for particular social networking sites simply by granting Facebook, Twitter, or LinkedIn temporary access to our email address book, so that those sites can automatically check the email addresses of our contacts against their lists of existing users. If five of our email buddies are already Twitter users, Twitter can let us know. So far, so good. The problem is that one can do the same operation with one's enemies as well. Email addresses can be added to address books manually, without ever having to email that person. Thus, just by knowing a person's email address, it might be possible to find her accounts on all social networking sites, even if she doesn't use those sites under her real name.

A 2010 study by Eurecom, a French research institute, sought to investigate the security vulnerabilities that such ease of use creates for the user. First, the researchers found 10.4 million email addresses on the Web; then they imported them into their address books; and, finally, they developed a simple script to automatically check with each of the popular social networking sites whether it had any users corresponding to those emails. As a result, they identified more than 876,941 emails linked to 1,228,644 profiles, with 199,161 emails having accounts on at least two sites, 55,660 on three, and so forth (11 people had their email accounts linked to seven social networking sites at once).

As was to be expected, some users who had accounts on multiple social networking sites provided different details to each (for example about their sexual orientation, location, or age). It's highly probable that quite a few of the people under investigation didn't want anyone to link the kind of frivolities they post to Twitter with their line of work, and yet researchers found at least 8,802 users who had accounts on both LinkedIn, a social network for professionals, and Twitter. If someone in that pool listed, say, "U.S. Department of Defense" as their employer on their LinkedIn profile, one could check what that person was tweeting about, even if the tweeting was done under a nickname.

Therefore, as long as social networking accounts are tied to one email address, it's also remarkably easy to tie them to a particular person, learn that person's name, and see what kind of hidden indiscretions that person may be engaging in, offline or online. The researchers, for example, found the profile of a married professor in his fifties who was also remarkably active on various dating sites. Similarly, activists who upload sensitive videos to YouTube thinking that no one could guess their real names from their usernames may be under much greater risks if they use the same email address to access Facebook and the secret police learns what that email address is.

Once alerted to such vulnerabilities, many social networking sites slightly tweaked their operating procedures, making it hard to do such checks in bulk. Nevertheless, it's still possible to find multiple online identities for individual emails through manual checking. This is not the kind of feature that is going to disappear soon, if only because it allows social networking sites to expand their user base.

Corporations are already taking advantage of the increasingly social nature of the Web. Hotels now use locations, dates, and usernames that appear on sites like TripAdvisor or Yelp to triangulate a guest's identity. If they find a likely match and the review happens to be positive, the review is added to a hotel's guest preference records. If it's negative, the travelers might be given a voucher to compensate for the inconvenience or, in the worst scenario, to be marked as "problem guests." Barry Hurd, the CEO of Seattle-based 123 Social Media, a reputation management company that works with more than five hundred hotels, believes that "technology is evolving so fast that in the future, every hotel representative could have a toolbar on his or her computer that reveals everything about a guest at the click of a mouse—every review, guest preference and even the likelihood that you'll be positively or negatively inclined toward your stay."

Of course, hotels are not authoritarian governments—they won't imprison guests in their rooms for expressing dissenting views—but if they can learn the real identities behind imaginary online nicknames, so can the secret police. Moreover, the corporate quest for de-anonymizing user identities can soon fuel a market in tools that can automate the process, and those tools can then be easily used in more ominous contexts. Intelligence agencies in the United States have already profited from data-gathering technology created on Wall Street. TextMiner, one such platform developed by Exegy, a firm that works with both intelligence agencies and Wall Street banks, can search through flight manifests, shipping schedules, and phone records as well as patterns that might form Social Security numbers or email accounts. "What was taking this one particular agency one hour to do, they can now do in one second," says Ron Indeck, Exegy's chief technology officer, in a phrase that sounds remarkably similar to the glee of the Chinese contractors at TRS Solutions. Thus, an entire year's worth of news articles from one organization can be searched and organized in "a couple of seconds." The private sector will surely continue churning out innovations that can benefit secret police everywhere. Without finding ways to block the transfer of such technologies to authoritarian states or, even more important, the kind of limits that should be imposed on such technologies everywhere, the West is indirectly abetting the work of the secret police in China and Iran.

But even in the absence of such tools, creative hacks will do the job just fine. A 2010 collaborative project between researchers at the Vienna University of Technology, the University of California at Santa Barbara, and Eurecom found an interesting way of de-anonymizing users of Xing, a popular German social networking site akin to Facebook and LinkedIn. Since most of us belong to a number of different social networking groups that vary according to our passions, life history, and lifestyle—for example, Save the Earth, Feed the Children of Africa, Alumni of the Best University in the World, Vegetarians of the World Unite—the probability that you and your friends belong to exactly the same groups is small (having attended the same liberal arts college in New England, your best friend may also want to save the earth and feed the children of Africa but also love Texas barbecue ribs).

Social networking sites do not usually hide lists of group members from nonmembers, so as not to erect too many communication barriers. It is thus possible to produce a nearly unique identifier, a "group fingerprint"—think of this as a list of all Facebook groups that a given user belongs to—for each of us. And the most obvious place to look for a matching fingerprint would be in our web browsers' history, for this is where a record of all the groups—and, of course, of all other websites we visit—is kept. All it takes to steal our browser history is to have us click on a malicious link, like the one mysteriously added to RSF's email petition, and everything we have been browsing in the last few days will no longer be private knowledge.

According to the 2010 report, producing a matching "group fingerprint" required the checking of 92,000 URLs, which took less than a minute. The researchers managed to correctly guess the identity of their target 42 percent of the time. In other words, if someone knows your Web history and you happen to be an avid user of social networking sites, she has a good chance of deducing your name. Soon, the secret police will just be able to look at the log from your favorite Internet café and learn who you are, even without asking for a copy of your passport (although that latter option is also increasingly common in authoritarian governments).

It's hardly surprising that the secret police in authoritarian regimes are excited about exploiting such vulnerabilities to fill in gaps in their databases. They may, for example, know email addresses of government opponents but not their identities. To learn their names, they could send the opponents fake emails containing malicious links that aim to steal their browsers' histories. In just a few minutes, they'll be able to attach names (as well as photos, contact details, and information about related connections) to their rather sparse database entries. Another problem is that social networking sites like Facebook don't thoroughly screen external developers—those who work on all those online games, quizzes, and applications—for trustworthiness. (Until very recently, they also did not impose clear limits on how much user data such applications could have access to, regardless of their actual needs.) This means, in essence, that a smart authoritarian regime can just put together a funny quiz about Hollywood movies and use it to gather sensitive information about its opponents. This is a nightmarish scenario for activists who struggle to keep their connections hidden from authorities; obviously, if the government knows all the Facebook friends of its fiercest political opponents, it would be silly not to pay close attention to their online activities, too, as there is always a good chance they also pose a threat.

Nor does it help that in their ill-conceived quest for innovation, technology companies utterly disregard the contexts in which many of their users operate, while significantly underestimating the consequences of getting things wrong. In early 2010, when Google launched Google Buzz, its Twitter-like service, they did not take appropriate care in protecting the identities of many of their users, disclosing their contact lists in the erroneous belief that no one would mind such intrusions into their privacy (even Andrew McLaughlin, Google's former senior executive and the deputy chief technology officer in the Obama administration, was trapped in the Buzz trap, as many of his former Google colleagues appeared in his contact list). Though Google executives downplayed the significance of the accident by claiming that no one got seriously hurt in the debacle, in truth we don't know how many new names and connections were added to the KGB's databases as a result. The real costs of Google's misjudgment cannot be immediately calculated.

Think, Search, Cough

Every time we post a greeting to our friend's Facebook wall, Google the name of our favorite celebrity, or leave a disapproving comment on the website of our favorite newspaper, we leave a public trail somewhere on the Internet. Many of these trails, like the comment on the newspaper's site, are visible to everyone. Some, like our Google searches, are only visible to us (and, of course, Google). Most, like that odd comment on the Facebook wall, fall somewhere in between.

Fortunately, we are not alone on the Internet—at least one billion other users are also blogging, Googling, Facebooking, and tweeting—and most of our information is simply lost in the endless ocean of digital ephemera produced by others. This is what privacy scholars call "security by obscurity." In most cases, obscurity still works, even though there are more and more exceptions to this rule. Ask anyone who has difficulty finding a job or renting an apartment because something embarrassing about him or her appears in Google searches or on Facebook. Nevertheless, aggregating these tiny digital trails into one big data set—sometimes across entire populations—could produce illuminating insights into human behavior, point to new trends, and help predict public reaction to particular political or social developments. Marketing and advertising companies understood the power of information a long time ago. The more they know about demographics, consumer habits, and preferences of particular customer types, the more they can tailor their product offerings, and the more they can make in sales as a result.

The digital world is no different. The history of our Internet search says more about our information habits than our patron files in the local library. The ability to identify and glean "intent" from a mere Internet search, matching advertisers with customers looking for their offerings, has allowed Google to turn the advertising business on its head. Thus, in addition to running the world's most successful advertising agency, Google also runs the most powerful marketing intelligence firm. This is because Google knows how to relate Internet searches to demographics and other searching and purchasing decisions of its customers (e.g., what percent of New Yorkers who searched for "digital camera" in the past twelve months ended up searching for "deals on iPhones").

But we're not just looking for better iPods and new deals on plasma TVs. We are also seeking information about people and places in the news ("has Michael Jackson died?"), about broader cultural trends ("what are the best novels of the decade?"), and, of course, about solving problems—mostly trivial but some important—that constantly pop up in our lives ("how to repair a broken washing machine").

There are many seasonal variations to how often we search for particular items (searches for "stuffed turkey" predictably increase before Thanksgiving), but the frequency of queries for most items is usually fairly consistent. Thus, whenever there is a sudden spike in the number of Google queries for a given term, it probably indicates that something extraordinary has just happened; the likelihood is even higher if the search spike is limited to a particular geographic area only.

For example, when an unusually high number of Internet users in Mexico began Googling terms like "flu" and "cold" in mid-April 2009, it signaled the outbreak of swine flu. In fact, Google Flu Trends, a dedicated Google service built especially for the purpose of tracking how often people search for flu-related items, identified the spike on April 20, before the swine flu became a cause célèbre with many in the media. And even though several scientific studies by health researchers found that Google's data is not always as accurate as other ways of tracking the spread of influenza, even they acknowledged how cheap and quick Google's system is. Besides, in fields that are not as data-intensive as disease control, Google does a much better job than the alternatives— if those exist at all.

Search engines have inadvertently become extremely powerful players in the business of gathering intelligence and predicting the future. The temptation—which Google executives, to their credit, have resisted so far—is to monetize the vast quantity of this trends-related information beyond just ad sales.

Technically, Google does know how often Russian Internet users search for the words "bribes," "opposition," and "corruption"; it even knows how such queries are distributed geographically and what else such potential troublemakers are searching for. It does not take a Nostradamus to interpret a sudden spike of Internet searches for words like "cars," "import," "protests," and "Vladivostok" as a sign of growing social tensions over increases in car tariffs brewing in Vladivostok, Russia's major outpost in the Far East.

This is the kind of data that Russian secret services would literally kill for. Such knowledge may, of course, make authoritarianism more responsive and inject at least a modicum of democracy into the process. But it's also possible that governments would use this knowledge to crack down on dissenters in a more effective and timely manner.

Internet search engines offer an excellent way to harness the curiosity of the crowds to inform the authorities of impending threats. Monitoring an Internet search could produce even more valuable intelligence than monitoring Internet speech, because speech is usually directed at somebody and is full of innuendo, while an Internet search is a simple and neutral conversation between the user and the search engine.

The intelligence value of search engines is not lost on the Internet gurus consulting authoritarian governments. In March 2010, speaking about the Kremlin's ambitions to establish its own search engine, Igor Ashmanov, one of the pioneers of the Russian Internet and someone who had consulted for the Kremlin about their national search plan in the past, was direct: "Whoever dominates the search market in the country knows what people are searching for; they know the stream of search queries. This is completely unique information, which one can't get anywhere else." If one assumes that authoritarian governments usually fall by surprise—if they are not surprised, they are probably committing suicide (e.g., the case of the Soviet Union)—then we also have to assume that, given how much data on the Internet can be harvested, analyzed, and investigated, surprises may become rarer.

But even if the governments' attempts to control—directly or indirectly—the world of Internet search would not bring immediate results, the Internet could boost their intelligence-gathering apparatus in other ways. The advent of social media has made most Internet users increasingly comfortable with the idea of sharing their thoughts and deeds with the world at large. It may not seem obvious, but trolling through all those blog posts, Twitter updates, photos, and videos posted to Facebook and YouTube could yield quite a lot of useful information for intelligence services—and not just about individual habits, as in the Belarusian KGB case, but also about broad social trends and the public mood as a whole. Analyzing social networks could offer even better insights than monitoring online searches, as one could correlate information coming from particular individuals (whether it's opinions or facts) in the light of what else could be known about these individuals from their social networking profile (how often they travel, what kind of online groups or causes they embrace, what movies they like, who else is in their network, etc.).

An authoritarian government, for example, may pay special attention to the opinions of those who are between twenty and thirty-five years old, frequently travel abroad, and have advanced degrees. One simply needs to spend some time browsing relevant Facebook groups (e.g., "Harvard class of 1998" or "I love traveling in the Middle East") to zero in on the right characters. In a sense, the world of social networking obviates the need for focus groups; finding smart ways to cluster existing online groups and opinions could be more effective. And they don't have to collect this data on their own. Plenty of private companies are already collecting data—mostly for marketing purposes—that governments, both authoritarian and democratic ones, would find extremely useful. Thus, while the KGB may no longer exist in 2020, its functions may still be performed by a smattering of private companies specializing in one particular aspect of information work.

Today governments can learn quite a lot about the prospects of political unrest in a particular country simply by paying particular attention to the most popular adjectives used by the digerati. Are they "happy" or "concerned"? Do they feel "threatened" or "empowered"?

What if one controls for religion? Do self-professed secular bloggers feel more satisfied than the religious ones?

Just imagine how useful it might be for the Iranian government to track how often Iranians use the word "democracy" in their public online conversations and how such mentions are spread across the country. (For example, are there any regions of Iran that are more democratically inclined and unhappy with the current regime than others?)

If proper controls for statistical bias are in place, such technology is often superior to opinion polls, which take time to develop and, when done in authoritarian countries, always carry the risk of people misrepresenting their views to avoid punishment. Such aggregated information may not be fully representative of the entire population, but it helps to keep the tab on the most troublesome groups. Thus, the fact that authoritarian governments can now learn more about the public mood in real time may only add to their longevity. They are less likely to misjudge the public reaction.

What's worse is that social media activity is not always a bad proxy for judging the relative importance of antigovernment activists. If tweets of a particular user are retweeted more often than average, it's a good idea for the government to start watching that individual closely and learn more about his or her social network. The viral culture of social media may at least indirectly help solve the problem of information overload that has affected censorship as well. It's the "online marketplace of ideas" that tells secret police whom to watch. From the perspective of the secret police, people who are unpopular probably don't even deserve to be censored; left to their own devices and nearly zero readers, they will run out of blogging energy in a month or so.

The Myth of an Overprotected Activist

Despite the terrifying efficiencies in the practice of surveillance that were introduced by digital technology, not all is lost. It would be disingenuous to suggest that the digital realm has nothing in store for dissidents; it has greatly enhanced many of their activities as well. One great intellectual challenge facing any scholar of today's Internet is being able to see the risks inherent in new technologies while not discarding the numerous security-enhancing opportunities that they offer. The only way to come up with a satisfying answer to the question of whether the Internet has eroded or strengthened the surveillance and control apparatus of authoritarian governments is to examine all major technologies one by one, in their specific contexts.

But first it may help to examine the ways in which the Internet has helped dissidents to conceal antigovernment activities. First, sensitive data can now be encrypted on the cheap, adding an extra level of protection to conversations between dissidents. Even though decryption is possible, it can eat a lot of government resources. This is particularly true when it comes to voice communications. While it was relatively easy to bug a phone line, this is not such an easy option with voice-over-the-Internet technology like Skype. (The inability to eavesdrop on Skype conversations bothers Western governments, too: In early 2009 the U.S. National Security Agency was reported to have offered a sizeable cash bounty to anyone who could help them break Skype's encrypted communications; to date no winners have been announced.)

Second, there is so much data being produced online that authorities cannot possibly process and analyze all of it. Comparable estimates for the developing world are lacking, but according to a 2009 study by researchers at the University of California at San Diego, by 2008 the information consumption of an average American reached thirty-four gigabytes of data per day, an increase of 350 percent compared to 1980. The secret police have no choice but to discriminate; otherwise, they may develop a severe case of attention deficit disorder, getting bogged down in reading millions of blogs and Twitter updates and failing to see the big picture. Thanks to this data deluge, it may take a few months before authorities discover the new hideout of activists, who thus gain a few months of unsupervised online collaboration. The authorities are much better informed about the parameters of the haystack, but the needle is still quite hard to find.

Third, technologies like Tor now make it possible to better protect one's privacy while surfing the Internet. A popular tool that was initially funded by the U.S. Navy but eventually became a successful independent project, Tor allows users to hide what it is they are browsing by first connecting to a random "proxy" node on the volunteer Tor network and then using that node's Internet connection to connect to the desired website. Interestingly, as users of the Saudi site Tomaar found out, tools like Tor also help to circumvent government filtering of the Internet, for, from the government's perspective, the user is not browsing banned websites but is simply connecting to some unknown computer. This is why once the Iranian government found out the proxies used by its opponents during the 2009 protests, many of them publicized by unsuspecting Westerners on Twitter, it immediately began blocking access to them.

But Tor's primary function remains guaranteeing its users' anonymity. Think of this as surfing the Internet using an anonymous network of helpers who fetch all the websites you need and thus ensure that you yourself are not directly exposed. As long as the government doesn't know these helpers by name, the helpers don't know each other, and you frequent enough other networks not to attract attention to the helpers, you can get away with browsing whatever you want.

But how many activists actually bother to read the fine print that is invariably attached to all modern technologies? Most probably ignore it. If the Soviet dissidents had to memorize the manuals to their smuggled photocopiers before distributing any samizdat, their output might have been considerably less impressive. And a lot of the tools are easy to misunderstand. Many users, including those in the most secretive government outfits, mistakenly believe that Tor, for example, is more secure than it actually is. Swedish researcher Dan Egerstad set up five Tor nodes of his own—that is, he became one of the final stage helpers—to learn more about data that passed through them. (The "helper" who finds herself as the final node on the network—that is, it helps to gain access to the desired target site rather than simply redirect the request to another "helper"—can see what websites it is actually "helping" to access, even though it won't know who is trying to access them.) Egerstad, who was arrested as a result of his little scholarly experiment, found that 95 percent of the traffic that passed through his experimental Tor connections—including government documents, diplomatic memos, and intelligence estimates—was not encrypted. Think of intercepting an envelope that doesn't have a return address. Would you be able to guess who wrote it? Sure, if you look inside: The letterhead may tell you everything you need to know. TOR is excellent at removing the sender's address from the envelope, but it doesn't destroy the letterhead, let alone the rest of the letter. There are, of course, plenty of other encryption technologies that can do this, but Tor is simply not one of them. That so many users exchanging sensitive information online— including activists and dissidents—do not have a firm understanding of the technologies they use is cause for serious concern. Eventually it puts them at completely unnecessary and easily avoidable risk.

Besides, even complete mastery of technology is often not enough. Your security is only as good as that of the computer you are working on; the more people have access to it, the more likely it is that someone could turn your computer into a spying machine. Given that a lot of Internet activism takes place on public computers, security compromises abound. For many antigovernment activists, cybercafés have become the new (and often the only) offices, as authorities keep a close eye on their home and office Internet connections. However, few Internet cafés allow their patrons to install new software or even use browsers other than Internet Explorer, which puts most innovative tools for secure communication out of easy reach.

Rainy Days of Cloud Computing

Some observers see many security-enhancing benefits to the Internet. For example, dissidents and NGOs can now use multifunctional online working environments to execute all their work remotely—"in the cloud"—without having to install any software or even store any data on poorly protected computers. All one needs is a secure browser and an Internet connection; there's no need to download any files or carry a portable copy of your favorite word processor on a USB thumb drive.

"Cloud activism" may, indeed, seem like something of a godsend, an ideal solution to data security concerns faced by many NGOs and activists. Take the case of Memorial, a brave Russian NGO that has gained worldwide recognition for its unyielding commitment to the documenting of human rights abuses and crimes committed in the country, from Stalin's rule to the more recent wars in Chechnya.

On November 4, 2008, only a day before an edgy conference on Stalin's role in modern Russia co-organized by Memorial, the Russian police raided its offices in Saint Petersburg and confiscated twelve hard drives containing the entire digital archives of atrocities under Stalin, including hours of audio histories and video evidence of mass graves. It was an institutional disaster. Not only did Memorial lose possession of (even if temporarily) twenty years of important work, but Russian authorities were supplied with potentially damning evidence against the organization. Given that historical memory—especially of the Stalin period—is a sensitive issue in Russia, finding fault with Memorial, which happens to be a staunch critic of the Kremlin, wouldn't be so hard. Russian police are notorious for finding fault with the most innocuous of documents or, worse, software and operating systems. (Quite a few Russian NGOs use illegal software in their offices, often without even realizing it until it is too late; on more than one occasion, the war on pirated software, which the West expects Moscow to fight with all its vigor, has been a good excuse to exert more pressure on dissenting NGOs.)

Fortunately, the courts concluded that the search had been conducted in violation of legal due process, and Memorial's hard drives were returned in May 2009. Nevertheless, the fact that authorities had simply walked in and confiscated twenty years of work posed a lot of questions about how activists might make digital data more secure.

Fans of "cloud activism" would point out that one way to avoid disasters like Memorial's is to shift all data into the cloud, away from local hard drives and onto the Internet, thus making it impossible for the authorities to confiscate anything. To get access to such documents, authorities would need a password, which, in most countries, they would not be able to obtain without a court order. (Of course, this would not work in countries that have absolutely no respect for the rule of law; one can learn the password by torturing the system administrator without having to go through the courts.)

The possibility of using online word-processing services like Google Docs and dumping all important data on the Internet may, indeed, seem like an improvement over storing data on easily damaged, insecure hard drives lying around NGOs' dusty offices. After all, the data could be stored on a remote server somewhere in California or Iowa, completely out of immediate reach of authoritarian governments, if only because it ensures that the latter cannot legally and physically get to the services storing it (or not immediately, at any rate).

While there is much to admire about this new cloud-based model, it also comes with tremendous costs, which could sometimes outweigh the benefits. One major shortcoming of producing and accessing documents in the cloud is that it requires a constant transmission of data between a computer and a server where the information is stored. This transmission is often done "in the open" (without proper encryption), which creates numerous security compromises.

Until very recently, many of Google's online offerings—including such popular services as Google Docs and Google Calendar—did not offer encryption as the default option. This meant that users connecting to Google Docs through, say, insecure Wi-Fi networks were playing with fire: virtually anyone could see what they were sending to Google's servers. Fortunately, the company altered its encryption policy after several high-profile security experts wrote a letter to its CEO, where they highlighted the unnecessary and easily avoidable risks to Google users. But Google is not the only player in this space—and where Google has the resources to spend on extra encryption, others may not. Making encryption the default setting may slow down the service for other users and impose new costs on the company's operations. Such improvements are not completely out of the question, however. A strong argument can be made—hopefully, by lawmakers on both sides of the Atlantic—that forcing Internet companies to enhance the security of their services makes a lot of sense from the perspective of consumer protection regulation. Instead of giving such companies a free pass because they are now the key players in the fight for Internet freedom, Western governments should continue looking for ways in which their services could be made extremely secure, for anything less than that would, in the long run, endanger too many people.

But other insecurities abound, too. The fact that many activists and NGOs now conduct all their business activities out of a single online system, most commonly Google—with calendar, email, documents, and budgets all easily available from just one account—means that should their password be compromised, they would lose control over all of their online activities. Running all those operations on their own laptops was not much safer, but at least a laptop could be locked in a safe. The centralization of information under one roof—as often happens in the case of Google—can do wonders from the perspective of productivity, but from the perspective of security it often only increases the risks.

On Mobile Phones That Limit Your Mobility

Much like cloud computing, the mobile phone is another activist tool that has not been subjected to thorough security analysis. While it has been rightly heralded as the key tool for organizing, especially in countries where access to the Internet and computers is prohibitively expensive, little has been said about the risks inherent to most "mobile activism."

The advantages of such activism are undeniable. Unlike blogging and tweeting, which require an Internet connection, text messaging is cheap and ubiquitous, and it doesn't require much training. Protesters using mobile phones to organize public rallies have become the true darlings of the international media. Protesters in the Philippines, Indonesia, and Ukraine have all taken advantage of mobile technology to organize and challenge their governments. This technology is not without its shortcomings and vulnerabilities, however.

First and foremost, authorities can shut down mobile networks whenever they find it politically expedient. And they do not have to cut off the entire country; it's possible to disconnect particular geographic regions or even parts of the city. For example, during the unsuccessful color revolution in Belarus in 2006, the authorities turned off mobile coverage in the public square where protesters were gathering, curbing their ability to communicate with each other and the outside world (the authorities claimed that there were simply too many people using mobiles on the square and the mobile networks couldn't cope with the overload). The Moldovan authorities made a similar move in spring 2009, when they turned off mobile networks in the central square of Chisinau, Moldova's capital, thus greatly hampering the communication capacity of those leading the local edition of the Twitter revolution. Such shutdowns can also be on a larger, national scale and last longer. In 2007 the government of Cambodia declared a "tranquility period," during which all three mobile operators agreed to turn off text messaging for two days (one of the official explanations was that it would help keep voters from being flooded with campaign messages).

Many authorities have mastered the art of keyword filtering, whereby text messages containing certain words are never delivered to their intended recipients. Or they may be delivered, but the authorities will take every step to monitor or punish their authors. In 2009 police in Azerbaijan reprimanded forty-three people who voted for an Armenian performer (Armenia and Azerbaijan are at war over the disputed Nagorno-Karabach territory) in the popular Eurovision contest, summoning some of them to police headquarters, where they were accused of undermining national security, and forced to write official explanations. The votes were cast by SMS. In January 2010, *China Daily*, China's official English-language newspaper, reported that mobile phone companies in Beijing and Shanghai began suspending services to cellphone users who were found to have sent messages with "illegal or unhealthy" content, which is the Chinese government's favorite euphemism for "smut."

This means that China's mobile operators would now be comparing all text messages sent by their users to a list of banned words and blocking users who send messages containing banned words. That's a lot of messages to go through: China Mobile, one of China's biggest mobile operators, processes 1.6 billion text messages per day. Even though the campaign officially claims to be fighting pornography, similar technology can be easily used to prevent the distribution of text messages on any topic; it all depends on the list of banned words. Not surprisingly, this list of "unhealthy words" comes from China's police. But there is also plenty of traffic in the other direction—that is, from companies to the state. Wang Jianzhou, China Mobile's CEO, stunned the attendees of the World Economic Forum in Davos in 2008 by claiming that his company provides data on its users to the government whenever the government demands it.

What's worse, Western companies are always happy to provide authoritarian governments with technology that can make filtering of text messages easier. In early 2010, as American senators were busy praising Google for withdrawing from China, another American technology giant, IBM, struck a deal with China Mobile to provide it with technology for tracking social networks (of the human, not virtual variety) and individuals' messaging habits: who sends what messages to whom and to how many people. (IBM, of course, was quick to point out that such technology is meant for helping Chinese mobile operators cut down on spam, but none can vouch that the same operators won't use it to curb political speech.)

Any technologies based on keyword filtering can, of course, be easily tricked. One can deliberately misspell or even substitute most sensitive words in a text message to fool the censors. But even if activists resort to misspelling certain words or using metaphors, governments could still make the most popular of such messages disappear. In fact, it's not the actual content of the messages that worries the government—no one has yet expressed cogent government criticism in a hundred forty characters or less—but the fact that such messages could go viral and be seen by millions of people. Regardless of the content being shared, such viral dissemination of information makes authoritarian governments feel extremely uneasy, as it testifies to how much their grasp on information has been eroded. In the most extreme cases, they won't hesitate to use the nuclear option and block most popular messages, without paying much attention to their content.

What is even more dangerous about using mobile phones for activism is that they allow others to identify the exact location of their owners. Mobile phones have to connect to local base stations; once a user has connected to three bases, it is possible to triangulate the person's position. In an online demonstration to its current and potential customers, ThorpeGlen, a U.K.-based firm, boasts that it can track "a specific target through ALL his electronic communications.... We can detect change of SIM and change of handset after identifying one suspect.... We can even detect that profile again even if the phone AND SIM are changed." This means that once you've used a cellphone, you are trapped. To clinch their marketing pitch, ThorpeGlen attached an online map of Indonesia that depicted the movements of numerous dots—millions of Indonesians with their cellphones; it allowed a viewer to zoom in on any particular sector. But it is hardly the only company offering such services; more and more start-ups cater to the vibrant consumer market in cellphone surveillance. For just $99.97 a year, Americans can load a little program called MobileSpy onto someone's cellphone and track that phone's location whenever they want.

Monitoring the geographic location of phone owners may enable the government to guess where big public actions might be happening next. For example, if the owners of the hundred most dangerous cellphone numbers are all seen heading to a particular public square, there is a good chance that an antigovernment demonstration will soon ensue. Furthermore, mobile companies have strong economic incentives to improve their location-identification technology, as it would allow them to sell geographically targeted advertising, such as prompts to check out the café next door. If anything, determining a person's location by tracing his or her mobile phone is poised to get easier in the future. While ThorpeGlen markets its services to law enforcement and intelligence firms in the West, it's not clear if any restrictions would prohibit the export of such technology elsewhere.

Many activists are, of course, aware of such vulnerabilities and are doing their best to avoid easy detection; however, their most favorite loopholes may soon be closed. One way to stay off the grid has been to buy special, unbranded models of mobile phones that do not carry unique identifiers present in most phones, which could make such devices virtually untraceable. Such models, however, also appeal to terrorists, so it's hardly surprising that governments have started outlawing them (for example, in the wake of 2008 attacks in Mumbai, India banned the export of such phones from China). The frequent use of new technologies by terrorists, criminals, and other extreme elements presents a constant challenge to Western governments who would like to both empower democratic activists and disempower many of the sinister non-state groups that are undermining the process of democratization.

Another favorite low-tech solution, disposable prepaid SIM cards, which allow activists to change their phone numbers on a daily basis, may not stay around for much longer either, as buying them is becoming more difficult in many parts of the developing world. Russia and Belarus, for example, require retailers to obtain a copy of the customer's passport when someone buys a prepaid card, which essentially eliminates the desired anonymity. In early 2010 Nigeria passed a similar law, and other African states are expected to follow. Since American policymakers fret about Al-Qaeda jihadists using prepaid SIM cards to coordinate terrorist acts, it's quite likely that similar measures will soon pass in the United States as well. In 2010, with the entire country abuzz with the Times Square terror threat, FBI Director Robert Mueller endorsed anti-terrorism legislation that would require prepaid cellphone sellers to keep records of buyers' identities.

As useful as mobile technology could be for countering the power of authoritarian states, it comes with numerous limitations. This is not to say that activists should not be harnessing its communications power. They should, but only after fully familiarizing themselves with all the risks involved in the process.

As the Web becomes more social, we are poised to share more data about ourselves, often forgetting about the risks involved. Most disturbingly, we do so voluntarily, not least because we often find such sharing beneficial. Thus, sharing our geographical location may alert our friends to our whereabouts and facilitate a meeting that may not have happened otherwise. What we often overlook is that by saying where we *are*, we are also saying where we are *not*. Obviously, this is a boon for burglars; privacy activists even set up a dedicated site provocatively called "Please Rob Me" to raise public awareness about such risks. Such a wealth of data is also of great value to authoritarian states. Today's digitized, nimble, and highly social surveillance has little in common with the methods practiced by Stasi and KGB in 1989. The fact that there are more ways to produce and disseminate data has not overloaded the censorship apparatus, which has simply adapted to this new age by profiting from the same techniques—customization, decentralization, and smart aggregation—that have propelled the growth of the Internet. The ability to speak and make connections comes with costs, and those costs may not always be worth the benefits.

Denying that greater information flows, combined with advanced technologies like face or voice recognition, can result in the overall strengthening of authoritarian regimes is a dangerous path to take, if only because it numbs us to potential regulatory interventions and the need to rein in our own Western corporate excesses. It's not a given that IBM should be selling SMS-filtering technology to authoritarian states; that services like Google Buzz should be launched with minimum respect for the privacy of its users; that researchers at public universities like the University of California should be accepting funding from the Chinese government to work on better video surveillance technology; or that Facebook should be abdicating their responsibility to thoroughly screen developers of its third-party applications. All of these developments are the result of either excessive utopianism, unwillingness to investigate how technology is being used in non-Western contexts, or unquenchable thirst for innovation with complete disregard for its political consequences. While the Internet by itself may not be liberating those living in authoritarian states, Western governments should not be making it easier to use in suppressing dissent.

chapter seven

Why Kierkegaard Hates Slacktivism

•   •   ■

I f you've been to Copenhagen, you've probably seen the Stork Fountain, one of the city's most famous sights. The fountain was made
even more famous thanks to a quirky Facebook experiment. In spring
2009 Anders Colding-Jorgensen, a Danish psychologist who studies how
ideas spread online, put the famous fountain at the center of his research
project. He started a Facebook group that implied—but never said so
explicitly—that the city authorities were about to demolish the fountain.
This threat was completely fictitious; Colding-Jorgensen himself had
dreamed it up. He publicized the group to 125 of his Facebook friends,
who joined the cause in a matter of hours. It was not long before their
friends joined, too, and the imaginary Facebook campaign against
Copenhagen's city council went viral. At the peak of its online success,
the group had two new members joining every minute. When the count
reached 27,500, Colding-Jorgensen decided it was time to end his little
experiment.

There are two strikingly different ways to make sense of the Stork
Fountain experiment. Cynics might say that the campaign took off
simply because Colding-Jorgensen looked like a respected activist

academic—just the kind of guy to start a petition about saving a fountain on Facebook. His online friends were likely to share his concern for the preservation of Denmark's cultural heritage, and since joining the group did not require anything other than clicking a few buttons, they eagerly lent their names to Colding-Jorgensen's online campaign. If that request had come from some unknown entity with few historically conscious contacts, or if joining in required performing a number of challenging chores, chances are the success of that crusade would have been far less spectacular. Or perhaps the campaign received so much attention because it was noticed and further advertised by some prominent blogger or a newspaper, thus giving it exposure it might never have earned on its own. On this rather skeptical reading, the success of online political and social causes is hard to predict, let alone engineer. Policymakers, therefore, should not pay much attention to Facebook-based activism. While Facebook-based mobilization will occasionally lead to genuine social and political change, this is mostly accidental, a statistical certainty rather than a genuine achievement. With millions of groups, at least one or two of them are poised to take off. But since it's impossible to predict which causes will work and which ones won't, Western policymakers and donors who seek to support or even prioritize Facebook-based activism are placing a wild bet.

Another, more optimistic way to assess the growth of activism on social networks is to celebrate the ease and speed with which Facebook groups can grow and go viral. From this perspective, Colding-Jorgensen's experiment has shown that when communication costs are low, groups can easily spring into action—a phenomenon the Internet guru Clay Shirky dubbed "ridiculously easy group forming." (Shirky acknowledges that some "bad groups"—for example, anorexic girls seeking to impress each other with their sacrifices—can be formed ridiculously easily as well.) Proponents of this view argue that Facebook is to group formation what Red Bull is to productivity. If a nonexistent or poorly documented cause could garner the attention of 28,000 people, more important, well-documented causes—genocide in Darfur, Tibetan independence, abuses of human rights in Iran—can certainly rally millions behind them (and they do). While there are still no universal benchmarks for evaluating the effectiveness of such groups, the fact that they exist—pushing updates to their members, pestering them with fund-raising requests, urging them to sign a petition or two—suggests that, despite occasional embarrassing gaffes, Facebook could be a valuable resource that political activists and their Western supporters need to master. That they may not know how to do this is a poor excuse for not getting engaged.

Digital Natives of the World, Unite!

And engaged many of them already are. When in 2008 the streets of Colombia got filled with up to a million angry protesters against the guerillas of the FARC movement, which has been terrorizing the country for decades, it was a Facebook group called No Más FARC (No More FARC) that got credited for this unprecedented mobilization. (In 2008 FARC dominated Colombian news with a series of high-profile kidnappings.) Launched by Oscar Morales, a thirty-three-year-old unemployed computer technician, the group quickly gained members and became a focal point for spreading information about the protests, earning the support of the Colombian government in the process.

The American government was just a Facebook request away as well. Morales, who later became a fellow at the George W. Bush Institute, got a note from U.S. State Department's Jared Cohen, the American bureaucrat who one year later sent the infamous email request to Twitter. Cohen wanted to come to Colombia to study the details of Morales' impressive online operation. Morales didn't seem to mind.

Cohen's visit to Colombia must have been inspirational, for just a few months later the State Department soft-launched an international organization called the Alliance of Youth Movements (AYM), built on the assumption that cases like Colombia's are going to be more widespread and that the U.S. government needs to be an early player in this field, doing its share to facilitate networking among such "digital revolutionaries." A series of high-profile summits of youth movements—one was even moderated by that staunch defender of Internet freedom Whoopi Goldberg—duly followed.

In its brief history, AYM has emerged as something of a digital-era equivalent of the Congress of Cultural Freedom, a supposedly independent artistic movement that in reality was created and funded by the CIA to cultivate anticommunist intellectuals during the early stages of the Cold War. (Unfortunately, AYM's literary output is nowhere as prodigious.) Now that the battle for ideas has shifted into cyberspace, it is bloggers rather than intellectuals that the U.S. government wants to court.

George W. Bush Institute's James Glassman, then the undersecretary of state for public diplomacy and public affairs, kicked off AYM's first summit in New York, explaining that the meeting's purpose was to "bring about two dozen groups together with top technologists from the United States and produce a manual . . . [to help] other organizations that want the information and technological knowledge to be able to organize their own anti-violence groups."

Companies like Facebook, Google, YouTube, MTV, and AT&T attended the New York summit, along with groups like the Burma Global Action Network, Genocide Intervention Network, and Save Darfur Coalition. (A representative from Balatarin, a prominent Iranian social news site, was present at AYM's second summit in Mexico.) The gathering was meant to send yet another powerful message that American companies, perhaps with a gentle push from the U.S. government, were playing an important role in facilitating democratization and that digital technologies—above all, social networking—were instrumental in pushing back against oppressors. "Any combination of these [digital] tools allows for a greater chance of civil society organizations coming to fruition regardless of how challenging the environment," proclaimed Jared Cohen, giving perhaps one of the sharpest articulations of both cyber-utopianism and Internet-centrism to date.

Impressed by the success of the Colombian group, American officials decided to embrace social networking sites as viable platforms for breeding and mobilizing dissent, expressing their willingness to fund the creation of new sites if necessary. Thus, in 2009 the State Department ran a $5 million grant competition in the Middle East, soliciting funding requests for projects that would "develop or leverage existing social networking platforms to emphasize priorities of civic engagement, youth outreach, political participation, tolerance, economic entrepreneurship, women's empowerment, or nonviolent conflict-resolution." (Apparently, there is no problem that social networking can't solve.) Most likely American officials would have dismissed the Stork Fountain experiment as just a minor embarrassment, the cost of doing business in this new digital environment, but hardly a good reason to stop harvesting the tremendous energy of social networking. But could it be that in their pursuit of short-term and instrumental mobilization goals, they may have overlooked the long-term impact of social networking on the political cultures of repressive societies?

To even begin answering that question, we may need to reconsider the lessons of the Danish fountain. Both interpretations of the Stork Fountain experiment—the one slamming it as an oddity and the one worshiping it as a powerful example of the power of the Internet to mobilize—suffer from several analytical deficiencies. Neither offers a good account of what membership in such networked causes does to the members themselves. Surely most of them are not just mindless activist robots, pressing whatever buttons required of them by their online overlords, without ever grappling with the meaning of what it is they are doing and trying to figure out how their participation in such communities might affect their views on the meaning of democracy and the importance of dissent. Nor do these two competing interpretations indicate what kind of effect such online campaigns may have on the effectiveness and popularity of other offline and individual activist efforts. While it's tempting to forget this in an era of social networking, the fight for democracy and human rights is fought offline as well, by decades-old NGOs and even by some brave lonely warriors unaffiliated with any organizations. Before policymakers embrace digital activism as an effective way of pushing against authoritarian governments, they are well-advised to fully investigate its impact both on its practitioners and on the overall tempo of democratization.

Poking Kierkegaard

Ironically, to get a more critical view on the meaning of the Colding-Jorgensen's Copenhagen experiment, we need to turn to another Dane: Søren Kierkegaard (1813–1855). Considered the father of existentialism, he lived in interesting times not entirely unlike our own. In the first part of the nineteenth century, the social and political consequences of both the Industrial Revolution and the age of Enlightenment were beginning to manifest themselves in full force. The European "public sphere" expanded at unprecedented rates; newspapers, magazines, and coffee houses rapidly emerged as influential cultural institutions that gave rise to a broad and vocal public opinion.

But whereas the majority of contemporary philosophers and commentators lauded this great leveling as a sign of democratization, Kierkegaard thought that it might result in a decline of social cohesion, a feast of endless and disinterested reflection, and a triumph of infinite but shallow intellectual curiosity that might prevent deep, meaningful, and spiritual engagement with a particular issue. "Not a single one of those who belong to the public has an essential engagement in anything," Kierkegaard bitterly observed in his journal. All of a sudden, people were getting interested in everything and nothing at the same time; all subjects, no matter how ridiculous or sublime, were getting equalized in such a way that nothing mattered enough to want to die for. The world was getting flat, and Kierkegaard hated it. As far as he was concerned, all the chatter produced in coffee houses only led to the "abolition of the passionate distinction between remaining silent and speaking." And silence for Kierkegaard was important, for "only the person who is essentially capable of remaining silent is capable of speaking essentially."

For Kierkegaard, the problem with the growing chatter—epitomized by the "absolutely demoralizing existence of the daily press"—was that it lay outside of political structures and exerted very little influence on them. The press forced people to develop strong opinions on everything but rarely cultivated the urge to act on them; often people were so overwhelmed with opinions and information that they would indefinitely postpone any important decisions. Lack of commitment, caused by the multiplicity of possibilities and the easy availability of quick spiritual and intellectual fixes, was the real target of Kierkegaard's critique. He believed that only by making risky, deep, and authentic—one of Kierkegaard's favorite terms—commitments, by discriminating between different causes, by dealing with both triumphs and disappointments of such choices, and by learning from the resulting experiences, do people acquire wisdom and fill their lives with meaning. "If you are capable of being a man, then danger and the harsh judgment of existence on your thoughtlessness will help you become one" is how he summed up the philosophy that would come to be known as existentialism.

It's not hard to guess what Kierkegaard would have made of today's Internet culture, dominated by the 24/7 cycle of punditry and fluid engagement with ideas and relationships. "What Kierkegaard envisaged as a consequence of the press's irresponsible and uncommitted coverage is now fully realized on the World Wide Web," writes Hubert Dreyfus, a philosopher at the University of California at Berkeley. A world where professing one's commitment to social justice requires nothing more than penning a socially conscious Facebook status would have greatly rankled Kierkegaard. His Twitter account would surely be hard to find. It's safe to assume that sites like RentAFriend.com, where you can "rent a friend to go to an event or party with you, teach you a new skill or hobby, help you meet new people, show you around town" by choosing from more than 100,000 members registered on the site, would not be much to Kierkegaard's liking. Ukrainian Web entrepreneurs have adapted RentAFriend's model to the protest needs of their country's numerous political movements by setting up a website that allows anyone organizing a rally to "shop" for registered users, mostly students, who, at just $4 hour per hour, are eager to chant political slogans of any ideology. The entrepreneurs would not be among Kierkegaard's Facebook friends either.

And yet the Dane's philosophy is useful in grasping the ethical and political problems associated with digital activism, especially in the context of authoritarian states. It's one thing for existing and committed activists who are risking their lives on a daily basis in opposition to the regime to embrace Facebook and Twitter and use those platforms to further their existing ends. They might be overestimating the overall effectiveness of digital campaigns or underestimating their risks, but their commitment is "authentic." It's a completely different thing when individuals who may have only cursory interest in a given issue (or, for that matter, have no interest at all and support a particular cause only out of peer pressure) come together and start campaigning to save the world.

This is the kind of shallow commitment that Kierkegaard detested and saw as corrupting the human soul. Such high-minded moralizing may seem out of place today, but then no one has yet toppled an authoritarian government by assuming the posture of a clown and cracking jokes about the guillotine. Even when structural conditions favor democratization, an opposition movement composed of meek and characterless individuals will most likely fail to capitalize on such openings.

The problem with political activism facilitated by social networking sites is that much of it happens for reasons that have nothing to do with one's commitment to ideas and politics in general, but rather to impress one's friends. This is not a problem caused by the Internet. For many people, impressing one's peers by pursuing highly ambitious causes like saving the Earth and ending another genocide may have been the key reason for joining various student clubs in college, but this time one can proudly wear the proof of one's membership in public. Explaining the Stork Fountain experiment to the Washington Post, Colding-Jorgensen said, "Just like we need stuff to furnish our homes to show who we are, on Facebook we need cultural objects that put together a version of me that I would like to present to the public."

Research by Sherri Grasmuck, a sociologist at Temple University, confirms Colding-Jorgensen's hunch, revealing that Facebook users shape their online identities in implicit rather than explicit ways. That is, they believe that the kinds of Facebook campaigns and groups they join reveal more about them than whatever they put in the dull "about me" pages. Thus, many of them join Facebook groups not only or not so much because they support particular causes but because they believe it's important to be seen by their online friends to care about such causes. In the past convincing themselves and, more important, their friends that they were indeed socially conscious enough to be changing the world required (at a minimum) getting off their sofas. Today, aspiring digital revolutionaries can stay on their sofas forever—or until their iPads' batteries run out—and still be seen as heroes. In this world, it doesn't really matter if the cause they are fighting for is real or not; as long as it is easy to find, join, and interpret, that's enough. And if it impresses their friends, it's a true gem.

Not surprisingly, psychologists have also noticed a correlation between the use of social networking and narcissism. A 2009 national poll of 1,068 U.S. college students conducted by researchers at San Diego State University (SDSU) found that 57 percent of them believe that their generation uses social networking sites for self-promotion, narcissism, and attention seeking, while almost 40 percent agreed with the statement that "being self-promoting, narcissistic, overconfident, and attention-seeking is helpful for succeeding in a competitive world." Jean Twenge, an associate professor of psychology at SDSU who conducted the study and also author of *The Narcissism Epidemic: Living in the Age of Entitlement*, believes that the very structure of social networking sites "rewards the skills of the narcissist, such as self-promotion, selecting flattering photographs of oneself, and having the most friends." There's nothing wrong with self-promotion per se, but it seems quite unlikely that such narcissistic campaigners would be able to develop true feelings of empathy or be prepared to make sacrifices that political life, especially political life in authoritarian states, requires.

Kandinsky and Vonnegut Are Now Friends!

Given how easy groups can form online, it is easy to mistake quantity for quality. Facebook is already facilitating the processes that do not really require much social glue to begin with. The truth is that it's natural for people to form groups. Social psychologists have long understood that while it doesn't take much to make a group of people feel they have a common identity, it is considerably harder to make them act in the interests of that community or make individual sacrifices in its name.

Beginning in the early 1970s much research in social psychology was dedicated to the so-called Minimal Group Paradigm, the minimal conditions that can foster a sense of group identity among complete strangers. It turns out that the fact of categorizing people into groups—using completely random, coin-tossing methods—already produces a strong feeling of group identity, enough to start discriminating against those who are not members of the group. This was first confirmed by a group of British researchers who showed a group of schoolboys pairs of highly abstract paintings by two artists, Paul Klee and Wassily Kandinsky, without identifying the authorship of each painting in the pair. Having solicited the boys' preferences, they used this information to form two groups, the Klee lovers and the Kandinsky lovers, although some children were told they were assigned to a group randomly rather than based on their preferences. Each boy was then given a fixed amount of money and was asked to allocate it among the other boys. Much to the surprise of the researchers, the children allocated more money to members of their own group, even though they had no prior shared experiences and no obvious future as a group, and it was highly unlikely they felt strongly about either Kandinsky or Klee (in fact, in some cases the researchers showed pairs painted by just one of them without telling the students).

On first sight, this only seems to bolster the case of Internet enthusiasts who celebrate the ease with which online groups can form. But as any tax collector would know, dividing a small pot of other people's money in a scientific experiment is not the same as agreeing to cofund a Kandinsky exhibition out of one's own pocket. Obviously, the weaker the common denominator among the members of a particular group, the less likely they would be inclined to act as a coherent whole and make sacrifices in the name of the common good. It's little wonder that members of most Facebook groups proudly flaunt their membership cards—but only until someone asks for hefty membership fees. Since there are no sacrifices to make on joining such groups, they attract all kinds of adventurists and narcissists. Notes the Canadian writer Tom Slee, "Sure, it's easier to sign up to a Facebook group than if you have to actually go and meet someone, but if signing up is so easy it's not likely to be much of a group, just as an automated phone apology that 'all our agents are busy right now' is cheap, and so is not much of an apology."

The widespread tendency to misread meaningless mutual associations, both offline and online, as something much deeper and politically significant is what Kurt Vonnegut was ridiculing in his 1963 novel *Cat's Cradle*, in which he wrote of the "granfalloon": groups of people who outwardly choose or claim to have a shared identity or purpose based on rather imaginary premises. "The Communist Party, the Daughters of the American Revolution, the General Electric Company, the Independent Order of Odd Fellows—and any nation, anytime, anywhere" were Vonnegut's most prominent examples. For Vonnegut, the granfalloon was based on little but air or, as he put it, whatever is "hiding under the skin of a toy balloon."

The Internet, with its promise of fostering "virtual communities" on the cheap and widely advertised by the earlier generation of cyber-utopians as something of a panacea to many of modern democracy's ills, has driven the costs of joining such groups to zero. But it's hard to imagine how it could, all by itself, help cultivate a deep commitment to serious causes. This, at least for the foreseeable future, would be the task of educators, intellectuals, and, in some exceptional cases, visionary politicians. Not much has changed in that regard since 1997, when Oxford University's Alan Ryan wrote that "the Internet is good at reassuring people that they are not alone, and not much good at creating a political community out of the fragmented people that we have become."

Killing the Slacktivist in You

Alas, those charmed by the promise of digital activism often have a hard time distinguishing it from "slacktivism," its more dangerous digital sibling, which all too often leads to civic promiscuity—usually the result of a mad shopping binge in the online identity supermarket that is Facebook—that makes online activists feel useful and important while having preciously little political impact.

Take a popular Facebook cause, Saving the Children of Africa. At first sight, it does look impressive, with over 1.7 million members, until you discover that they have raised about $12,000 (less than one-hundredth of a penny per person). In a perfect world, this shouldn't even be considered a problem: It's better to donate one-hundredth of a penny than do nothing at all. But attention is limited, and most people only have a few hours a month (perhaps an optimistic estimate) to spend on improving the common good. Thanks to its granularity, digital activism provides too many easy ways out. Lots of people are rooting for the least painful sacrifice, deciding to donate a penny where they may otherwise donate a dollar. While the social science jury is still out on how exactly online campaigning may cannibalize its offline brethren, it seems reasonable to assume that the effects are not always positive. Furthermore, if psychologists are right and most people support political causes simply because it makes them feel happier, then it's quite unfortunate that joining Facebook groups makes them as happy as writing letters to their elected representatives or organizing rallies without triggering any of the effects that might benefit society at large.

A good way to tell whether a digital campaign is serious or "slacktivist" is to look at what it aspires to achieve. Campaigns of the latter kind seem to be premised on the assumption that, given enough tweets, the world's problems are solvable; in the language of computer geeks, given enough eyeballs, all bugs are shallow. This is precisely what propels so many of these campaigns into gathering signatures, adding new members to their Facebook pages, and asking everyone involved to link to the campaign on blogs and Twitter. This works for some issues, especially those that are geography bound (e.g., performing group community service at a local soup kitchen, campaigning against a resolution passed by a local town council, etc.). But with global issues, whether it's genocide in Darfur or climate change, there are diminishing returns to awareness raising. At some point one must convert awareness into action, and this is where tools like Twitter and Facebook prove much less successful.

Not surprisingly, many of these Facebook groups find themselves in a "waiting for Godot" predicament: Now that the group has been formed, what comes next? In most cases, what comes next is spam. Most of these campaigns—remember many of them, like the anti-FARC campaign in Colombia, pop up spontaneously without any carefully planned course of action—do not have clear goals beyond awareness raising. Thus, what they settle on is fund-raising. But it's quite obvious that not every problem can be solved with an injection of funds. If the plight of sub-Saharan Africa or even Afghanistan is anything to judge by, money can only breed more trouble unless endemic political and social problems are sorted out first.

The fact that the Web has made raising money easy may result in making it the primary focus of one's campaigning, when the real problems lie elsewhere. Asking for money—and receiving it—may also undermine one's efforts to engage group members in more meaningful real-life activities. The fact that they have already donated some money, no matter how little, makes them feel as if they have already done their bit and should be left alone. Some grassroots campaigns are beginning to realize this. For example, the website of Free Monem, a 2007 pan-Arab initiative to free an Egyptian blogger from jail, featured the message "DON'T DONATE; Take action" and had logos of Visa and MasterCard in a crossed red circle in the background. According to Sami Ben Gharbia, a Tunisian Internet activist and one of the Free Monem organizers, the message on the site was a way to show that their campaign needed more than money as well as to shame numerous local and international NGOs that like to raise money without having any meaningful impact on the situation. In other words, the fact that the technology for raising money is so superb these days may push some movements to pursue monetary objectives when what they really need to be doing is politics and advocacy instead (granted, those cost money, too).

On the Increased Productivity of Lonely Warriors, or Why Some Crowds Are Wise but Lazy

That said, the meager fund-raising results of the Saving the Children of Africa campaign—assuming that money is what they are after, as their "about" page states ("This group need [sic] financial support to be able to help deprived children in all African nations")—still look quite puzzling. Were one to evaluate the effectiveness of this group's efforts based on its immense potential (having access to 1.7 million people who self-identify as interested in helping the cause) the assessment wouldn't be too kind. Surely even a dozen people working on their own would be able to raise more than $12,000 in the few years that have passed since the group's founding. Is there a danger that the popularity of Facebook might nudge activists toward embracing some kind of "group fetishism," in which they opt for a group solution to problems that could be solved much faster and better by solo artists? Now that almost every problem could be tackled by collective action rather than individually, is there a risk that a collective urge might also delay the solution?

Before the Silicon Valley crowd went gung-ho about the wisdom of crowds, social psychologists and management experts were already studying conditions under which individuals who work as groups may be less effective than the same individuals working solo. One of the first people to discover and theorize this discrepancy was the French agricultural engineer Max Ringelmann.

In 1882 Ringelmann conducted an experiment in which he asked four individuals to pull on a rope, first alone and then in groups, and then compared the results. The rope was attached to a strain gauge so it was possible to measure the pull force. To Ringelmann's surprise, the total pull force of the group pull was consistently less than the sum of the individual pull forces, even as he adjusted the number of individuals participating in the experiment. What has become known as the Ringelmann Effect is thus the opposite of synergy.

In the century that has passed since Ringelmann's original experiment, plenty of other tests have proven that we usually put much less effort into a task when other people are also doing it alongside us. In fact, calling it the Ringelmann Effect is only adding theoretical luster to what we already knew intuitively. We don't have to make fools of ourselves by singing "Happy Birthday" at the top of our lungs; others will do the job just fine. Nor do we always clap our hands as loudly as we could—much to the disappointment of performers. The logic is clear: When everyone in the group performs the same mundane tasks, it's impossible to evaluate individual contributions, and people inevitably begin slacking off (it's for this reason that another name for this phenomenon is "social loafing"). Increasing the number of participants diminishes the relative social pressure on each and often results in inferior outputs.

Hearing of Ringelmann's experiments today, one can't help noticing the parallels to much of today's Facebook activism. With the power of Facebook and Twitter at their fingertips, many activists may choose to tackle a problem collectively when tackling it individually would make more strategic sense. But just as "the madness of crowds" gives rise to "the wisdom of crowds" only under certain, carefully delineated social conditions, "social loafing" leads to synergy only once certain conditions are met (it's possible to monitor and evaluate individual contributions, and the group members are aware that such evaluation is going on; tasks to be performed are unique and difficult, etc.) When such conditions are missing, pursuing a political end collectively rather than individually is no more desirable than choosing what to have for breakfast by polling one's neighbors. It's certainly possible for a group to meet all the conditions, but it often takes a lot of effort, leadership, and ingenuity. This is why effective social movements don't spring up in a day.

But Facebook simply does not provide for the kind of flexibility that this requires. Once we join a Facebook group to fight for a cause, we move at the group's own pace, even though we could be much more effective fighting on our own. Our contributions to achieving the group's stated objectives are hard to verify, so we may join as many as we want without fearing that we might get reprimanded. It's not Facebook's fault, of course. Most popular social networking sites were not set up for activists by activists; they were created for the purposes of entertainment and attract activists not because they offer unique services but because they are hard to block.

Even though Facebook activism offers only a limited vision of what is really possible in the digital space, the network effect—the fact that so many people and organizations are already on Facebook—makes it hard to think outside the box. Activists can easily set up a website with better privacy defaults and a gazillion more functions, but why should they bother building it if it may fail to attract any visitors? Most campaigns have no choice but to conform to the shallowness and limitations of Facebook communications; in a tradeoff between scale and functionality, most of them choose the former. Thus, for many such campaigns, the supposed gains of digital activism are nothing but illusory: Whatever they save through their newly found ability to recruit new members, they lose in trying to make these new members act as a group—and, preferably, without giving in to social loafing. Facebook may have made finding volunteers easier, but only at the cost of having to spend more time getting those volunteers to do any work.

Furthermore, the increasingly social nature of information consumption in the digital age may result in certain causes (like those having to do with the immediate environment, one's friends, alma mater, and so on) gaining a disproportionally higher place on one's immediate agenda. Often this is a useful change in the focus of political activism. While many students are wasting their energy on "saving" Darfur by joining Facebook groups, their own universities are run without the scrutiny they deserve from the student body. Some kind of a balance between the global and the local is desirable, but as social networking sites inundate their members with information and suggestions carefully selected based on their demographics, location, and existing networks, it may well be that global activism is once again the aristocratic privilege of the widely traveled and widely read upper classes.

Everybody Can't Be Che Guevara

The decentralization of political organizing may have wonderful implications for knowledge creation—Wikipedia is one example—but the reality is that decentralization itself is not a sufficient condition for successful political reform. In most cases, it's not even a desired condition.

When every node on the network can send a message to all other nodes, confusion is the new default equilibrium condition. This is becoming obvious to anyone managing political campaigns—giving volunteers a chance to spam everyone on the list can paralyze the entire effort. One academic observer of how campaigns worked during the Iowa Presidential Primary in 2008 was shocked by the amount of mis- and overcommunication experienced by their staff, noting that much of it was uncalled for and harmful and remarking that "it would have taken very peculiar priorities in activist groups or campaign organizations to generate 45 phone calls or letters in a few days to a single Goldwater supporter in 1964 or a McGovern supporter in 1972." Not so today: It's distressingly easy to send 450 or 4,500 emails with a click of a button.

The ease with which supporters can now be mobilized online may eventually block the campaigners' imagination and preclude them from experimenting with more costly—but also potentially more effective—strategies. As the *New Yorker*'s Malcolm Gladwell, in a rather Kierkegaardian train of thought, asked the audience at F5 Expo, a Canadian technology conference, in 2010: "What would have happened to Castro if he had had Twitter and Facebook? Would he have gone to the trouble of putting together an extraordinary network that allowed him to defeat Batista?" What Gladwell seems to be saying is that, despite Facebook and Twitter's superb ability to mobilize millions of people in a matter of minutes, it's not such mobilization but rather the ability to organize and wisely expend one's *resources* (it helps if they include a hundred or more fearless gun-totting bearded guerillas) that makes or breaks a revolution. But since Twitter and Facebook are within much easier reach, it may be tempting to start one's quest for a revolution in the digital rather than the physical realm. This may have worked if activists campaigns were all like Wikipedia and other open-source projects, where tasks are granular, risk-free, and well-defined, and the timeline is extremely short. But you can't simply join a revolution any time you want, contribute a comma to a random revolutionary decree, rephrase the guillotine manual, and then slack off for months. Revolutions prize centralization and require fully committed leaders, strict discipline, absolute dedication, and strong relationships based on trust.

The unthinking glorification of digital activism makes its practitioners confuse priorities with capabilities. Getting people onto the streets, which may indeed become easier with modern communication tools, is usually the last stage of a protest movement, in both democracies and autocracies. One cannot start with protests and think of political demands and further steps later on. There are real dangers to substituting strategic and long-term action with spontaneous street marches. Angela Davis, a controversial activist in the civil rights movement, knows a thing or two about organizing. Davis, who used to associate with the Black Panthers in the early 1970s, gradually emerged as one of the most talented organizers on the left, having played an important role in the struggle for civil rights. Today she is concerned with the long-term effects of the growing ease of mobilization on the effectiveness of social movements. "It seems to me that mobilization has displaced organization, so that in the contemporary moment, when we think about organizing movements, we think about bringing masses of people into the streets," writes Davis in her 2005 book *Abolition Democracy: Beyond Empire, Prisons, and Torture.*

The dangers of this development are obvious. The newly gained ability to mobilize may distract us from developing a more effective capacity to organize. As Davis remarks, "it is difficult to encourage people to think about protracted struggles, protracted movements that require very careful organizing interventions that don't always depend on our capacity to mobilize demonstrations." Just because you can mobilize a hundred million people on Twitter, in other words, does not mean you should; it may only make it harder to accomplish more strategic objectives at some point in the future. Or as Davies herself puts it: "The Internet is an incredible tool, but it may also encourage us to think that we can produce instantaneous movements, movements modeled after fast food delivery."

It seems that Iran's Green Movement may have been much more successful in 2009 had they heeded Davis's advice. While the unique decentralized nature of Internet communications allowed the protesting Iranians to effectively bypass censorship and broadcast information outside of Iran, it also prevented the movement from acting in a strategic thought-out fashion or, at least, speaking with one voice. When the time came to act in unison, thousands of Facebook groups couldn't collect themselves into a coherent whole. Iran's Twitter Revolution may have drowned in its own tweets: There was just too much digital cacophony for anyone to take decisive action and lead the crowds. As one Iranian commentator bitterly remarked on his blog: "A protest movement without a proper relationship with its own leaders is not a movement. It is no more than a blind rebellion in the streets which will vanish sooner than you can imagine." Social media only further added to the confusion, for, while information seemed to be coming from everywhere, it was not obvious that anyone was in control. "Cell phone cameras, Facebook, Twitter . . . seem . . . to be making everything happen much faster. There's no time to argue what it all means—what the protesters want, if they're ready to die. The movement rolls forward, gathering speed, and no one really knows where it's going," writes a young Iranian who participated in the 2009 protests, got arrested, and penned a book about all those experiences under the pseudonym Afsaneh Moqadam.

Just because the Internet allows everyone to lead doesn't mean that nobody should follow. It's not so hard to imagine how any protest movement might be overstretched by the ease of communications. When everyone can send a tweet or a Facebook message, it's safe to assume that they will. That those numerous messages would only increase the communication overload and may slow down everyone who receives them seems to be lost on those touting the virtues of online organizing.

Dissidents Without Dissent

The problem with most dominant interpretations of digital activism is that they are too utilitarian in spirit—asking questions like "how many more clicks / eyeballs / signatures on Facebook can I garner if I invest more money, time, staff?"—and often overlook a softer cultural side of activism. The fact that Facebook allows us to achieve a certain objective X looks less impressive if achieving that objective by means of Facebook also displaces an activity Y that may be more important—depending on the context—in the longer run. For example, even though microwaves and frozen foods can help minimize the time spent on cooking food, few of us rush to this solution when we entertain a dinner party, not least because there are other more important qualities to cooking, eating, and socializing than just time or cost savings. The calculus of measuring quality of life demands a few more steps than simply adding all the efficiencies and subtracting all the inefficiencies; it also requires a good understanding of what particular values are important in a particular context of human relations.

If an authoritarian regime can crumble under the pressure of a Facebook group, whether its members are protesting online or in the streets, it's not much of an authoritarian regime. The real effects of digital activism would thus most likely be felt only in the long term rather than immediately. Over the long haul, the availability of such mobilization opportunities begins to influence deeply rooted political structures and established political processes of a particular society, authoritarian or not. The challenge for anyone analyzing how the Internet may affect the overall effectiveness of political activism is, first, to determine the kind of qualities and activities that are essential to the success of the democratic struggle in a particular country or context and, second, to understand how a particular medium of campaigning or facilitating collective action affects those qualities and activities.

For example, it's safe to assume that in most countries toppling a powerful authoritarian regime would demand dissidents who are strategic, well-organized, but above all brave and ready to die or go to prison if the circumstances so require. Obviously, only a small share of any country's population would be eager to make such sacrifices; this is why there is still such a heroic aura around the word "dissident." Such people may not be terrifically successful in undermining the power of the regime, but they might (one thinks of Gandhi) be setting an important moral example that could nudge the rest of their fellow citizens.

Significant political change requires an embrace not only of conventional politics but of its most hyperactive and brutal elements: arrests, intimidation, torture, and expulsions from universities. Solzhenitsyn and Sakharov may have been more effective communicators if they had access to the Internet, but it's not certain they would have been more effective dissidents. It was not what they said (or, for that matter, how they said it) that awoke Russians from seven decades of a political coma, but rather what they did—bravely defied the authorities, spoke their minds, and faced the consequences. Dissidents were much more than hubs for the gathering and dissemination of information; their movement, likewise, was more than the sum of such hubs, with dissident culture enabling certain kinds of risky behavior that helped to punch holes in the once-solid structure of authoritarianism. What mattered the most about the dissidents was not what they produced but what their activities enabled them to achieve on other fronts.

The nurturing of dissent has always depended on the ability of existing dissidents to cultivate certain myths around their activities, if only to encourage others to follow suit. Many in the Russian dissident community still fondly remember how Sakharov and his wife, Elena Bonner, would secretly meet in the park—a radio their sole companion—to listen to and transcribe foreign radio broadcasts. Or how the Czech and Polish dissidents would secretly meet in the mountains on the Czech-Polish border, sit next to each other pretending to be resting, only to pick up each other's bag on leaving, thus facilitating cross-border exchange of samizdat. Such tales, whether true or not, helped to cultivate a certain image of a renegade dissident; such a distinct cultural phenomenon must have had enormous political repercussions, if only in terms of inciting romantic youths to join the movement.

Since most successful dissident groups that operated in closed societies before the Internet did not embed visiting anthropologists who could hang around them for a year and study how they came to be who they were, we only have a cursory knowledge of what gave rise to their dissent and courage. Take the issue of censorship, which, on first sight, may look unconnected to dissent. Will people be more likely to become dissidents if they regularly run into government censorship? Will it help if the censorship is visible and intrusive—think radio jamming, with all of its noises and crackles—as opposed to the more silent, internalized, and mostly invisible censorship of newspapers? At least one scholar of Cold War history argued that radio jamming could indirectly breed dissenting attitudes, for it "excites listener curiosity about programs being jammed, increases suspicion of the authorities' motives for jamming, and supports the people's faith in what Radio Liberty has to say." This is not to argue that censorship is good but rather to suggest that most people who chose to oppose their governments during communism did not just wake up one day fashioning themselves as dissidents; their politicization was a slow and complicated process that we are only beginning to understand.

The kind of oppositional politics made possible by the Internet—where all communication is assumed to be protected (even when it is not), where anonymity is the default rather than the exception, where there is a "long tail" of political causes an activist may be involved in, where it's easy to achieve tactical but mostly marginal victories over the state—is not likely to produce the next Václav Havel. Someone still has to go to prison. And many bloggers do just that. But they are still predominantly lone wolves, all too often by their own choice, who operate without much in the way of popular appeal. Instead of building sustainable political movements on the ground, they spend their time receiving honorary awards at Western conferences and providing trenchant critique of their governments in interviews with Western media. Yoani Sánchez, a prominent Cuban blogger hailed by *Time* magazine as "one of the world's most influential people," is far better known outside of Cuba than inside, which, of course, is not for lack of trying, itself an act of heroism given Cuba's restrictive system of media controls. Still, in their ability to guide the moral outlook for an entire generation, Sánchez's numerous blog posts, poignant as they are, hardly amount to a single play by Havel. Becoming Cuba's Havel may not be the goal that Sánchez has set for herself, but this is not how most of her Western supporters, who confuse blogging with samizdat, see it.

Likewise, it's certainly remarkable that gays in Nigeria can now form an online Bible study class, as the *Economist* approvingly reported in one of its recent issues, for they might be beaten up if they showed up at homophobic Nigerian churches instead. But let's be honest: We simply don't know if the availability of such virtual meeting space is going to help the long-term prospects of gay rights in Nigeria. After all, changing social attitudes on such charged issues would require a series of painful political, legal, and social reforms and sacrifices, which may or may not have been made easier by the Internet. Sometimes the best way to launch an effective social movement is to put an oppressed group into a corner that leaves no other option but dissent and civil disobedience. The danger is that the temporary false comfort of the digital world may result in that group never quite feeling the corner as forcefully.

No Such Thing as Virtual Politics

The danger that "slacktivism" poses in the context of authoritarian states is that it may give young people living there the wrong impression that another kind of politics—digital in nature but leading to real-world political change and the one underpinned entirely by virtual campaigns, online petitions, funny Photoshopped political cartoons, and angry tweets—is not only feasible but actually preferable to the ineffective, boring, risky, and, in most cases, outdated kind of politics practiced by the conventional oppositional movements in their countries. But despite one or two exceptions, this is hardly the case at all. If anything, the entertainment void filled by the Internet—the ability to escape the gruesome and boring political reality of authoritarianism—would make the next generation of protesters less likely to become part of traditional oppositional politics. The urge to leave the old ways of doing politics behind is particularly strong in countries that have weak, ineffectual, and disorganized opposition movements; often the impotence of such movements in their fight against the governments generates more anger among the young people than the governments' misdeeds.

But whether we like it or not, such movements are often the only hope that such societies have. Young people have no other choice but to join in and try to improve them. Denouncing their governments and applying for permanent residence in Twitter-land is not an option likely to reinvigorate the moribund political process in many of these countries.

"In terms of their impact [on the Arab world, new media] seem more like a stress reliever than a mechanism for political change," writes Rami Khouri, editor-at-large of Lebanon's the *Daily Star*, who fears that the overall impact of such technologies on political dissent in the Middle East might be negative. "Blogging, reading politically racy Websites, or passing around provocative text messages by cellphone is . . . satisfying for many youth. Such activities, though, essentially shift the individual from the realm of participant to the realm of spectator, and transform what would otherwise be an act of political activism, mobilizing, demonstrating or voting into an act of passive, harmless personal entertainment." Mr. Khouri may be slightly overstating the case—digital activists in the Middle East can boast of quite a few accomplishments, particularly when it comes to documenting police brutality—but his overall concern about the long-term effects of digital activism on politics at large is well-justified.

Seeing how the worlds of offline and online politics collide in the case of Belarus, my home country, I do detect a certain triumphalism about online politics among the younger generation. Many young people, frustrated by the inability of the opposition to mount a challenge to the country's tough ruler, are beginning to wonder why they should even bother with poorly attended town halls, rigged elections, exorbitant fines, and inevitable jail time if the Internet allows doing politics remotely, anonymously, and on the cheap. But this has proved no more than utopian dreaming: No angry tweets or text messages, no matter how eloquent, have been able to rekindle the democratic spirit of the masses, who, to a large extent, have drowned in a bottomless reservoir of spin and hedonism, created by a government that has read its Huxley. When most Belarusians use the Web to gorge on free entertainment on YouTube and LiveJournal, seeking solace from the dreadful political realities of the real world, politicizing requires more than just sending them requests to join antigovernment Facebook groups, no matter how persuasive those are.

While policymakers shouldn't ignore the multiple successes of activists who have used the Internet and social media to their advantage around the globe, from campaigning against Pervez Musharaf in Pakistan to shaming Shell about its dubious activities in Nigeria to fighting misogynist fundamentalists in India, they should also remember that, even when successful, such campaigns always come with hidden social, cultural, and political costs. This is even more relevant if they target a powerful authoritarian state. One of the main reasons why the anti-FARC protests were so successful in Colombia was because they were opposing a group much hated by the Colombian government. When the same group of activists used their Facebook know-how to launch similar anti-Chavez protests in Venezuela in September 2009, expecting up to sixty million people to join the protests all over the globe, only a few thousand actually showed up (that Chavez launched a smart propaganda campaign and countered with "grassroots" protests of his own did not help either). Whenever Hillary Clinton touts the power of social networking to change the world and David Miliband, her former British counterpart, speaks of "civilian surge" and muses on how new media can help "fuel the drive for social justice," one should scrutinize those claims closely. While it may be true that new forms of activism are emerging, they may be eroding rather than augmenting older, more effective forms of activism and organizing.

chapter eight

Open Networks, Narrow Minds:
Cultural Contradictions
of Internet Freedom

● ● ■ ■

F or all the praise that American diplomats heaped on Twitter for its prominent role in the Iranian protests of 2009, one extremely ironic development has gone largely unnoticed: by allowing Iranians to share photos and videos from the streets of Tehran, Twitter's executives may have violated U.S. law. Few of Twitter's cheerleaders in the media paid any attention to the fact that the tough sanctions imposed by the U.S. government on Iran extend to American technology companies, including those offering Internet services to ordinary Iranians.

In fact, these American sanctions, administered predominantly by the Department of Treasury and Department of Commerce, far away from the cyber-utopian offices of the State Department, have hurt the development of the Iranian Internet as much as the brutal crackdowns of the Iranian police. Until March 2010, almost a year after the protests, Iranians could not legally download Google's Chrome browser, place calls on Skype, or chat via MSN messenger. All these services (and many more) were subject to a rather byzantine set of restrictions imposed by the U.S. government. Some could have been overcome, but most American companies chose not to bother; such fights would be too expensive to mount, while the profits they would make by selling online advertising to Iranians didn't seem lucrative enough.

Most Iranian opposition groups have a hard time finding an American company willing to host their websites; those who have more luck with European or Asian firms cannot easily pay for them, as U.S.-based online payment systems like PayPal do not offer their services to Iranians. Even more bizarrely, those who want to pierce Iran's numerous firewalls by circumventing government blocks of sites like Twitter cannot easily do so, as the export of anticensorship tools also falls within the scope of the sanctions regime. Furthermore, most technologies that feature encryption are subject to a complex set of special export regulations, waivers, and licenses. Ironically, various nonprofit groups funded by the U.S. government keep training Iranian activists to use many of those tools despite the sanctions; in a sense, American taxpayers are funding the training of Iranians in using the tools that the U.S. government doesn't allow them to use.

American diplomats eventually realized that the current sanctions regime "is having an unintended chilling effect on the ability of companies such as Microsoft and Google to continue providing essential communications tools to ordinary Iranians," as a State Department representative put it in a letter sent to the Senate six months after the Iranian protests. In March 2010 the Treasury agreed to make limited amendments to its regulations, allowing the export of "publicly-available mass market online services . . . incident to the exchange of personal communications over the Internet" to Iran (Cuban and Sudanese sanctions were also amended accordingly). These amendments, however, don't extend to the export of most censorship-circumvention software, meaning that Iranians still face legal hurtles when they want to break through the firewalls that American policymakers so badly want them to pierce.

A Dollar in a Haystack

This doesn't mean that it's impossible to get such tools to Iran legally: One can ask the U.S. government for an export license. Not surprisingly, some of the tools that make it to Iran are backed by people and organizations that are more successful in pressuring the U.S. government into granting them such a license. Thus, it's those who have the best resources—lawyers, publicists, and lobbyists—rather than the best technologists who are most likely to see their products used by Iranians.

In March 2010, after a publicity-heavy campaign in the media, a technology called Haystack was granted one such license. Haystack came out of nowhere during the Iranian protests in 2009. It was founded by two twenty-something American techies with no ties to Iran. They got mesmerized by the pictures coming out of Tehran and wanted to help by finding a way to give Iranians access to banned websites. Thus, they designed Haystack—a technology that would not only pierce through the firewalls but would also make it seem as if its users were browsing innocuous sites like weather.com. Austin Heap, the public face of Haystack who quickly became a media darling, kept boasting that his software was not only very effective but also perfectly secure.

Such claims were impossible to verify, as no one could evaluate Haystack: Its website contained several "Donate!" buttons but had no link to download the actual software. Haystack's founders claimed this was on purpose: They simply did not want to let the Iranian government reverse-engineer their software before anyone in Iran actually got a chance to use it.

This seemed like a good enough explanation to the media—and Haystack continued receiving glowing coverage in the *International Herald Tribune*, NPR, *Christian Science Monitor*, and BBC News (in March 2010 *The Guardian* even proclaimed Austin Heap to be "Innovator of the Year"). In August 2010 *Newsweek* published an approving profile of Haystack in which its founder got most of the limelight. "Tomorrow I meet with [Sens. John] McCain, [Bob] Casey, maybe [Carl] Levin, but I don't know if I will have enough time," Austin Heap told *Newsweek*.

Many experts in the technology community were getting increasingly skeptical: If Haystack was, indeed, so good, why wouldn't its founders let anyone find flaws in it? Given that it was meant to be used by Iranian dissidents, such concerns were justified.

Soon enough, the outlandish claims made by Haystack's founders angered some of the Iranians that had originally been recruited to test Haystack in the country—and they leaked a copy of Haystack to independent third-party testers. A few hours of testing revealed that using Haystack in Iran was extremely unsafe, as the software left digital traces that the Iranian government could use to identify its users. As it turned out, Haystack only had a few dozen testers in Iran—a far cry from 5,000 users that Austin Heap claimed in March 2010. By mid-September 2010 testing of Haystack in Iran was halted, while everyone working on the Haystack project—including its high-profile advisory board—resigned.

While little is known of any Haystack-related arrests in Iran, it's not so hard to imagine what would have happened if Haystack did go into mass distribution in the country. How many Iranians would learn that Haystack violated many of the good practices that an ideal anticensorship tool should aspire to when the U.S. government happily granted it an export license and Secretary Clinton even mentioned it in passing in one of her interviews?

Many Iranian users may be no more sophisticated about such matters than American journalists who chose to pen admiring reviews of Haystack. Differences between various circumvention technologies are difficult to grasp for nontechnologists; many may mistakenly believe that Haystack is safer than it really is simply because the U.S. government thought it a worthwhile export. Not surprisingly, the endorsement by the U.S. government only raised Haystack's profile, resulting in even more admiration by the media.

Clearly, the way in which the review process works currently is flawed, and some may even say dangerous. While many have called for the sanctions to be scrapped altogether, there are surely less radical ways in which the process could be driven by the assessment of the actual features of a technology rather than the ability of its founders to generate buzz in the media and among policymakers.

Mugabe Blogs Here

If this could be of any consolation to Iranians, they are not alone in their predicament; countries like Belarus, Cuba, North Korea, Syria, Zimbabwe, and certain areas of Sudan also face various sanctions imposed by the U.S. government. Fortunately, some of them are highly targeted, as is the case of Belarus and Zimbabwe, where they cover only dozens of so-called specially designated nationals, mostly current and former government officials (but also entire organizations) who have been known to engage in outrageous abuses of power.

In theory, such targeting sounds like a great way to prevent the effects of sanctions extending to innocent individuals. But the reality is much more complex. Unfortunately, many American Internet companies would rather not take the chance that some crony of Robert Mugabe has covertly or overtly taken virtual residence on their site, as this could lead to fines and even jail time for their executives. The only way to avoid such risk is to extensively vet all new users from Zimbabwe, a practice so expensive and time-consuming that many companies, particularly those that don't have large compliance budgets, prefer simply to ban all Zimbabwean nationals and even specify that in their terms of service. (The fact that Zimbabwe is not an important profit center also means that such a decision is relatively easy to make.) Besides, it takes a lot of optimism to believe that American Internet companies would all work to address the differences in the nature of sanctions imposed on different countries. Finding how the differences in sanctions imposed on Cuba and Syria should translate into the provision of specific services to their nationals is a task that often requires nothing short of an Ivy League law degree. Most companies simply opt for the lowest common denominator—a blanket ban on all nationals of those countries.

Often this results in rather surreal situations, in which an American company would cite the regulations of the U.S. government to stop providing Internet services to the entities and individuals that enjoy the moral or financial support of the U.S. government. Consider what BlueHost, one of the largest Internet hosting providers in the United States, did to the websites of the Belarusian-American Studies Association, a DC-based nonprofit entity that is frequently consulted by the U.S. State Department on Belarus policy, and Kubatana, one of the leading anti-Mugabe civil society organizations in Zimbabwe, which also enjoys extensive contacts with the American government. On discovering that both entities are run by individuals who are citizens of Belarus and Zimbabwe, BlueHost simply terminated their contracts and threatened to delete all their content from its servers, since its terms of service (the fine print all of us have to scroll through searching for that "Next" button) specified that no deals with the nationals of Belarus and Zimbabwe were allowed, supposedly because of U.S. sanctions— a gross misinterpretation of a highly targeted policy. BlueHost's CEO wasn't swayed even when the U.S. ambassador to Zimbabwe wrote to him to confirm Kubatana's impeccable anti-Mugabe credentials.

It took a letter from the U.S. Treasury Department to convince Blue-Host to change its practices. Similar overcompliance is still common among Internet companies. In April 2009 the popular social networking site LinkedIn decided to ban all Syrian users from its site, citing U.S. sanctions. After its CEO saw that such a move led to a lot of LinkedIn bashing in the blogosphere, he reversed the decision, explaining that it was the result of overzealous interpretation of existing regulations.

That most sanctions imposed by the United States on unruly governments fail to accomplish their goals is, of course, an open secret in Washington and beyond. But the futility of such sanctions in regulating technology is even more apparent. To assume that the leaders of Belarus or Zimbabwe would be bothering to purchase services from American hosting companies while they can easily get them from their domestic (and often state-controlled) firms is simply ridiculous. The chasm between the U.S. government's rhetoric on Internet freedom and the reality of their own restrictions on the exports of technology has not been lost even on authoritarian states, which, on multiple occasions, have used it to bolster their own propaganda that Washington doesn't mean what it says (in 2009, a government newspaper in China deplored the U.S. government for not allowing the downloads of MSN messenger in Cuba). But there are many other reasons why such sanctions need to go; whatever its real role and significance in the Iranian protests, had Twitter complied with the letter of the law in the summer of 2009, it might have deprived Americans of an important channel of information. (Some lawyers speculate that this would have bordered closely on what they call "prior restraint" and may have even violated the First Amendment.)

The ineffectiveness of sanctions has, of course, rarely stopped American leaders from embarking on quixotic adventures. Still, it would be disingenuous not to acknowledge that a campaign to promote Internet freedom around the globe loses much of its allure when the U.S. government itself creates so many hurdles for people who want but are unable to take full advantage of the Internet. One danger of making Internet freedom into a guiding orientation for the Western impetus to promote democracy is that it diverts attention away from the misdeeds and poor policies of Western governments themselves, focusing almost exclusively on the draconian Internet controls of authoritarian governments. As the situation in Iran so aptly demonstrated, with the U.S. State Department asking a company to continue providing the services it shouldn't have been providing in the first place, even American officials can get lost in their own policies and sanctions. Until those are simplified and purged of unnecessary hurdles, the Internet is only working at half of its fully democratizing capacity.

A Doll with Censored Nipples

In 2008 the Moroccan government's fear of the Internet made international headlines after it jailed Fouad Mourtada, a Moroccan engineer who had supposedly set up a fake Facebook profile for Prince Moulay Rachid, one of the country's rulers. It never became clear how the Moroccan government traced Mourtada; some commentators even accused Facebook of turning him in. Given Facebook's reputation with the government, the Moroccan activist Kacem El Ghazzali must have accepted the possibility that his government might find a way to ban access to his innocently named Facebook group (Youth for the Separation Between Religion and Education). El Ghazzali wants to establish a clearer dividing line between religion and education in his country. He may not be calling for regime change, but given the rather soporific pace of Moroccan politics, even such supposedly apolitical campaigns draw ire from the country's rulers.

El Ghazzali's case is not unique; there is a rapidly growing network of other occasional single-issue activists working on reforming Morocco. Thanks to the Internet, many of them are able to register their disagreement with various policies of their government and find like-minded individuals inside the country, in the diaspora, and elsewhere in the Arabic-speaking world who may assist in campaigning. Predictably, the government is not thrilled, making every effort to obstruct such activism, especially if it involves profanity or humor. The proliferation of such online initiatives may not always be terrifically effective from a policy-planning perspective—everything else being equal, charges of slacktivism are inevitable—but the real contribution of Facebook groups to the democratization of Morocco may lie in pushing the boundaries of what can and cannot be said in this conservative society rather than mobilizing street protests. (To label any Facebook activity as extremely useless or extremely useful just because it takes place on Facebook would thus be an obvious case of Internet-centrism; what may be destructive in the context of Belarus, a society with a much more open, even if still state-manipulated, public sphere, may actually be quite useful in the context of the more socially conservative Morocco.)

One day in 2010 El Ghazzali logged onto Facebook and discovered that his group was gone, along with the list of its more than 1,000 members. It was not obvious how the Moroccan government could be involved—unless, that is, they had friends in Palo Alto, California, where Facebook's headquarters are located. It turned out that Facebook itself had deleted his group, and it did not bother to give El Ghazzali any explanation or even warn him about what was coming.

When he emailed the company demanding an explanation, his own profile on the site was deleted as well. When a few days later several prominent international bloggers stood up for El Ghazzali and the story got media attention in the West, Facebook restored the banned group but still did not explain its motivation for the original blocking. El Ghazzali himself was not so lucky: He had to create a brand-new Facebook account for himself, since his original account was not restored.

Such (relatively) happy endings are rare; most similar cases do not attract the kind of attention to push Facebook and other intermediaries to rein in their bureaucratic excesses. Had it not been for the international attention that El Ghazzali's case had received, he, like many other activists before him, would have had to rebuild his online campaign from scratch. One can't easily accuse Facebook of any legal wrongdoing—after all, it is a private company and can do whatever it wants to. Perhaps, for reasons of their own, Facebook executives did not want to be seen as taking sides in Morocco's secularism debate; alternatively, the deletion of El Ghazzali's group was simply the result of a human error, of someone mistaking his group for a quasi-revolutionary collective asking for the overthrow of the Moroccan government (since Facebook didn't issue a press release, we will never know). What is clear is that, contrary to the expectations of many Western policymakers, Facebook is hardly ideal for promoting democracy; its own logic, driven by profits or ignorance of the increasingly global context in which it operates, is, at times, extremely antidemocratic.

Were Kafka to pen his novel *The Trial*—in which the protagonist is arrested and tried for reasons that are never explained to him—today, El Ghazzali's case could certainly serve as inspiration. That much of digital activism is mediated by commercial intermediaries who operate on similar Kafkaesque principles is cause for concern, if only because it introduces too much unnecessary uncertainty into the activist chain. Imagine that El Ghazzali's group was planning a public protest on the very day that its page got deleted: The protest could have easily been derailed. Until there is complete certainty that a Facebook group won't be removed at the most unfortunate moment, many dissident groups will shy away from making it their primary channel of communication.

In reality, there is no reason why Facebook should even bother with defending freedom of expression in Morocco, which is not an appealing market to its advertisers, and even if it were, it would surely be much easier to make money there without crossing swords with the country's rulers. We do not know how heavily Facebook polices sensitive political activity on its site, but we do know of many cases similar to El Ghazzali's. In February 2010, for example, Facebook was heavily criticized by its critics in Asia for removing the pages of a group with 84,298 members that had been formed to oppose the Democratic Alliance for the Betterment and Progress of Hong Kong, the pro-establishment and pro-Beijing party. According to the group's administrator, the ban was triggered by opponents flagging the group as "abusive" on Facebook.

This was not the first time that Facebook constrained the work of such groups. In the run-up to the Olympic torch relay passing through Hong Kong in 2008, it shut down several groups, while many pro-Tibetan activists had their accounts deactivated for "persistent misuse of the site." It's not just politics: Facebook is notoriously zealous in policing other types of content as well. In July 2010 it sent multiple warnings to an Australian jeweler for posting photos of her exquisite porcelain doll, which revealed the doll's nipples. Facebook's founders may be young, but they are apparently puritans.

Many other intermediaries are not exactly unbending defenders of political expression either. Twitter has been accused of silencing online tribute to the 2008 Gaza War. Apple has been bashed for blocking Dalai Lama–related iPhone apps from its App Store for China (an application related to Rebiya Kadeer, the exiled leader of the Uighur minority, was banned as well). Google, which owns Orkut, a social network that is surprisingly popular in India, has been accused of being too zealous in removing potentially controversial content that may be interpreted as calling for religious and ethnic violence against both Hindus and Muslims. Moreover, a 2009 study found that Microsoft has been censoring what users in the United Arab Emirates, Syria, Algeria, and Jordan could find through its Bing search engine much more heavily than the governments of those countries. Anyone visiting websites that contain words like "sex" or "porn" in their URLs in a country like Jordan will be able to access them; however, were Jordanians to search for anything containing those terms on Bing, they would simply see a warning from Microsoft.

Dangerous Intermediaries

Is there a secret plot by the world's largest technology companies to restrict global freedom of expression? Probably not. The sheer amount of content uploaded to all these sites makes it impossible to administer them without making mistakes. The border between a video promoting violence and a video documenting human rights abuses is rather blurry and often impossible to determine without an intimate knowledge of the context in which the video was made. Google, for example, has been accused of removing from YouTube a series of videos from Egypt that depicted police brutality on the basis of their being too violent. (Google later acknowledged that it had done so in error.) But knowing that a video captures an act of police brutality rather than a scene from a horror movie requires knowing the context, and this is not so easy, given that twenty-four hours of video is uploaded to YouTube *every minute.* The only way to completely avoid making mistakes on this front is to hire panels of human rights lawyers and pair them up with regional experts to review every piece of controversial content that is found on sites like YouTube and Facebook. To its credit, YouTube is much more open about its content removal strategies than Facebook; one may disagree with the principles it employs—and especially with the way in which its video-analysis technology "recommends" videos that need special attention by humans—but at least the company is transparent about it, thus making it easier for activists to make educated guesses.

Some companies have tried to address this issue by introducing ways in which users themselves can report videos that they find offensive, somehow alleviating the burden on their own internal police. So far, however, such features have triggered a disturbing surge in cyber-vigilantism. For example, a well-coordinated group of two hundred culturally conservative users in Saudi Arabia, known as "Saudi Flagger," regularly monitor all Saudi Arabia–related videos uploaded to YouTube. En masse, they complain about the videos they do not like—most of them critical about Islam or Saudi's rulers—"flagging" them for YouTube's administrators as inappropriate and misleading. (The group's members have a more philosophical take on their work: "All we do is to perform our duty towards our religion and homeland," Mazen Al Ali, one of the group's volunteers, told the Saudi daily *Al Riyadh* in 2009.) Good judgment, as it turns out, cannot be crowdsourced, if only because special interests always steer the process to suit their own objectives.

Perhaps it is only natural that in its quest for more and better eyeballs, digital activism breeds a culture of dependence on large intermediaries, where those with a dissenting viewpoint have to read pages of fine print before sharing their subversive thoughts online. What's worse, the fine print is often ambiguous and inconclusive. (Who would have guessed Facebook frowns on dolls' nipples?) Even those who master it in full can never be sure that they are not breaching some arcane rule. While activists can minimize their exposure to intermediaries by setting up their own independent sites, chances are that their efforts may not receive the kind of global attention that they might on Facebook or YouTube. Faced with the painful choice between scale and control, activists usually choose the former, surrendering full control over their chosen platform.

None of the popular Web 2.0 sites have handled such issues with consistency. Some clearly activist content is deemed offensive and removed; some stays on and attracts millions of views. The ensuing uncertainty works against digital activists. Who wants to invest time, money, and effort into building an antigovernment Facebook group only to have it deleted by the site's administrators? As a result, supporting structures that could have provided fertile foundations for building social capital online never solidify quite fully.

There are no simple remedies for such problems. This is not a fight against almighty Chinese censors; it's a fight against well-meaning technology types in the Bay Area who, not wanting to turn their sites into playgrounds for terrorists, sadists, or some dangerous fringe movements, tend to overcensor or adopt one-size-fits-all censorship policies that don't try too hard to study what it is they are censoring. Of course, no one expects Facebook to stop making money and turn their site into a colony of revolutionary cells, but the least they can do is to remove any ambiguity from their censorship process, for it's the ambiguity that confuses so many activists.

Ultimately, the rapidly growing role of Western intermediaries is yet further proof that the battle for Internet freedom, however ill-conceived it may be, should also be fought in the spacious meeting rooms of Silicon Valley. Winning the battles in Moscow, Beijing, or Tehran won't automatically turn the Facebooks and Googles of this world into responsible global citizens. Unfortunately, there was little acknowledgment of that fact in Hillary Clinton's seminal speech on the subject, even thought this is an area where Western policymakers could accomplish the most simply by means of legislation. Nascent industry-wide initiatives like the Global Network Initiative—which Facebook didn't join, claiming that, as a young company, it did not have the resources to pay the $250,000 membership fee—that aspire to make technology companies pledge their commitment to a set of values are, in principle, a worthy undertaking. But ensuring compliance to the very principles that companies pledge allegiance to may require a strong push from governments in North America and Europe. Microsoft, for example, is a member of GNI, and yet the way its Bing search engine works in the Middle East does not fully adhere to the spirit of the initiative. Unless technology companies are somehow made to deliver on their own pledges, initiatives like GNI will be little but publicity stunts, meant to assure policymakers that the companies joining it are responsible global citizens.

Most unfortunately, it seems that the relatively short-lived quest for Internet freedom has already been corrupted by that old Washington problem: the tight embrace between policymakers and the industry. Two of the high-profile State Department appointees who spearheaded much of the work on Internet freedom, including establishing a close partnership with Silicon Valley firms, left Washington to work for those very firms. One, Katie Jacobs Stanton, adviser to the Office of Innovation, left to work for Twitter as its head of international strategy; the other, Jared Cohen, went to Google to head its new think tank. Of course, such turnover is nothing new for Washington, but it hardly provides an effective foundation for promoting Internet freedom or taking a critical view of the practices of technology firms, which are also in desperate need of suave executives with government experience.

The Beam in Thine Own Cyberspace

Even though the emerging public debate about Internet freedom inevitably ends with loud calls to oppose greater control of the Internet by authoritarian governments, Western policymakers should not let such rhetoric get the better of their common sense. Otherwise it would become all too easy to ignore the problems and debates about Internet regulations in their own backyard. The reality is that in their statements and actions most policymakers already acknowledge that a free Internet unburdened by regulation is likely to be as conducive to democratization as a government unburdened by the rule of law.

As the Internet gains in importance and penetrates more and more walks of public life, Western governments are poised to feel—and many of them are already feeling—growing pressure to regulate it. Some of that pressure will inevitably have illegitimate, harmful, and undemocratic origins; much of it won't. The way forward is to acknowledge that the public pressure to regulate the Web is growing and that not all of the ensuing regulation should be resisted because the Internet is the sacred cow of the libertarian movement. The only way to get it right is to avoid holding on to some abstract absolute truths—for example, that the Internet is a revolutionary force that should be spared any regulation whatsoever—but rather to invest one's energy into seeking broad public agreement on what acceptable, transparent, just, and democratic procedures by which such regulation is to occur should look like.

While the discussion of those ideal procedural principles merits a book of its own, anyone designing them should be aware of some major inconsistencies between the strong antiregulation impetus of Western foreign policy and the equally strong pro-regulation impetus of Western domestic policy. For while American diplomats are preaching the virtues of a free and open Internet abroad, an Internet unburdened by police, court orders, and censorship, their counterparts in domestic law enforcement, security, and military agencies are preaching—and some are already pursuing—policies informed by a diametrically opposite assessment of those virtues.

The quixotic quest to promote and defend Internet freedom may be a doomed enterprise from the beginning, for its ambitious objectives are programmed to collide with equally ambitious domestic objectives. To stay oblivious to the inevitability of this collision is to give false hope to activists and dissidents in authoritarian states, who may be naïvely hoping that the West will stick to its promises.

That the government needs to be brought into cyberspace or else cyberspace may lead to lawlessness in the real world is a view rapidly gaining traction among Western policymakers. "Cyberspace is increasingly Hobbesian, and the belief of the pioneers that a 'social contract' would emerge naturally from the self-organizing internet community without the intervention of the state has proven to be either wrong or moving at a pace so slow that it threatens security," writes James Lewis, a senior fellow at the Center for Strategic and International Studies in Washington and one of the authors of the report *Securing Cyberspace for the 44th Presidency*, something of a cybersecurity blueprint for the Obama administration. "The emancipatory aspirations for a libertarian cyberspace that would, to unparalleled extents, privilege social freedom over regulation, may end up in a socio-technical regime that largely undermines and reverses the freedom it once enabled," concurs Jeanette Hofmann of the London School of Economics.

It is, then, hardly surprising that some Western governments—Australia leads the pack here—are constantly flirting with censorship schemes that bear an eerie resemblance to those of China. For several years now European governments have been trying to pass hard-hitting legislation aimed at curbing illegal file-sharing, which may result in more aggressive tracking of users by their ISPs. The U.S. government, under immense pressure from the corporate sector and various activist groups, may soon be pushing to control the Internet on several fronts at once. Military and law enforcement agencies are the most aggressive in their push toward more Internet control. The Obama administration has been lobbying to allow the FBI to get access to more Internet records—like email addresses and browsing history—without seeking court orders. The White House's logic in this case rests on a particular and rather aggressive interpretation of existing rules about phone records, and many privacy activists take exception to establishing functional identity between phone numbers and email addresses. Whatever the legal merits of the administration's argument, it is obvious that once such measures go into effect in the United States, it would be impossible to stop other governments from expanding their own legal provisions to match the American standard.

Just like their peers in China and Iran, law enforcement professionals in the West are beginning to troll social networking sites, searching for details of their cases or just looking for new threats. On a purely rhetorical level, it's hypocritical for democratic governments to criticize authoritarian governments for employing the same tactics. While the world's attention was fixed on the young people arrested in Iran, most Western observers paid no attention when the New York Police Department went after and arrested Elliot Madison, a forty-one-year-old American activist from Queens who used Twitter to help protesters against the G20 summit in Pittsburgh evade the police. Nor did the world see much public outcry when in early 2010 two members of Philadelphia's city council considered legal action against Facebook, Twitter, and MySpace after those sites had been used to organize violent snowball fights in the city. When democratically elected politicians in the West champion the organizing power of social media while the police arrest citizens who take advantage of that power, how can the West expect to hold high moral ground over China and Iran? If American lawmakers are willing to punish popular Internet sites for facilitating snowball fights, it's hard not to expect the Iranian government to punish them for facilitating street protests.

Plenty of American decision makers in the defense and intelligence communities are pushing to reengineer the Internet to better protect the country from cyberwar by making it easier to track cyber-attacks, hardly good news for anyone concerned about privacy. When the director of the FBI publicly admits that he doesn't bank online out of security concerns, it's a sure bet that more control and regulation of the Internet are on their way. Cybercrime has been one of the few consistently growing activities on the Internet, and its future prospects look bright. Now, with the rapidly expanding trade in virtual goods on social networking sites and other websites, crimes that target such goods have spiked as well (in 2009, the fraud rate for merchants selling virtual goods was 1.9 percent, compared to 1.1 percent for those selling physical goods online).

That so many online transactions are anonymous is believed to be the chief reason behind rapidly growing rates of cybercrime. Not surprisingly, many governments are attempting to link our online actions to our real names. Speaking at a April 2010 cybersecurity conference, Stewart Baker, former counsel of NSA, was just expressing a popular view in intelligence circles when he said that "anonymity is the fundamental problem we face in cyberspace." In his much-discussed 2010 book about cyberwarfare, Richard Clarke, a senior national security official in many administrations, proposed that more ISPs should engage in "deep packet inspection," a practice that would allow them to better analyze the information sent and received by their customers, thus identifying cyber-threats and dealing with them at an early stage. Clarke's is a legitimate proposal that deserves debate and scrutiny by the public, but it's important to remember that it's through deep packet inspection—and using equipment bought from European companies—that Iran manages to keep such a tight hold on the Web. Very little can be done about Iran's use of the technology: Nokia-Siemens, one of the companies that supplied Iran with the inspection equipment, rightfully points out that this is the same equipment used by Western governments, even if they may not engage in such practices as aggressively as the Iranians. As deep packet inspection becomes even more widespread in the West, it will become nearly impossible to hold Nokia-Siemens—never mind Iran—responsible for its actions. The public may decide that it wants more deep packet inspections to address the threats posed by cybercrime or terrorism; they'll just need to remember it will have rather chilling effects on the business of promoting democracy abroad.

More junior and more tech-friendly military staffers have also been trying to figure out how to tame the Internet. In "Sovereignty in Cyberspace," a 2010 article published in *Air Force Law Review*, Lieutenant Colonel Patrick Franzese, who is with the U.S. Strategic Command, proposed that "[American] users wanting to access the Internet globally could be required to use a biometric scanner before continuing." Franzese's justification for establishing a tighter control over the Internet is common in military circles: "Cyberspace provides states and non-state actors the opportunity to negate the United States' conventional military advantage, circumvent its natural boundaries, and directly attack critical infrastructure inside the United States." It certainly helps that reining in cyberspace seems considerably easier than other domains. The notion of an Internet kill-switch is probably just an urban myth, but little about the infrastructure of today's Internet precludes some kind of a biometric scanner standing between users and the Internet. (Indeed, many laptops today already carry fingerprint-scanning devices.)

It is not just the military folks who are concerned about controlling the Web. Parental associations want to make it easier to track online pedophile activities and protect their children. Hollywood, music studios, and publishing companies are pushing for better ways to track and delete unauthorized exchange of copyright-protected content. Banks want stricter identity controls to minimize online fraud. That more and more people in the developing world are getting online is not seen as a precursor to a truly global conversation but rather as a precursor to global hell populated by Nigerian email scammers. Back in 1997, Eli Noam, a professor of communications at Columbia University, rightly observed that a free Internet—an Internet without barriers of any sort, where governments cannot erect barriers to protect their citizens against practices and services that they deem illegal—is not what the American public wants, so Americans should stop being in denial. "For all the rhetoric of an Internet 'free trade zone,' will the United States readily accept an Internet that includes Thai child pornography, Albanian tele-doctors, Cayman Island tax dodges, Monaco gambling, Nigerian blue sky stock schemes, Cuban mail-order catalogues?" asked Noam on the pages of the *New York Times*. The answer was no in 1997, and it's a much more resounding no today.

And the conversation gets even weirder once we get out of the United States and look at other Western democracies. When South Korean lawmakers want their government to be more effective in banning any South Koreans from visiting any North Korean websites, it's hard to imagine how a common Western position on Internet freedom may ever emerge. Such cacophony is not lost on authoritarian governments, who take every opportunity to introduce their own Internet controls and justify them based on greater regulation of the Web by their peers in the West. In February 2006, when confronted with criticism that there is too much Internet control in China, Liu Zhengrong, who then supervised Internet affairs for the information office of China's State Council, quoted the American experience with the USA PATRIOT Act and asked why China cannot be allowed to do the same. "It is clear that any country's legal authorities closely monitor the spread of illegal information. We have noted that the U.S. is doing a good job on this front. . . . So why should China not be entitled to do so?" So far Western democracies have not come up with a satisfying answer.

Western policymakers' nearly exclusive attention to problems like cybercrime and censorship may have crowded out serious debate on arguably more important issues like privacy. Lawmakers in most countries—with the possible exceptions of Germany, Switzerland, and Canada—have found themselves too overwhelmed to regulate social networks, essentially giving carte blanche to sites like Facebook. Furthermore, most cheerleaders of Web 2.0 believe that the calls for more privacy are unjustified and we as a society need to adjust to a world where everything is transparent. "We will simply become much more accepting of indiscretions over time. The point is, we don't really care about privacy anymore. And Facebook is just giving us exactly what we want," writes Michael Arrington of the popular technology blog TechCrunch. "I'd rather have entrepreneurs making high-profile mistakes about [privacy] boundaries, and then correcting them, than silently avoiding controversy . . . or avoiding a potentially contentious area of innovation because they are afraid of backlash," says Tim O'Reilly, the iconic publisher of technology books.

Such a stance is seriously problematic, for it has dire implications for users in authoritarian states. While many of us in the developed world can maybe survive the demise of privacy as long as other legal institutions are working well (and that's a very big "maybe"), it might easily have disastrous consequences elsewhere. Developing countries, where most citizens do not have bankable credit cards and are thus of little interest to online advertisers, hardly matter for Silicon Valley. No one is going to design a more secure version of their social network for them, even if the political situation in their countries requires a more careful attitude toward sharing personal data. A laissez-faire regulatory approach that glosses over high-profile mistakes in the name of innovation may eventually give us a shiny portable guide to the best frappuccinos in the neighborhood, but it may also inadvertently compromise the security of Iranian bloggers, who won't be treated to many frappuccinos in Tehran's Evin Prison.

As long as Western governments regulate the Internet out of concerns for terrorism or crime, as they currently aspire to, they also legitimize similar efforts—but this time done primarily for political reasons—undertaken by authoritarian governments. Even worse, in areas like cybercrime, the military and intelligence communities on both sides of the Atlantic would actually be quite happy to see Russian and Chinese governments establish stronger control over their respective national Internets. The West's own desire to have those governments do something about uncontrollable, even if hardly devastating, cyber-attacks that are regularly unleashed by their hacker populations trumps the impetus to promote abstract goods like Internet freedom simply because the security of America's own trade secrets always comes before the security of foreigners' social networking profiles.

To top it all off, officials tasked with U.S. domestic Internet policy—above all, the Federal Communications Commission—are also fond of the term "Internet freedom," by which they refer, primarily, to the issue of network neutrality, that is, ensuring that all types of content are treated equally and are not discriminated against by ISPs. The landmark net neutrality legislation proposed by FCC bears the name "Internet Freedom Act of 2010." It may be the case that drawing some parallels between the foreign and the domestic uses of the term may help both diplomats and technology policy wonks bring greater media attention to their cause, but, most likely, it would make both meanings extremely fuzzy, creating rhetorical traps for the U.S. government. In late 2009, while speaking on the subject of network neutrality, Andrew McLaughlin, the deputy chief technology officer of the U.S. government, said that "if it bothers you that the Chinese government [censors the Web], it should bother you when your cable company does it." He thus unwillingly supplied authoritarian governments with one more potential opportunity to chide the United States for not sticking to the principles it wants to promote abroad. Should the FCC's own aspirations to promote net neutrality be undercut by Congress, the Chinese and Iranian governments would score some major propaganda points by simply pointing out that American lawmakers, too, are regularly impinging on Internet freedom. Such is the cost of building government policy around highly ambiguous terms and then choosing to use them in completely different contexts.

Cyberwar Can Be Good for You

But foreign policy challenges and contradictions would also make the defense of Internet freedom hard to mount in the long term. As it becomes easier to organize targeted and well-contained attacks (i.e., without any unintended side effects) on sites of, say, Islamic extremists, there will be more calls to simply disable them, if only to stop future terrorist attacks. Of course, such sites also present immense intelligence value, which may explain why so many of them are allowed to operate. But this choice between attacking and spying on sites that the West hates or fears does not sound like a good way to burnish its credentials as a defender of Internet freedom.

Before the West makes an unconditional commitment to keeping the Internet free at all costs and in all situations, it also needs to consider that such policy is likely to clash with its own need to control and disrupt flows of information under pressing circumstances. Back in the 1990s it was quite fashionable to talk of "information intervention." Jamie Metzl, then a U.S. State Department official who emerged as a leading advocate of the policy of information intervention, persuasively argued that "the time has come to develop, refine, and institutionalize information-based responses to incendiary mass communications." By that Metzl meant primarily the ability to jam broadcasts inciting people to genocide.

Adjusting this concept to the Internet era raises many interesting questions. Would Western powers allow foreign radio stations broadcasting ethnic prejudice and hatred over the Internet to continue operating if there was a possibility of another genocidal war? Such was the unfortunate role of the radio in Rwanda and Yugoslavia in the 1990's, only the media hadn't yet shifted online. The liberal interventionists in the West probably would want to retain that capability; as Metzl rightly pointed out in 1997, "the free flow provisions of international telecommunications law hardly trump the revisions of the genocide convention that make inciting genocide illegal under international law." The lack of a quick "off" button that could simply shut down most Internet-based communications in a given region would become apparent the moment a large-scale genocide struck. If anything, the West wants to promote Internet freedom with a few giant asterisks, but the asterisks somehow get lost in translation.

This may seem like an overblown concern—ISPs may simply be down during the next genocide—but we have to remember that Western governments, concerned as they are about terrorism, will always want to preserve the ability to turn parts of the Internet off, if only temporarily or if only to a few foreign websites. Few sane policymakers would endorse a foreign policy that doesn't provide for such capability. Actually, such temporary Internet shutdowns happen all the time, even when no genocide is taking place. In 2008 the U.S. military launched cyber-attacks against an Islamist Internet forum in Saudi Arabia—ironically, itself first set up by the CIA to learn more about the jihadists' plans—to prevent the jihadists from collaborating and launching joint attacks on American targets in Iraq.

Cyber-attacks present us with an intellectually complex case that deserves a much more rigorous treatment than is allowed by inherently reductionist concepts like Internet freedom. When Hillary Clinton proclaimed that "countries or individuals that engage in cyber attacks should face consequences and international condemnation," she forgot to mention that American hackers, too, regularly launch cyber-attacks on the websites of governments they do not like. Most recently, this happened during the Iranian protests, when many Americans and Europeans eagerly joined an extremely well-publicized—mostly over Twitter—campaign to launch cyber-attacks on the websites of the Iranian government and thus thwart their ability to spread lies and propaganda. "The public's ability to strike back is something that every government should be reminded of from time to time" is how Matthew Burton, a former analyst with the U.S. Defense Intelligence Agency who participated in the attacks, justified his involvement. But this was not such a good idea after all: The attacks slowed down the Iranian Internet, making it harder to upload photos and videos from the street protests.

The most interesting part about this cyber-campaign was that American authorities did not react to it. The problem with such a seemingly cool stance is that when similar attacks were launched against the governments of Estonia and Georgia—supposedly by Russian nationalists—many officials on both sides of the Atlantic were quick to demand that Russia should stop tolerating its hackers and prosecute them. It sounded like a credible admonishment, but America's inaction in Iran meant that the United States, at least, ceded that moral ground. It's hard to avoid accusations of duplicity when America's own citizens— including former spies like Burton—openly spearhead attacks on the websites of a sovereign country that they happen to dislike. Despite Hillary Clinton's unambiguous proclamations to the contrary, Western policymakers simply do not yet have a coherent policy on cyber-attacks, nor do they know what that policy should look like. Instead of banning them outright, they should try to come up with a more sophisticated approach that may accept that some such attacks are inevitable and, potentially, even desirable.

Many cyber-attacks—especially those of the DDoS variety—may simply be construed as acts of civil disobedience, equivalent to demonstrations in the streets. It's not obvious that a campaign to limit the public's ability to practice those would abet the cause of democratization. If society tolerates organizing sit-ins in university offices and temporarily halting their work, there is nothing wrong—at least, in principle—with allowing students to organize DDoS attacks on university websites. In fact, this is already happening, with various degrees of success. In March 2010 Ricardo Dominguez, a professor at University of California at San Diego, called on his students to launch DDoS attacks on the website of the university's president to protest more than $900 million in budget cuts (the university administrators disconnected the professor's own server in retaliation). Some European courts have already ruled on that matter—in favor of DDoS as a means of dissent. In 2001 a German activist launched a series of DDoS attacks on the websites of Lufthansa to protest the fact that the airline allowed the German police to use its planes to deport asylum seekers. He argued that his campaign amounted to a virtual sit-in, and a German appeals court agreed.

The morality and legality of such cases have to be judged on a case-by-case basis. Clearly it would be inappropriate to outlaw all cyber-attacks across the board or proclaim them to be immoral. Imagine that pro-democracy activists in some authoritarian country governed by a ruler friendly to the United States—say Egypt or Azerbaijan—use Twitter and Facebook to launch or publicize a series of cyber-attacks on their government's websites and get arrested as a result. What should the U.S. government do in the face of such a Sartrean predicament? Speaking up on behalf of those activists would mean condoning cyber-attacks as a legitimate means of expressing dissent and would thus risk triggering a cascade; staying silent would mean reneging on core principles of Internet freedom, further entrenching authoritarian rule, and inviting even more cyber-attacks. It's a tricky situation that cannot be resolved in the abstract; what is clear, though, is that it is a bit premature to make major political commitments that would force Western policymakers to choose one over the other regardless of the context in which such cyber-attacks happen.

You Can't Be a "Little Bit Free" on the Internet

Perhaps Western governments harbor no ambitions of promoting Twitter revolutions. It's possible they just want to chide authoritarian governments for excessive Internet censorship and unexplained cyber-attacks. Maybe all they want is to promote freedom *of* the Internet rather than freedom *via* the Internet. Nevertheless, it's not the Western governments' original intentions that shape responses from their authoritarian adversaries; it's the perceptions of those intentions. There is such a long-running suspicion about the motives of the United States in many parts of the world that John Mearsheimer, a prominent scholar of international relations at the University of Chicago, justifiably concludes, "It should be obvious to intelligent observers that the United States speaks one way and acts another." Nowhere is this chasm more obvious than in what the State Department says about Internet freedom and what the Department of Defense does about Internet control.

Even Western policymakers cannot agree on the extent to which the power of the Internet should be harnessed to engender democratic change around the world. "The problem is that in Washington, the phrase 'global Internet freedom' is like a Rorschach test, in which different people look at the same ink splotch and see very different things," writes Rebecca MacKinnon, who, as a leading expert on China's Internet, had the privilege of giving several congressional testimonies and thus studying the Internet freedom zeitgeist on Capitol Hill. MacKinnon is quick to add that such a lack of clarity is also the main reason why "there is [still] no political consensus whatsoever on how to coordinate the conflicting interests and policy goals."

Nevertheless, as discussions on the subject advance, it's already possible to outline various schools of thought. One must distinguish between the weak form of Internet freedom promoted by the Obama administration and foreign policy liberals and its strong form, which is embraced by those who favor a more assertive, neoconservative foreign policy (its adherents are scattered across numerous think tanks like the George W. Bush Institute, the Hudson Institute, Freedom House, many of which were present at the Bush Institute gathering in Texas).

Whereas the weak form implies an almost exclusive focus on defending online freedom of expression—freedom *of* the Internet—the strong version, eagerly embraced by the cyber-cons, seeks to promote freedom *via* the Internet and envisions the Internet as an enabler of some kind of 1989-inspired bottom-up revolt, with tweets replacing faxes. To use Isaiah Berlin's famous distinction, while the weak form of Internet freedom is preoccupied mostly with promoting negative liberty (i.e., freedom *from* something: government online surveillance, censorship, DDoS attacks), the strong form of Internet freedom is more concerned with advancing the causes of positive liberty (i.e., freedom *to* do something: mobilize, organize, protest).

The strong agenda operates with the plain old rhetoric of "regime change" but spruced up with the libertarian language of Palo Alto. The weak agenda, it seems, aspires for little else but the preservation of the Internet as it is today, and it is ultimately rooted in the defense of freedom of expression, as codified in Article 19 of the UN Declaration of Human Rights ("Everyone has the right to freedom of opinion and expression; this right includes freedom to hold opinions without interference and to seek, receive and impart information and ideas through any media and regardless of frontiers"). The vision that underpins the struggle to build a world with few limits on speech does not necessarily forgo democracy promotion as one of its objectives; rather it takes a much longer view. The cyber-cons, of course, wouldn't mind preserving a free Internet either, but for them it's mostly an instrument to enable democratic rebellions in Belarus, Burma, and Iran.

Those in the weak agenda camp—most of them self-proclaimed fans of liberalism and international institutions—are walking into a trap of their own making, for most nonexperts, at least judging by the irrational exuberance over Iran's Twitter Revolution, interpret the term in its strong form, characterized by a much more aggressive use of the Internet to overthrow authoritarian regimes. The first image that comes to mind when one hears the words "Internet freedom" is that of the dying Neda Agha-Soltan surrounded by Iranian youngsters with mobile phones, not the picturesque conference room of the International Telecommunications Union in Geneva hosting a debate about the future of Internet governance. The problem is that if this more aggressive interpretation sticks around—and so far all the indicators show that it will—liberals' ability to protect the free flow of information on the Internet as well as to actually promote freedom *without* the Internet (i.e., through more conventional offline means) could be severely compromised.

That there are two different kinds of Internet freedom is lost on most media commentators in America, who believe that it's one of the few truly bipartisan issues facing the nation. Commenting on the George Bush cyber-dissidents conference in Dallas, Barrett Sheridan, a staff reporter for *Newsweek*, admired the fact that "there aren't many ideas that unite former U.S. president George W. Bush and his successor, Barack Obama, but one safe topic for conversation would be Internet freedom and the power of technology to foment democratic revolutions." Why Obama should even be having this conversation is not obvious. Elsewhere in his foreign policy, he did everything he could to dispel the myth that he wants to follow his predecessor in "fomenting democratic revolutions." The ambiguity over the meaning of Internet freedom, however, risks canceling out whatever other measures he has taken to present his foreign policy as the opposite of George Bush's.

On the same day that Hillary Clinton delivered her seminal Internet freedom speech, James Glassman and Michael Doran, Glassman's hawkish former colleague in the George W. Bush administration, published an op-ed in the *Wall Street Journal* in which they hinted at how they would go about harnessing the power of the Internet in the case of Iran. They called on the U.S. government to use technology to provide moral and educational support, increase communications within Iran and between Iran and the outside world, and refute Iranian propaganda. Here was a clear example of a "strong" agenda being put on the table, and, most probably, parts of the U.S. foreign policy establishment would be active in making it happen.

Marc Lynch, a prominent scholar of Middle Eastern politics, was quick to notice how easy it would be to twist Clinton's speech—which aspired to little other than defending online freedom of expression—for more sinister ends. For hawks like Glassman and Doran, wrote Lynch on his *Foreign Policy* blog, "Internet freedom, which Clinton presents as an abstract universal good, is clearly and unapologetically a weapon to be wielded against the Iranian regime. . . . Most of the world probably assumes that Clinton has the same goal in mind as Glassman and Doran, even if she doesn't say so."

But Clinton's speech was itself not particularly clear as to why Internet freedom is worth defending. On the one hand, she did acknowledge America's commitment to the weak agenda by saying that "we stand for a single internet where all of humanity has equal access to knowledge and ideas." But Clinton also hinted that the reasons for such a wide embrace of Internet freedom are more pragmatic: "The internet can help humanity push back against those who promote violence and crime and extremism. In Iran and Moldova and other countries, online organizing has been a critical tool for advancing democracy and enabling citizens to protest suspicious election results." In essence, what she really said is this: We'd like to promote Internet freedom so that everyone can express and read anything they want, but we also hope that this will essentially lead to a number of democratic revolutions.

This scenario, of course, is unlikely to come true, let alone help to promote democracy, if only because there is not enough space to maneuver in existing American policies, tied as they are to long-running concerns over terrorism, energy supply, and the politics of military bases. Technologists, in their typical streak of Internet-centrism, can talk all they want about the "Internet freedom agenda"—it makes them feel important, after all—but it is not going to alter what motivates the United States to behave as it does in the Middle East or Central Asia any more than its overall concerns with human rights and freedom of expression. Concerns over getting oil out of Azerbaijan won't give way to concerns over getting tweets from the Azeri opposition anytime soon, if only because Washington has long made a strategic decision not to undermine the friendly Azeri regime.

This is not to say that Clinton wouldn't chide the country's government for cracking down on bloggers, as she did during a June 2010 visit to Azerbaijan. This is not the kind of criticism, however, that could seriously threaten the relationship between the two countries. Rather, it is the kind of criticism that assists American officials in presenting themselves as holding democracy above their own energy needs. While this may certainly help them cope with the often cynical nature of their work, the impact of such posturing on Azeri authorities is zero. The greater danger here is that the supposed presence of a new pillar of American foreign policy—and this is how Internet freedom is often presented by senior American diplomats—will detract the public from asking the tough questions about the older, far more influential pillars, some of which are clearly beginning to crumble. As such, it would become much harder to evaluate the continuity of American policies, to view and criticize them in their totality. Since the plight of bloggers makes for far better copy than the plight of human rights campaigners, some observers might mistakenly believe that the U.S. government is, in fact, extremely critical of its allies.

As Rami Khouri, of Lebanon's *Daily Star*, so poignantly remarked on the gap between America's highly idealistic rhetoric on Internet freedom and its rather cynical actions in the rest of its foreign policy: "One cannot take seriously the United States or any other Western government that funds [online] political activism by young Arabs while it simultaneously provides funds and guns that help cement the power of the very same Arab governments the young social and political activists target for change." But Khouri may have underestimated American diplomats' own capacity for self-delusion. They take themselves seriously, and it's quite possible that they would be the first to believe that a fight for a free Internet—fought, for some reasons, only abroad—could somehow compensate for the lack of any serious changes elsewhere in American foreign policy. Unfortunately, virtually nothing about the current situation suggests that American foreign policy can muster enough decency and idealism to erect this new shiny pillar of Internet freedom; in its current incorporation, the Internet freedom agenda looks more like a marketing ploy.

Recent developments indicate that Washington's newly declared commitment to Internet freedom will be shaped by pre-Internet policies and alliances. Thus, even though a week before Clinton's seminal speech Jordan, America's staunchest ally in the Middle East, announced a new harsh Internet censorship law, she never referred to it (Clinton mentioned many other countries, like Uzbekistan, Vietnam, and Tunisia nevertheless).

The biggest tragedy of the Obama administration's Internet freedom agenda, even in its weakest form, is the unleashing of a conceptual monster so ambiguous as to greatly impede the administration's ability to accomplish other objectives. China and Iran, for example, want to keep tight control over the Internet, not only because they fear that their citizens may discover the real state of affairs in their countries, but also because they believe that the Internet is America's favorite tool of starting antigovernment rebellions. And the stronger such beliefs, the more challenging it would be for liberals to keep the Internet unregulated and hope that gradually it will help foster a strong demand for democracy. Marc Lynch hit the nail on the head when he wrote, "When the U.S. says to Iran or to other adversarial regimes that it should respect 'freedom of internet expression' or 'freedom of internet connectivity,' those regimes will assume that it is really trying to use those as a rhetorical cover for hostile actions." Translated into policies, the very concept of Internet freedom, much like "the war on terror" before it, leads to intellectual mush in the heads of its promoters and breeds excessive paranoia in the heads of their adversaries. This is hardly the kind of change that American foreign policy needs in the age of Obama.

The End of the American Internet

The interpretation of Internet freedom as a cover-up for regime change might seem ridiculous if it weren't so widely shared by some of America's most powerful movers and shakers. Such cyber-jingoism is poised to backfire, however, on American companies, which have been exporting Internet freedom, perhaps in its weakest form, for years.

Before all this talk about Internet freedom began in earnest, no political leader would think of Twitter users as a serious political force to contend with. They were seen as just a bunch of bored hipsters, who had an irresistible urge to share their breakfast plans. Suddenly, almost overnight, these tweeting bohemians became the Che Guevaras of the Internet. And which dictator, we might ask, wants a battalion of iPad-armed revolutionaries drifting through his sushi bars in search of fellow conspirators?

Web 2.0 has moved from the periphery of politics in authoritarian states to its very center—not because it has gained in importance or has acquired new abilities to topple governments, but because both leaders and media in the West grossly overstated its role, alerting the dictators to its future significance. But the significance of the Internet, at least when it comes to fostering new public spaces conducive to democratic norms, will only be felt in the long term—and only if the governments are hapless enough to stay out of the process of shaping these spaces according to their own agendas. There is nothing to celebrate here: Seemingly innocuous digital spaces that may have otherwise been left free of government supervision are now watched with more rigor and intensity than antigovernment gatherings in physical spaces. As Carlos Pascual, the U.S. ambassador to Mexico and a career diplomat with decades of valuable experience in international politics, told the *New York Times Magazine*: "If and when in a particular country . . . there's a perception that Twitter or Facebook is a tool of the U.S. government . . . that becomes dangerous for the company, and it becomes dangerous for people who are using that tool. It doesn't matter what the reality is. . . . There is some sort of a line there, and we [in the U.S. government] have to respect that line."

As the world learns about the mysterious but never substantiated role that Twitter played in Iran, about equally mysterious collaboration between Google and the National Security Agency, and about foreign trips that the U.S. State Department organizes for Silicon Valley executives (so far, they have taken such trips to India, Iraq, Mexico, Syria, and Russia), many authoritarian governments are beginning to feel uneasy, even though most of the Internet activities pursued by their citizens are still as silly as they used to be. The only difference is that now the Web is being perceived as some kind of a "made in America" digital missile that could undermine authoritarian stability. When it comes to such sensitive services as email, this is not an entirely irrational reaction. How might the American government react if it learned that the vast majority of its citizens had their email managed by a Chinese company that had extensive contacts with the People's Liberation Army? A government does not need to be authoritarian to feel threatened when its citizens store all their secrets on foreign servers.

Many governments are only now beginning to realize how tightly their own communication systems are tied to American infrastructure. "The dominance of American companies in the software and hardware industries as well as in web-based services affords US government agencies huge advantages in monitoring what is happening out in cyberspace," notes the political commentator Misha Glenny. It's logical that more governments will try to challenge that dominance. Even though the notion of "information sovereignty"—the idea that governments might have legitimate concerns over the nationalities and allegiances of those who mediate their information markets—has been somewhat discredited by the fact that so many Chinese and Cuba propaganda officials like to invoke it in their speeches, it is poised to rise in importance in proportion to the role of the Internet in international politics. (Judging by its nervous response to transnational information powerhouses like WikiLeaks, the U.S. government is increasingly concerned about its information sovereignty as well.)

Given the amount of research and technology money coming out of America's defense and intelligence communities, it's hard to find a technology company that does *not* have a connection to the CIA or some other three-lettered agency. Even though Google does not publicize this widely, Keyhole, the predecessor to Google Earth, which Google bought in 2005, was funded through In-Q-Tel, which is the CIA's for-profit investment arm. That Google Earth is somehow a CIA-funded vehicle for destroying the world is a recurring theme in rare comments given by those working in security agencies of other countries. Lt. Gen. Leonid Sazhin of the Russian Federal Security Service was not just speaking for Russia when he expressed his frustration in 2005: "Terrorists don't need to reconnoitre their target. Now an American company is working for them." It doesn't help that In-Q-Tel has recently invested in a company that monitors buzz on Twitter, supposedly to give the intelligence community "early-warning detection on how issues are playing internationally," as its spokesperson put it. In July 2010 both In-Q-Tel and Google jointly invested in the same social media monitoring company, triggering even more rumors and conspiracy theories. Whatever the motives, the perception it creates is that there is a tight connection between the much-feared CIA and the buzz created on social media; many authoritarian leaders haven't forgotten that Radio Free Europe was initially funded by the CIA and that the original name of Radio Liberty was Radio Liberation. So the fear that the CIA is in the business of funding revolutionary media is not entirely unfounded.

The stronger the perception that American companies are tools for accomplishing the objectives of the U.S. government, the more resistant foreign governments will be to those companies doing business in their countries. This will inevitably push these governments to invest in their own equivalent of popular American online services or find ways to discriminate against foreign firms to bolster domestic industry. In late 2009 Turkey announced a plan to give a government-run email account to every citizen as well as to launch a search engine that would better respond to "Turkish sensibilities." Iran promptly followed Turkey's path in February 2010, announcing a similar national email plan after banning Gmail; the Iranian authorities also announced their plans for a national search engine in the summer of 2010. A month later, the Russian government announced that it, too, was considering giving every citizen a government-run email account, if only to make it easier to identify them when they deal with the increasingly electronic government. As already noted, Russian politicians have also been seriously considering creating a government-run search engine, to challenge Google's rapid growth in the country; according to Russian media, $100 million has been disbursed for that purpose.

John Perry Barlow, a cyber-utopian former lyricist of the Grateful Dead, who in 1996 wrote "A Declaration of the Independence of Cyberspace," a libertarian manifesto for the digital age, likes to point out that "in cyberspace, the First Amendment is a local ordinance." This, however, may have been just a temporary equilibrium that could soon go away as other foreign governments discover that they would rather not have America own key parts of the infrastructure of the information society. The moment Western—and, in this case, predominantly American—policymakers start talking about embracing the geopolitical potential of the Internet, everyone else reconsiders the wisdom of letting the Americans keep the Internet to themselves, in terms of both Washington's dominant role in Internet governance and Silicon Valley's market leadership.

Just as important is the fact that local Chinese and Russian Internet companies may offer far better and more useful web services by the sheer virtue of knowing the demands of their respective Internet cultures. As such, they have proved successful at attracting local audiences and, more important, complying with the censorship requests of their own governments. The politicization of Web 2.0 services is likely to amplify the role of local clones of global sites. "There's this hubris [among Americans] that drives the belief that what matters in China is Twitter and Facebook and YouTube. Ultimately, that's not what matters. . . . It's Weibo, Kaixin or RenRen, and Youku and Tudou," says Kaiser Kuo, a popular Chinese American blogger, remarking on the far greater popularity of China's domestic services.

From the perspective of freedom of expression, the inevitable end of the American Internet doesn't look like good news. As bad and unresponsive as Facebook and YouTube might be as intermediaries, they would probably still do a better job defending liberty and self-expression than most Russian or Chinese companies (if only because the latter are more easily pressured by their own governments). The impressive gains in influencing foreign audiences on American-run Internet platforms that have been achieved in the last five years of universal ecstasy over Web 2.0 are easy to lose, especially if Western policymakers don't acknowledge that the role American Internet companies play is increasingly seen as political. It is becoming more and more difficult to convince the world that Google and Twitter are not just the digital-age equivalents of Halliburton and Exxon Mobile.

On the Dubious Virtues of Exporting Damaged Goods

When the Obama administration decided to apply some of its Internet savvy to help improve American democracy, it did not expect to face many problems. But when Obama's uber-geeks tried to crowdsource the process of agenda setting and ask Internet users for questions they thought the administration should be answering, they were faced by the brutal reality of Internet democracy. The most popular question was about the decriminalization of marijuana. "There was one question that was voted on that ranked fairly high and that was whether legalizing marijuana would improve the economy and job creation, and I don't know what this says about the online audience," said Obama, responding to some of the submitted questions. His answer to the question was no, and the incident somewhat dampened Obama's enthusiasm for consulting the public—if only because in cyberspace "the public" happens to be whoever has the most Facebook friends and can direct them to a given poll.

Online town halls aside, there is a vibrant debate about the Internet's impact on the health of democratic institutions. In most contexts, transparency is helpful, but, at the same time, it can also be quite costly. Even seasoned advocates for Internet-based transparency like Lawrence Lessig are beginning to sound much more cautious on the subject. Letting voters rank various government services may inadvertently foster an even greater cynicism and become a political liability. As Archon Fung, a professor of government at Harvard's Kennedy School, notes, one unfortunate consequence of excessive government transparency may be simply to "further de-legitimize government, because what the transparency system is doing is helping people catch government making mistakes. . . . [It] is like creating a big Amazon rating system for government that only allows one- or two-star ratings."

Politicians, on the other hand, may find it harder to make independent decisions without thinking about what would happen when all their memos and lunch schedules make it online. But even if they do, voters may draw the wrong conclusions, as Lessig pointed out in a trenchant 2009 article in the *New Republic*: that a senator has had a lunch meeting with a CEO does not mean that the senator's vote that indirectly benefits the interests of that CEO was not driven by public interest. Of course, lobbyists and special interests are still usurping online spaces. Special interests have successfully explored the Internet to plant their own messages, micro-tailoring them to the newly segmented audiences and prompting the political commentator Robert Wright to complain that "technology has subverted the original idea of America," adding that "the new information technology doesn't just create generation-3.0 special interests; it arms them with precision-guided munitions."

The list of unanswered questions about the relationship between the Internet and democracy is infinite. Will the Internet foster political polarization and promote what Cass Sunstein called "enclave extremism"? Will it further widen the gap between news junkies and those who avoid political news at all costs? Will it decrease the overall amount of political learning, as young people learn news from social networks? Will it prevent our future politicians from making any risky statements— now stored for posterity—in their prepolitical careers so as not to become unelectable? Will it allow genuinely new voices to be heard as opposed to just raised? Or are the critics of digital democracy like George Washington University's Matthew Hindman, who believes that "putting up a website is like hosting a talk show on public access television at 3:30 in the morning," justified in concluding that the digital public sphere is driven by elitism, being "a de facto aristocracy dominated by those skilled in the high deliberative arts"?

The brightest minds in both the government and the academy simply don't have good answers to most of these questions. But if they are so unsure about the Internet's impact on the health of our own democracy, how confident could they be that the Web can foster democracy in countries that are running short on it already? Is it really reasonable to believe that Internet users in authoritarian countries, many of whom have little experience with democratic governance, will suddenly start wearing Thomas Jefferson's avatar in cyberspace? Isn't it a bit premature to start touting the benefits of a medium the West itself does not yet know how to comfortably embed into its own political institutions? After all, one can't be calling for imposing more restrictions on sites like WikiLeaks, as many American policymakers did in the summer of 2010, and be disparaging China and Iran for similar impulses.

If it turns out that the Internet does help to stifle dissent, amplify existing inequalities in terms of access to the media, undermine representative democracy, promote mob mentality, erode privacy, and make us less informed, it is not at all obvious how exactly the promotion of so-called Internet freedom is also supposed to assist in the promotion of democracy. Of course, it may also be true that the Internet does none of those things; the important thing is to acknowledge that the debate about the Internet's effects on democracy isn't over and to avoid behaving as if the jury is already out.

The Hidden Charms of Digital Orientalism

Whatever analytical insights we in the West have acquired while thinking about the Internet in the democratic context are rarely invoked when we look at authoritarian states. Whenever the Chinese authorities crack down on unlicensed cybercafés, we have a tendency to see it as a sign of encroachment on democratic freedoms rather than of social concerns. It's as if we can't ever imagine that Chinese or Russian parents, too, might have some valid concerns about how their kids spend their free time.

In the same vein, as we are beginning to debate the impact of the Internet on how we think and learn—tolerating the possibility that it may actually impede rather than facilitate those processes—we rarely pose such questions in the authoritarian context. It's hard to imagine a mainstream American magazine running a cover story like "Is Google Making China Stupid?" (as the *Atlantic* did in 2008, only without the reference to China). Why? Because it's only Americans and Europeans Google is presumed to be making stupid; for everyone else, it's presumed to be a tool of enlightenment. While many in the West concede that the Internet has not solved and may have only aggravated many negative aspects of political culture—consider the rise of the "death panels"–kind of discourse—they are the first to proclaim that when it comes to authoritarian states, the Internet enables their citizens to see through the propaganda. Why so many of their own fellow citizens— living in a free country with no controls on freedom of expression— still believe extremely simplistic and misleading narratives when all the facts are just a Google search away is a question that Western observers should be asking more often.

Nowhere is such tendency to glorify the impact of the Internet abroad more obvious than in the speeches given by American politicians. On a 2009 visit to Shanghai, Barack Obama was all too happy to extol the virtues of the Internet, saying that "the more freely information flows, the stronger the society becomes, because then citizens of countries around the world can hold their own governments accountable. They can begin to think for themselves. That generates new ideas. It encourages creativity." In contrast, when he spoke to the graduates of Hampton University in Virginia less than six months later, Obama communicated almost a completely different message, complaining about "a 24/7 media environment that bombards us with all kinds of content and exposes us to all kinds of arguments, some of which don't always rank all that high on the truth meter. . . . With iPods and iPads and Xboxes and PlayStations . . . information becomes a distraction, a diversion, a form of entertainment, rather than a tool of empowerment, rather than the means of emancipation."

Hillary Clinton, the leading defender of Internet freedom, sounded much more cautious when she was a junior senator from New York and was accountable to her constituents. One of the few high-profile bills she sponsored while in the Senate was a 2005 bill (ironically, cosponsored with Sam Brownback, that other defender of Internet freedom) that authorized more government-funded research on "the effects of viewing and using electronic media, including television, computers, video games, and the Internet, on children's cognitive, social, physical, and psychological development." In a high-profile speech delivered in 2005, Clinton called the Internet "the biggest technological challenge facing parents and children today." She warned that "when unmonitored kids access the Internet, it can also be an instrument of enormous danger. Online, children are at greatly increased risk of exposure to pornography, identify theft, and of being exploited, if not abused or abducted, by strangers."

But such things, of course, only happen on the American Internet. Chinese and Russian parents would never worry about such a thing! Or ask their governments to do something about it! This does smack of a certain digital Orientalism, only that all of our biases and prejudices of the Orient have turned into its equally unquestioning admiration. While it may be the only psychological cure for the guilt of imperialism, idealizing the politics of authoritarian states would not be good for either their citizens or those of us who would like to see them eventually turn democratic. In 2006 Massage Milk and Milk Pig, two popular Chinese bloggers, had enough of such uncritical glorification of the Chinese Internet by foreign commentators. They posted a message on their blog—"Due to unavoidable reasons with which everyone is familiar, this blog is temporarily closed"—and began waiting for calls from the Western press. And those did follow. The BBC reported that one of the blogs was "closed down by the authorities," adding that the act had coincided with the annual session of the Chinese legislature. Reporters Without Borders, too, issued a statement condemning the censorship. Only, of course, there was no censorship.

Those schooled in the history of U.S. foreign policy might notice an unhealthy resemblance between today's efforts to enlist the Internet, especially Silicon Valley's godfathers, as cultural ambassadors and the U.S. State Department's efforts in the mid-1950s to recruit black jazz musicians to play the same role. Just the way everyone outside of the United States was asked to believe that jazz was apolitical, despite the fact that its leading American practitioners were regularly discriminated against in their own country, people are now being asked to believe that the Internet, too, is apolitical, even though the very technology CEOs enlisted to spread the gospel of Internet freedom around the world have to regularly comply with the growing demands of the U.S. National Security Agency. If history is anything to judge by, such sheer duplicity has rarely been of much help in accomplishing the objectives of American foreign policy. The point is not that the U.S. government should avoid exploiting the fact that so many popular Internet companies are American but that, in exploiting it, diplomats should not lose sight of the fact that these companies *are* American, with all the biases and expectations that such label elicits.

But another fallacy is often at play as well: By looking at authoritarian states, Western observers often notice (false) similarities to their own processes and problems. Mamoun Fandy, a U.S.-based, Saudi-born scholar of Middle Eastern politics, notes that "the main problem arises when we mistake the means and the processes for mirror images of Western structures accompanied by specific expected functions: in other words, seeing only what is familiar and selecting data on the basis of this familiarity." In other words, since policymakers believe that bloggers can make politicians more accountable in the context of democratic politics in the West, they tend to believe that such outcomes are unavoidable in other contexts. But this is not a given, and so many of the processes we see do not reflect the fundamental structural changes we take for granted. Or as Fandy so insightfully observes, "to see a debate similar to the American show *Crossfire* does not mean that freedom of speech in the Arab world is fully realized, any more than to see voting and ballot boxes means that democracy has taken hold."

Unfortunately, it does appear that whoever suggested that Secretary Clinton use the term "Internet freedom" must have been either short on metaphors or naïve about world politics. This is not to say that there are no threats to the openness of the Internet or the security of its users; it's just that there were probably much better, less political, and more intellectually coherent ways to bring all of those problems to global attention without instilling American's own politicians with a false sense of accomplishment.

chapter nine

Internet Freedoms and Their Consequences

• • ■ ■

The millions of tourists passing through India's Hindu temples every day probably have no idea they're being filmed. The cameras are there not to spy but to broadcast religious rituals over the Internet, allowing those who are unable to travel to the temple to participate remotely.

The digitization of India's religious practices has spawned a number of related online businesses. Saranam.com, for example, charges from $4 to $300 to perform a whole menu of religious ceremonies in virtually any temple in the country. If you are too busy to travel to the temple yourself, a small fee enables one of Saranam's "franchisees" to perform the required religious services on your behalf. E-Darshan.org, another Indian innovation that was probably inspired by YouTube, aggregates videos of rituals held in the most famous temples throughout the country and broadcasts some of the rituals in real time as well.

India is hardly unique here. In China, a start-up called china-tomb.cn is capitalizing on the rapidly growing demand for online mourning services. The Chinese have a custom of visiting their relatives' memorials and tombs during the Qingming festival; this tomb-sweeping tradition has been part of the Chinese tradition since the eighth century. Travel within China is often impractical. Thus a new generation of websites charges mourners a small fee (just a little bit over a dollar) to set up virtual alternatives and sweep them over the Web. Chinese Internet users also have a wide choice of online cemeteries, full of online memorial halls, e-tombstones, e-incenses, and e-flowers (some of the cemeteries were launched by the Chinese government in the early 2000s). These options for holding a virtual memorial service for the deceased are quite popular, especially in the diaspora.

Political religion is also profiting from technology. In early 2010 Egypt's Muslim Brotherhood, an Islamist organization that is surprisingly comfortable with modern technology, unveiled a Wiki-like site where anyone can help chart the movement's history (the site was launched with 1,700 articles). The idea is to both produce a historical document and introduce young people, especially those who do not have access to the movement's most important texts, to the movement's key ideas. Hosted on a server in the United States, the site is considerably harder for the Egyptian authorities to shut down (as they already did with an earlier Brotherhood site that was hosted in Egypt).

As it turns out, the Internet has been reviving many of the religious and cultural practices that globalization was supposed to erode, if not eliminate altogether. Consider the case of Gokce, a small Turkish village near the country's southern border with Syria. Even though polygamy was banned in Turkey in 1926, the practice still lives on in many rural areas, including Gokce. Until very recently Gokce's lonely hearts searching for yet another wife had to hop on a bus and travel to Syria. Today most such "romantic" endeavors start online, thanks to Gokce's first Internet café, which opened in 2008. "Everyone's coming to the internet cafe now to find a wife. Sometimes, there's no space to sit down," the cafe's owner told EurasiaNet. As a result, polygamy is on the rise, with brides from Morocco, who don't need a visa to enter Turkey, being particularly popular. Instead of turning Gokce's male residents into cosmopolitan defenders of women's rights, the Internet has only entrenched their status as cosmopolitan polygamists. Likewise, while it's tempting to believe that the introduction of mobile phones and text messaging technology into Saudi households may have afforded more privacy to women, in reality, the opposite may have happened, as husbands can now receive automated text messages whenever their wives leave the country.

Tweets will not dissolve all of our national, cultural, and religious differences; they may actually accentuate them. The cyber-utopian belief that the Internet would turn us into uber-tolerant citizens of the world, all too eager to put our vile prejudices on hold and open up our minds to what we see on our monitors, has proved to be unfounded. In most cases, the only people who still believe in the ideal of an electronic global village are those who would have become tolerant cosmopolitans even without the Internet: the globe-trotting intellectual elite. The regular folk don't read sites like Global Voices, an aggregator of the most interesting blog posts from all over the world; instead, they are much more likely to use the Internet to rediscover their own culture—and, dare we say it, their own national bigotry.

The good news is that we are not rushing toward a globalized nirvana where everyone eats at MacDonald's and watches the same Hollywood films, as feared by some early critics of globalization. The bad news is that, under the pressure of religious, nationalist, and cultural forces reignited by the Internet, global politics is poised to become even more complex, contentious, and fragmented. While many in the West view the Internet as offering an excellent opportunity to revive the least credible bits of modernization theory—the once popular belief that, with some assistance, all developing societies can reach a take-off point where they put their history, culture, and religion on hold and simply follow in the policy steps of more developed nations—such ideas don't have much basis in reality.

Egypt's Muslim Brotherhood certainly does not perceive the Internet to be a tool of hyper-modernization, because they reject the very project of hyper-modernization, at least as it is being marketed by the neoliberal institutions that are propping up the Mubarak regime they oppose. And although others have doubts about their vision for the future of Egypt and the Middle East in general, the Brothers have nothing against using modern tools like the Internet to achieve it. After all,

modern technologies abet all revolutions, not just those that are decid-
edly pro-Western in character. Even such a devout conservative as Ayatollah Khomeini did not shy away from using audiotapes to distribute his
sermons in the shah's Iran. "We are struggling against autocracy, for democracy, by means of xeroxracy," was one of the numerous technology-
worshiping slogans adopted by the anti-shah intelligentsia in the late
1970s. Had Twitter been around at the time, the anti-shah demonstra-
tors would surely be celebrating Twitterocracy. And even though the
Islamic Republic did embrace many elements of modernity—cloning,
a vibrant legal market in organ donations, string theory, to name just a
few areas where contemporary Iran is far ahead of its peers in the Middle East—its politics and public life are still shaped by religious discourse. It's quite likely that a large chunk of both the West's funds and
its attention will need to go toward mitigating the inevitable negative
effects that Internet-powered religion will have on world affairs. This
is not a moral evaluation of religion: It has proved to be good for democracy and freedom at some points in history, but history has also
shown how pernicious its influence can be.

A commitment to Internet freedom—or a combination of its various
elements—may be the right and inevitable moral choice the West needs
to make (albeit with a thousand footnotes), but the West must also understand that a freer Internet, by its very nature, may significantly
change the rest of the agenda, creating new problems and entrenching
old ones. This doesn't mean that the West should embark on an ambitious global censorship campaign against the Internet. Rather, different
countries require a different combination of policies, some of them
aimed at countering and mitigating the influence of religion and other
cultural forces and some of them amplifying their influence.

Smallpox Strikes Back

Nationalism, too, is going through a major revival on the Web. Members of displaced nations can find each other online, and existing nationalist movements can delve into the freshly digitized national
archives to produce their own version of history. New Internet services

often open up new venues for contesting history. Nations are now arguing about whether Google Earth renders their borders in accordance with their wishes. Syria and Israel continue battling about how the contested Golan Heights territory should be listed in Facebook's drop-down menus. Indian and Pakistan bloggers have been competing to mark parts of the contested territory of Kashmir as belonging to either of the two countries on Google Maps. The site had also been under attack for listing some Indian villages in the Arunachal Pradesh province, on the Indian-Chinese border, under Chinese names and as belonging to China. Cambodians, too, have been outraged by Google Earth's decision to mark eleventh-century Preah Vihear temple, ownership of which was awarded to Cambodia in a 1962 court ruling, as part of Thailand.

But such fights over the proper marking of digital assets aside, has the Internet reduced our prejudices against other nations? Was Nicholas Negroponte, one of the intellectual fathers of cyber-utopianism, correct when he predicted in 1995 that "[on the Internet] there will be no more room for nationalism than there is for smallpox"? The evidence for such sweeping claims is thin. In fact, quite the opposite may have happened. Now that South Koreans can observe their old enemies from Japan through a 24/7 digital panopticon, they are waging cyber-wars over such petty disputes as figure skating. Many of the deeply rooted national prejudices cannot be cured by increased transparency alone; if anything, greater exposure may only heighten them. Ask Nigerians how they feel about the entire world believing them to be a nation of scammers who only use the Internet to inform us that a Nigerian chieftain was kind enough to include us in his will. Perversely, it's Nigerians themselves who—often quite willingly—use the Internet to create and perpetuate stereotypes about their nation. Had Facebook and Twitter been around in the early 1990s, when Yugoslavia was rapidly descending into madness, cyber-utopians like Negroponte would have been surprised to see Facebook groups calling for Serbs, Croats, and Bosnians to be exterminated popping up all over the Web.

Perhaps, nationalism and the Internet are something of natural allies. Anyone eager to satisfy their nostalgia for the mighty Soviet, Eastern

German, or Yugoslavian past can do so easily on YouTube and eBay, basking in a plethora of historical memorabilia. But it's not just memorabilia; historical facts, too, can now be easily compiled and twisted to suit one's own interpretation of history. Fringe literature dealing with revisionist or outright racist interpretations of history used to be hard to find. Major publishers would never touch such contentious material, and the independent publishers that took the risk usually published only a handful of copies. That world of scarcity is no more: Even the most obscure nationalistic texts, which previously could only be found in select public libraries, have been digitized by their zealous fans and widely disseminated online. Thus, extreme Russian nationalists who believe that the Great Ukrainian Hunger of 1933 was a myth or, at any rate, does not deserve to be called a genocide, can now link to a number of always-available scanned texts, residing somewhere in the cloud, that look extremely persuasive, even if historically incorrect.

And it's not just myth-making based on frivolous interpretations of history that thrives online. The Internet also abets many national groups in formulating legitimate claims against the titular nation. Take the case of the Circassians, a once great nation scattered all over the Northern Caucasus. History was not kind to them: The Circassian nattion was broken into numerous ethnic pieces that were eventually crammed into Russia's vast possessions in the Caucasus. Today, the Circassians make up titular nations of three Russian federal subjects (Adygeya, Karachay-Cherkessia, and Kabardino-Balkaria) and, according to the 2002 Russian population census, number 720,000 people. During the Soviet era, the Kremlin's strategy was to suppress Circassian nationalism at all costs; thus most Circassians were separated into subgroups, depending on their dialect and place of residence, becoming Adygeys, Adygs, Cherkess, Kabards, and Shapsugs. For much of the twentieth century, Circassian nationalism lay dormant, in part because the Soviets banned any competing interpretations of what happened in the Russian-Circassian war in the nineteenth century. Today, however, most of the scholarly and journalistic materials related to the war have been scanned and uploaded to several Circassian websites, so that anyone can access them. Not surprisingly, Circassian nationalism has been quite assertive of late. In 2010 a dedicated website was set up to call on residents of the five nations to list themselves simply as "Circassians" in Russia's 2010 census, and an aggressive online campaign followed. "The internet seems to offer a lifeline to Circassian activists in terms of rejuvenating their mass appeal," notes Zeynel Abidin Besleney, an expert on Circassian nationalism at the School of Oriental and African Studies at the University of London.

Russia, with its eighty-nine federal subjects, has certainly more than one Circassian problem on its plate. Tatars, the largest national minority in Russia, for a long time had to suffer under the policy of Russification imposed by Moscow. Now their youngsters are turning to popular social networking sites to set up online groups that focus on the issue of national Tatar revival. Not only do they use such groups to watch new videos and share links to news and music, but they are also exposed to information, often missing from Russia's own media, about Tatar history and culture and the internal politics of Tatarstan.

As the Circassian and the Tatar cases illustrate, thanks to the Internet many of the Soviet (and even Tsarist) myths that seemed to tie the nation together no longer sound tenable, with many previously captive nations beginning to rediscover their national identities. How Russia will keep its territorial integrity in the long run—especially if more nations try to secede or at least to overcome artificial ethnic divisions of the Soviet era—is anyone's guess. Not surprisingly, the Kremlin's ideologists like Konstantin Rykov have begun emphasizing the need to use the Internet to tie the Russian nation together. At this point, it's impossible to tell what such increased contentiousness means for the future of democracy in Russia, but it would take a giant dose of optimism to assume that, somehow, modern-state Russia would simply choose to disintegrate as peacefully as the Soviet Union did and, more, that democracy would prevail in all of its new parts. Developing an opinion about the long-term impact of the Internet on Russian democracy would inevitably require asking—if not answering—tough questions about nationalism, separatism, center-periphery relations, and so forth. (And not just in Russia: Similar problems are also present in China and, to a lesser degree, Iran, which have sizable minorities of their own.)

The influence of the diasporas—many of whom are not always composed of progressive and democracy-loving individuals—is also poised to rise in an age when Skype facilitates so much of the cultural traffic. Will some of that influence be positive and conducive to democratization? Perhaps, but there will also surely be those who will try to stir things up or promote outdated norms and practices. Thomas Hylland Eriksen, an anthropologist at the University of Oslo, notes that "sometimes, the elites-in-waiting use the Net to coordinate their takeover plans; sometimes diasporas actively support militant and sometimes violent groups 'at home,' knowing that they themselves do not need to pay the price for an increase in violence, remaining as they do comfortably in the peaceful diaspora."

The problem with Internet freedom as a foundation for foreign policy is that in its simplification of complex forces, it may actually make policymakers overlook their own interests. To assume that it's in the American, German, or British interest to simply let all ethnic minorities use the Internet to carve out as much space as they can from the dominant nation, whether it is in Russia, China, or Iran (not to mention much more complicated cases like Georgia), would simply be to badly misread their current policies and objectives. One might argue that these are cunning policies, and it's an argument worth having. The first rather ambiguous articulation of Internet freedom policy by Hillary Clinton simply preferred to gloss over the issue altogether, as if, once armed with one of the most powerful tools on Earth, all nations would realize that, compared to YouTube, all those bloody wars they've been fighting for centuries have been a gigantic waste of time. Finding a way to grapple with the effects of new, digitally empowered nationalism is a formidable task for foreign policy professionals; one can only hope that they won't stop working on it even if the imperative to promote Internet freedom would divert their time and attention elsewhere.

Putting the *Nyet* in Networks

When in 2008 Evdins Snore, a young Latvian director, produced *The Soviet Story*, a documentary that made rather unpleasant comparisons between Stalinism and Nazism, Russian nationalists took to LiveJournal, the most popular blogging platform in the country, to debate the best response to the film. Alexander Djukov, the leader of Historical Memory, a nationalistic outlet, wrote an emotional appeal on his own blog, offering to publish a detailed study of how Snore manipulated the truth, if only the community could help raise funds to cover the printing costs.

Within hours, his post attracted more than seventy responses, with many people not only offering to contribute money but also sharing tips about other networks and organizations to tap into—including the contact details of the supposedly pro-Russian member of the European Parliament from Latvia. The money was duly raised—thanks to Yandex Money, Russia's own version of PayPal—and the book appeared in print less than six months later, to much fanfare in the Russian media and the blogosphere alike.

It's not obvious what one should make of such online efforts. Nothing illegal was done. It is also hard to deny that such a campaign would have been difficult to mount before the advent of Web 2.0, if only because many such nationalistic movements were too disorganized and geographically dispersed. Those who care about promoting freedom and democracy in Russia today now have to fight not just the state but also various nonstate actors that, thanks to the Internet, have suddenly become overmobilized.

The problem is that the West began its quest for Internet freedom based on the mostly untested cyber-utopian assumption that more connections and more networks necessarily lead to more freedom or more democracy. In her Internet freedom address, Hillary Clinton spoke of the importance of promoting what she dubbed a "freedom to connect," saying that it's "like the freedom of assembly, only in cyberspace. It allows individuals to get online, come together, and hopefully cooperate. Once you're on the internet, you don't need to be a tycoon or a rock star to have a huge impact on society." The U.S. State Department's Alec Ross, one of the chief architects of Clinton's Internet freedom policy, said that "the very existence of social networks is a net good."

But are social networks really goods to be treasured in themselves? After all, the mafia, prostitution and gambling rings, and youth gangs are social networks, too, but no one would claim that their existence in the physical world is a net good or that it shouldn't be regulated. Ever since Mitch Kapor, one of the founding fathers of cyber-utopianism, proclaimed that "life in cyberspace seems to be shaping up exactly like Thomas Jefferson would have wanted: founded on the primacy of individual liberty and a commitment to pluralism, diversity, and community" in 1993, many policymakers have been under the impression that the only networks to find homes online would be those promoting peace and prosperity. But Kapor hasn't read his Jefferson closely enough, for the latter was well aware of the antidemocratic spirit of many civil associations, writing that "the mobs of the great cities add just so much to the support of pure government as sores do to the strength of the human body." Jefferson, apparently, was not persuaded by the absolute goodness of the "smart mobs," a fancy term to describe social groups that have been organized spontaneously, usually with the help of technology.

As Luke Allnut, an editor with Radio Free Europe, points out, "where the techno-utopianists are limited in their vision is that in this great mass of Internet users all capable of great things in the name of democracy, they see only a mirror image of themselves: progressive, philanthropic, cosmopolitan. They don't see the neo-Nazis, pedophiles, or genocidal maniacs who have networked, grown, and prospered on the Internet." The problem of treating all networks as good in themselves is that it allows policymakers to ignore their political and social effects, delaying effective response to their otherwise harmful activities. "Cooperation," which seems to be the ultimate objective of Clinton's network building, is too ambiguous of a term to build meaningful policy around.

A brief look at history—for example, at the politics of Weimar Germany, where increased civic engagement helped to delegitimize parliamentary democracy—would reveal that an increase in civic activity does not necessarily deepen democracy. American history in the post-Tocqueville era offers plenty of similar cues as well. The Ku Klux Klan was also a social network, after all. As Ariel Armony, a political scientist at Colby College in Maine, puts it, "civic involvement may . . . be linked to undemocratic outcomes in state and society, the presence of a 'vital society' may fail to prevent outcomes inimical to democracy, or it may contribute to such results." It's political and economic factors, rather than the ease of forming associations, that primarily set the tone and the vector in which social networks contribute to democratization; one would be naïve to believe that such factors would always favor democracy. For example, if online social networking tools end up overempowering various nationalist elements within China, it is quite obvious that the latter's influence on the direction of China's foreign policy will increase as well. Given the rather peculiar relationship between nationalism, foreign policy, and government legitimacy in China, such developments may not necessarily be particularly conducive to democratization, especially if they lead to more confrontations with Taiwan or Japan.

Even Manuel Castells, a prominent Spanish sociologist and one of the most enthusiastic promoters of the information society, has not been sold on the idea of just "letting a thousand networks bloom." "The Internet is indeed a technology of freedom," writes Castells, "but it can make the powerful free to oppress the uninformed" and "lead to the exclusion of the devalued by the conquerors of value." Robert Putnam, the famed American political theorist who lamented the sad state of social capital in America in his best-selling *Bowling Alone*, also cautioned against the "kumbaja interpretation of social capital." "Networks and associated norms of reciprocity are generally good for those inside the network," he wrote, "but the external effects of social capital are by no means always positive." From the perspective of American foreign policy, social networks may, indeed, be net goods, but only as long as they don't include anyone hiding in the caves of Waziristan. When senator after senator deplores the fact that YouTube has become a second home to Islamic terrorists, they hardly sound like absolute believers in the inherent democratic nature of the networked world.

One can't just limit the freedom to connect to the pro-Western nodes of the Web, and everyone—including plenty of anti-Western nodes—stands to profit from the complex nature of the Internet. When it comes to democracy promotion, one major problem with a networked society is that it has also suddenly overempowered those who oppose the very process of democratization, be they the church, former communists, or fringe political movements. As a result, it has become difficult to focus on getting things done, for it's not immediately obvious if the new, networked threats to democracy are more ominous than the ones the West originally thought to fight. Have the nonstate enemies of democracy been empowered to a greater degree than the previous enemy (i.e., the monolith authoritarian state) has been disempowered? It certainly seems like a plausible scenario, at least in some cases; to assume anything otherwise is to cling to an outdated conception of power that is incompatible with the networked nature of the modern world. "People routinely praise the Internet for its decentralizing tendencies. Decentralization and diffusion of power, however, is not the same thing as less power exercised over human beings. Nor is it the same thing as democracy. . . . The fact that no one is in charge does not mean that everyone is free," writes Jack Balkin of Yale Law School. The authoritarian lion may be dead, but now there are hundreds of hungry hyenas swirling around the body.

Safe to Disconnect

Even worse, the supposed lawlessness and networked anarchy enabled by the Internet have resulted in greater social pressure to tame the Web. In a sense, the more important the Internet becomes, the greater the onus to rein in its externalities. Promoting the freedom to connect will be a tricky proposal to sell to voters, many of whom actually want the government to promote the freedom to disconnect—at least for particular political and social groups. If the last decade is anything to judge by, the pressure to regulate the Web is as likely to come from concerned parents, environmental groups, or various ethnic and social minorities as it is from authoritarian governments. The truth is that many of the opportunities created by a free-for-all anonymous Internet culture have been creatively exploited by people and networks that undermine de-

mocracy. For instance, it's almost certain that a Russian white suprema-
cist group that calls itself the Northern Brotherhood would have never
existed in the pre-Internet era. It has managed to set up an online game
in which participants—many of them leading a comfortable middle-
class existence—are asked to videotape their violent attacks on migrant
guest workers, share them on YouTube, and compete for cash awards.

Crime gangs in Mexico have also become big fans of the Internet.
Not only do they use YouTube to disseminate violent videos and promote a climate of fear, but they are also reportedly going through social
networking sites hunting for personal details of people to kidnap. It
doesn't help that the offspring of Mexico's upper classes are all interconnected on Facebook. Ghaleb Krame, a security expert at Alliant International University in Mexico City, points out that "criminals can
find out who are the family members of someone who has a high rank
in the police. Perhaps they don't have an account on Twitter or Facebook, but their children and close family probably do." It's hard to imagine Mexican police officers becoming braver as a result. And social
networking can also help to spread fear: In April 2010, a series of Facebook messages warning of impending gang wars paralyzed life in Cuer-
navaca, a popular resort, with only a few brave people daring to step
outside (it proved to be a false alarm).

The leaders of al-Shabab ("The Lads"), Somalia's most prominent
Islamist insurgency group, use text messaging to communicate with
their subordinates, avoiding any face-to-face communication and the
risks it entails. It's not a particularly contentious conclusion that they
have become more effective—and thus more of a menace—as a result.

Plenty of other less notorious (and less violent) cases of networked
harm barely receive any global attention. According to a 2010 report
from the Convention on International Trade in Endangered Species,
an international intergovernmental organization, the Internet has created a new market for trade in extinct species, allowing buyers and sellers to find each other more easily and trade more effectively. Kaiser's
spotted newt, found only in Iran, may be the first real victim of the
Twitter Revolution. According to reports in the *Independent*, more than

ten companies are selling wild-caught specimens over the Internet. Not surprisingly, the newt's population was reduced 80 percent between 2001 and 2005 alone.

Another informal market the Internet has boosted is organ trading. Desperate individuals in the developing world are bypassing any intermediaries and are offering their organs directly to those who are willing to pay up. Indonesians, for example, use a website called iklanoke.com, a local alternative to Craigslist, where their postings usually go unmonitored by police. A typical ad from Iklanoke reads, "16-year-old male selling a kidney for 350 million rupiah or in exchange for a Toyota Camry."

Text messaging has been used to spread hate in Africa, most recently in Muslim-Christian squabbles that erupted in the central Nigerian city of Jos in early 2010 that took the lives of more than three hundred people. Human rights activists working in Jos identified at least 145 such messages. Some instructed the recipients how to kill, dispose of, and burn bodies ("kill before they kill you. Dump them in a pit before they dump you"); others spread rumors that triggered even more violence. According to Agence France-Presse, one such message urged Christians to avoid food sold by Muslim hawkers, as it could have been poisoned; another message claimed political leaders were planning to cut water supplies to dehydrate members of one faith.

Two years earlier Kenya lived through an eerily similar tumultuous period. The political crisis that followed Kenya's disputed election that took place on December 27, 2007, showed that the networks fostered by mobile technology, far from being "net goods," could easily escalate into uncontrollable violence. "If your neighbor is kykuyu, throw him out of his house. No one will hold you responsible," said a typical message sent at the peak of the violence; another one, also targeting Kykuyus, said, "Let's wipe out the Mt. Kenya mafia," adding, "Kill 2, get 1 free." But there was also a more disturbing effort by some Kykuyus to use text messaging to first collect sensitive information about members of particular ethnic groups and then distribute that information to attack and intimidate them. "The blood of innocent Kykuyus will cease to flow! We will massacre them right here in the capital. In the name of justice put down the names of all the Luos and Kaleos you know from work, your property, anywhere in Nairobi, not forgetting where and how their children go to school. We will give you a number on where to text these messages," said one such message. At one point, the Kenyan authorities were considering shutting down mobile networks to avoid any further escalation of violence (between 800 and 1,500 people died, and up to 250,000 were displaced).

Even though text messaging also proved instrumental in setting up a system that helped to track how violence spread around Kenya—a success story that gained far more attention in the media—one can't just disregard the fact that text messaging also helped to mobilize hate. In fact, text messages full of hatred and highly intimidating death threats kept haunting witnesses who agreed to testify to the high-level Waki Commission set up to investigate the violence two years after the clashes. ("You are still a young man and you are not supposed to die, but you betrayed our leader, so what we shall do to you is just to kill you" was the text of a message received by one such witness.)

The bloody Uighur-Han clashes that took place in China's Xinjiang Province in the summer of 2009 and resulted in a ten-month ban on Internet communications appear to have been triggered by a provocative article posted to the Internet forum www.sg169.com. Written by an angry twenty-three-year-old who had been laid off by the Xuri Toy Factory in China's Guangdong Province, 3,000 miles from Xinjiang, the article asserted that "six Xinjiang boys raped two innocent girls at the Xuri Toy Factory." (China's official media stated that the rape accusations were fake, and foreign journalists could not find any evidence to substantiate such claims either.) Ten days later, the Uighur workers at the toy factory were attacked by a group of angry Han people (two Uighurs were killed, and over a hundred were injured). That confrontation, in turn, triggered even more rumors, many of which overstated the number of people who had been killed, and the situation got further out of control soon thereafter, with text messaging and phone calls helping to mobilize both sides (the authorities eventually turned off all phone communications soon thereafter). A gruesome video that showed several Uighur workers being beaten by a mob armed with metal pipes quickly went viral as well, only adding to the tensions.

Even countries with a long democratic tradition have not been sparred some of the SMS-terror. In 2005, many Australians received text messages urging attacks on their fellow citizens of Lebanese descent ("This Sunday every Fucking Aussie in the shire, get down to North Cronulla to help support Leb and wog bashing day.... Bring your mates down and let's show them this is our beach and they're never welcome back"), sparking major ethnic fights in an otherwise peaceful country. Ethnic Lebanese got similar messages, only calling for attacks on non-Lebanese Australians. More recently, right-wing extremists in the Czech Republic have been aggressively using text messaging to threaten local Roma communities. Of course, even if text messaging had never been invented, neo-Nazis would still hate the Roma with as much passion; to blame their racism on mobile phones would be yet another manifestation of focusing on technology at the expense of political and social factors. But the ease, scale, and speed of communications afforded by text messaging makes the brief and previously locally contained outbursts of neo-Nazi anger resonate in ways that they could have never resonated in an era marked by less connectedness.

Perhaps, the freedom to connect, at least in its current somewhat abstract interpretation, would be a great policy priority in a democratic paradise, where citizens have long forgotten about hate, culture wars, and ethnic prejudice. But such an oasis of tolerance simply does not exist. Even in Switzerland, commonly held up as a paragon of decentralized democratic decision making and mutual respect, the freedom to connect means that a rather small and marginalized fraction of the country's population managed to tap the power of the Internet to mobilize their fellow citizens to ban building new minarets in the country. The movement was spearheaded by right-wing blogs and various groups on social networking sites (many of them featuring extremely graphic posters—or "political Molotov cocktails," as Michael Kimmelman of the *New York Times* described them—suggesting Muslims are threatening Switzerland, including one that showed minarets rising from the Swiss flag like missiles), and even peace-loving Swiss voters could not resist succumbing to the populist networked discourse. Never underestimate the power of Twitter and Photoshop in the hands of people mobilized by prejudice.

While Internet enthusiasts like to quote the optimistic global village reductionism of Marshall McLuhan, whom *Wired* magazine has chosen as its patron saint, few of them have much use for McLuhan's darker reductionism, like this gem from 1964: "That Hitler came into political existence at all is directly owing to radio and the public-address system." As usual, McLuhan was overstating the case, but we certainly do not want to discover that our overly optimistic rhetoric about the freedom to connect has deprived us of the ability to fix the inevitable negative consequences that such freedom produces. Some networks are good; some are bad. But all networks require a thorough ethical investigation. Promoting Internet freedom must include measures to mitigate the negative side effects of increased interconnectedness.

Do Weak States Need Powerful Gadgets?

All the recent chatter about how the Internet is breaking down institutions, barriers, and intermediaries can make us oblivious to the fact that strong and well-functioning institutions, especially governments, are essential to the preservation of freedom. Even if we assume that the Internet may facilitate the toppling of authoritarian regimes, it does not necessarily follow that it would also facilitate the consolidation of democracy. If anything, the fact that various antidemocratic forces— including extremists, nationalists, and former elites—have suddenly gained a new platform to mobilize and spread their gospel suggests that the consolidation of democracy may become harder rather than easier.

Concerns that the information revolution will weaken the nation state are not new. Arthur Schlesinger Jr., the Pulitzer-winning historian who advised John F. Kennedy, foresaw where increasing computerization might lead, if left unchecked, when he wrote in 1997: "The computer turns the untrammeled market into a global juggernaut crashing across frontiers, enfeebling national powers of taxation and regulation, undercutting national management of interest rates and exchanges rates, widening disparities of wealth within and between nations, dragging down labor standards, degrading the environment, denying nations the shaping of their own economic destiny, accountable to no one, creating a world economy without a world polity." Fortunately, things haven't proved as dramatic as he expected, but Schlesinger's prophecy does point to the importance of thinking through what it is we value about state institutions in the context of democratization and ensuring that the Internet does not fully erode those qualities. While it's tempting to use the Internet to cut off all the heads of the authoritarian hydra, no one has yet succeeded in building a successful democracy with that dead hydra (the state apparatus) still lying there. It's then hardly surprising that those living in democracies may not appreciate the fact that without a strong state any kind of journalism—regardless of whether it's performed by mainstream media or bloggers—is impossible.

As Silvio Waisbord, a scholar of press freedom at the George Washington University, points out: "'the state' means functional mechanisms to institutionalize the rule of law, observe legislation to promote access to information, facilitate viable and diversified economies to support mixed media systems, ensure functional and independent tribunals that support 'the public's right to know,' control corruption inside and outside newsrooms, and stop violence against reporters, sources, and citizens." If the cyber-utopians believe their own rhetoric about crumbling institutions, they've got a major problem on their hands, and yet they repeatedly refuse to engage with it.

Nowhere is this more evident than in countries like Afghanistan, where an already weak government is made even weaker by various political, military, and social forces. We can continue celebrating the potential role of the mobile phone in empowering Afghan women, but the Taliban has terrorized many of the mobile networks into shutting down their services between certain hours of the day (avoiding compliance with the Taliban's demands is not an option; when some carriers tried that, the Taliban responded by attacking cellphone towers and murdering their staff). The Taliban doesn't want to shut the system down entirely—for they also use cellphones to communicate—but they still manage to show who is in control and to dictate how technology should be used. Without a strong Afghan government, the numerous empowerment opportunities associated with the mobile phone will never be realized.

Stephen Holmes, a professor of law at New York University, makes this point in an essay titled "How Weak States Threaten Freedom" published in the *American Prospect* in 1997. Commenting on how Russia was slowly disintegrating under the pressures of gangsterism and corrupt oligarchy, Holmes explains what many Westerners have overlooked: "Russia's politically disorganized society reminds us of liberalism's deep dependence on efficacious government. The idea that autonomous individuals can enjoy their private liberties if they are simply left unpestered by the public power dissolves before the disturbing realities of the new Russia."

Internet enthusiasts often forget that if a government, even an authoritarian one, loses the ability to exercise control over its population or territory, democracy is not necessarily inevitable. As tempting as it is to imagine all authoritarian states as soulless Stalinist wastelands, where every single thing the government does aims at restricting the freedom of the individual, this is a simplistic conception of politics. Were the Russian or Chinese state bureaucracies to collapse tomorrow, it would not be pure democracy that would replace them. It would probably be anarchy and possibly even ethnic strife. This does not mean that either country is unreformable, but reforms can't start by blowing up the state apparatus first.

This is yet another one of those instances in which the peaceful transition to democracy in postcommunist Eastern Europe was interpreted by the West as the ultimate proof that once a government run by authoritarian crooks is out, something inherently democratic would inevitably emerge in its place. The peaceful transition did happen, but it was the result of economic, cultural, and political forces that were rather unique both to the region and to that particular moment in history. The manner in which the transition took place was not predefined by some general law of nature, positing that people always want democracy and, once all barriers are removed, it will necessarily triumph over every single challenge. The utopian vision inherent in such views was on full display in 2003, when, after the statue of Saddam Hussein was toppled in Baghdad, nothing even remotely resembling Eastern European postcommunist democracy ever came to replace it. (Holmes, in another essay, vividly summed up such reductionist views as "remove the lid and out leaps democracy.")

The only thing worse than an authoritarian state is a failed one. The fact that the Internet empowers and amplifies so many forces that impinge on citizen's rights and hurt various minorities is likely to result in more aggressive public demands for a stronger state to protect citizens from the lawlessness of cyberspace. As child pornographers, criminal gangs, nationalists, and terrorists use the Internet to cause more and more harm, the public's patience will sooner or later run out. When Chinese netizens find themselves targets of attacks by "human flesh search engines," they are not waiting for Robin Hood to come and protect them. They expect their otherwise authoritarian government to draft adequate privacy laws and enforce them. Similarly, when corrupt Russian police officers leak databases containing the personal details of many citizens, including their passport details and cell phone numbers, and these databases resurface on commercial Internet sites, it hardly makes Russians celebrate the virtues of limited government. Multiply the power of the Internet by the incompetence of a weakened state, and what you get is a lot of anarchy and injustice. The reason why so many otherwise astute observers see democracy where there is none is that they confuse the democratization of access to tools with the democratization of society. But one does not necessarily lead to the other, especially in environments where governments are too weak, too distracted, or too unwilling to mitigate the consequences of the democratization of tool access. More and cheaper tools in the wrong hands can result in less, not more, democracy.

It's much like the perpetual debate about blogging versus journalism. Today anyone can blog because the tools for producing and disseminating information are cheap. Yet giving everyone a blog will not by itself increase the health of modern-day Western democracy; in fact, the possible side effects—the disappearance of watchdogs, the end of serendipitous news discovery, the further polarization of society—may not be the price worth paying for the still unclear virtues of the blogging revolution. (This does not mean, of course, that a set of smart policies—implemented by the government or private actors—won't help to address those problems.) Why should it be different with the Internet and politics? For all we know, many social ills may have become considerably worse since the dawn of social media. We need to start looking at the totality of side effects, not just at the fact that the costs of being a political activist have fallen so dramatically.

In debunking the "more access to technology = more democracy" fallacy, Gerald Doppelt, a professor of philosophy at the University of California at San Diego, suggests some further issues we need to ponder. "In order to evaluate the impact of any particular case of technical politics on the democratization of technology and society," writes Doppelt, "we need to ask who is this group of users challenging technology, where do they stand in society, what have they been denied, and what is the ethical significance of the technical change they seek for democratic ideals?" Without asking those questions, even the sharpest observers of technology will keep circling around the paradoxical conclusion that the blogging Al-Qaeda is good for democracy, because blogs have opened up new and cheap vistas for public participation. Any theory of democracy that doesn't go beyond the cost of mobilization as its only criteria of democratization is a theory that policymakers would be well-advised to avoid, even more so in the digital age, when many costs are plummeting across the board.

Not asking those questions would also prevent us from identifying the political consequences of such democratization of access to technology. Only irresponsible pundits would advocate democratizing access to guns in failed states. But the Internet, of course, has so many positive uses—some of which promote freedom of expression—that the gun analogy is rarely invoked by anyone. The good uses, however, do not always cancel out all the bad ones; if guns could also be used as megaphones, they would still make good targets for regulation. The danger is that the colorful banner of Internet freedom may further conceal the fact that the Internet is much more than the megaphone for democratic speech, that its other uses can be extremely antidemocratic in nature, and that without addressing those uses the very project of democracy promotion might be in great danger. The first prerequisite to getting Internet freedom policy right is convincing its greatest advocates that the Internet is more important and disruptive than they have previously theorized.

Why Rational Politics Doesn't Fit a Hundred Forty Characters

As the Internet mediates more and more of our foreign policy, we are poised to surrender more and more control over it. Of course, the era when diplomats could take the time to formulate deep and extremely careful responses to events was already over with the arrival of the telegraph, which all but killed the autonomy of the foreign corps. As far as thoughtful foreign policy is concerned, it's all been downhill from there. It's hardly surprising that John Herz, the noted theorist of international relations, observed in 1976 that "where formerly more leisurely but also cooler and more thoroughly thought-out action was possible, one now must act or react immediately."

The age of Internet politics deprives the diplomats of more than just autonomy. It's also the end of rational policymaking, as policymakers are bombarded with information they cannot process, while a digitally mobilized global public demands an immediate response. Let's not kid ourselves: Policymakers cannot craft effective policies under the influence of blood-curling videos of Iranian protesters dying on the pavement.

By 1992 George Kennan, the don of American diplomacy and author of the famous "Long Telegram" from Moscow, which shaped much of American thinking during the Cold War and helped to articulate the policy of containment, had come to believe that the media killed America's ability to develop rational foreign policy. Back then viral political videos were still the bread and butter of network television. After he watched the gruesome footage of several dead U.S. Army rangers being dragged through the streets of Mogadishu on CNN, Kennan made the following bitter note in his diary, soon republished as an op-ed in the

New York Times: "If American policy from here on out . . . is to be controlled by popular emotional impulses, and particularly ones invoked by the commercial television industry, then there is no place—not only for myself but for what have traditionally been regarded as the responsible deliberative organs of our government, in both executive and legislative branches." Kennan's words were soon seconded by Thomas Keenan, the director of the Human Rights Project at Bard College, who believes that "the rational consideration of information, with a view to grounding what one does in what one knows, now seems overtaken and displaced by emotion, and responses are now somehow controlled or, better, remote-controlled by television images."

Now that television images have been superseded by YouTube videos and angry tweets, the threshold of intervention has dropped even lower. All it took to get the U.S. State Department to ask Twitter to put off their maintenance was a high number of tweets of highly dubious provenance. When the whole world expects us to react immediately—and the tweets are piling up in the diplomats' in-boxes—we are not likely to rely on history, or even our own experiences and earlier mistakes, but instead decide that tweets + young Iranians holding mobile phones = a Twitter Revolution.

William Scheuerman, a political theorist who studies the role of speed in international affairs, is right to worry that "the historical amnesia engendered by a speed-obsessed society invites propagandistic and fictional retellings of the past, where political history is simply recounted to the direct advantage of presently dominant political and economic groups." Apparently, it's also fictional retellings of the most recent present that a speed-obsessed society should be concerned about. When facts no longer shape their reactions, policymakers are likely to produce wrong responses.

The viral aspect of today's Internet culture is hardly exerting a positive influence on diplomats' ability to think clearly. Back in the 1990s, many pundits and policymakers liked to denigrate (and a select few worship) the so-called "CNN effect," referring to the power of modern media to exert pressure on decision makers by streaming images from the scene of a conflict, eventually forcing them to make decisions they may not have otherwise made. CNN's supposed—but mostly unproven—influence on foreign policy in the 1990s could at least be justified by the fact that it was speaking on behalf of some idealistic and even humanistic position; we knew who was behind CNN, and we knew what their (mostly liberal) biases were.

The humanism of a bunch of Facebook groups is harder to verify. Who are these people, and what do they want? Why are they urging us to interfere or withdraw from a given conflict? Where the optimists see democratization of access, the realists may see the ultimate victory of special interests over agenda setting. Governments, of course, are not stupid. They are also taking advantage of this tremendous new opportunity to cover their own attempts to influence global public opinion in the cloth of vox populi, either directly or through the work of proxies. Take Megaphone, a technology developed by a private Israeli firm. It keeps track of various online polls and surveys, usually run by international newspapers and magazines, that ask their readers questions about the future of the Middle East, Palestine, the legitimacy of Israeli policies, etc. Whenever a new poll is found, the tool pings its users, urging them to head to a given URL and cast a pro-Israel vote. Similarly, the tool also offers to help mass-email articles favorable to Israel, with the objective of pushing such articles to the "most emailed" lists that are available on many newspaper websites.

But it's not only nimble guerilla-like Web experiments like Megaphone that are influencing global public opinion. The truth is that Russia and China have created their own CNNs, which aim to project their own take on the world news. Both have vibrant websites. As American and British news media are experimenting with paywalls to remain afloat, it's government-owned English-language media from Russia and China that stand to benefit the most. They would even pay people to read them!

For all intents and purposes, navigating the new "democratized" public spaces created by the Internet is extremely difficult. But it's even more difficult to judge whether the segments that we happen to see are representative of the entire population. It's never been easier to mistake a few extremely unrepresentative parts for the whole. This in part explains why our expectations about the transformative power of the Internet in authoritarian states are so inflated and skewed toward optimism: The people we usually hear from are those who are already on the frontlines of using new media to push for democratic change in authoritarian societies. Somehow, the Chinese bloggers who cover fashion, music, or pornography—even though those subjects are much more popular in the Chinese blogosphere than human rights or rule of law—never make it to congressional hearings in Washington.

The media is not helping either. Assuming they speak good English, those blogging for the Muslim Brotherhood in Egypt may simply have no intention of helping BBC or CNN to produce yet another report about the power of the blogosphere. That's why the only power Western media cover is usually secular, liberal, or pro-Western. Not surprisingly, they tell us what we wanted to hear all along: Bloggers are fighting for secularism, liberalism, and Western-style democracy. This is why so many Western politicians fall under the wrong impression that bloggers are natural allies, even harbingers, of democracy. "If it's true that there are more bloggers per head of population in Iran than any other country in the world, that makes me optimistic about the future of Iran," said then UK's foreign minister, David Miliband, while visiting Google's headquarters. Why this should be the case—given that Iran's conservative bloggers, who are often more hard-line than the government and are anything but a force for democracy, equality, and justice, are a formidable and rapidly expanding force in the Iranian blogosphere—is unclear. Chances are that Miliband's advisors simply never ventured beyond a handful of pro-Western Iranian blogs that dominate much of the media coverage of the country. It's hard to say what Miliband would make of certain groups of Chinese nationalists who, when they're not making anti-Western or anti-CNN videos, are busy translating books by Western philosophers like Leibniz and Husserl.

Things get worse when Western policymakers start listening to bloggers in exile. Such bloggers often have a grudge against their home country and are thus conditioned to portray all domestic politics as an extension of their own struggle. Their livelihoods and careers often depend on important power brokers in Washington, London, and Brussels making certain assumptions about the Internet. Many of them have joined various new media NGOs or even created a few of their own; should the mainstream assumptions about the power of blogging shift, many of these newly created NGOs are likely to go under.

Not surprisingly, people who get grants to harness the power of the Internet to fight dictators are not going to tell us that they are not succeeding. It's as if we've produced a few million clones of Ahmed Chalabi, that notoriously misinformed Iraqi exile who gave a highly inaccurate picture of Iraq to those who were willing to listen, and hired them to tell us how to fix their countries. Of course, the influence of exiles on foreign policy is a problem that most governments have had to deal with in the past, but bloggers, perhaps thanks to the inevitable comparisons to Soviet dissidents and the era of samizdat, are often not subjected to the level of scrutiny they deserve.

Why Some Data Need to Remain Foggy

It's not just emotion and speed that threaten the integrity of policy-making. It's also the growing availability and mobility of information. As mobile phones proliferate to the most remote corners of the world, data has become easier to gather. Previously disconnected local populations can now report on virtually anything, ranging from the effects of natural disasters to instances of human rights violations to election fraud. Suddenly, tragedy is more visible, and (or so the hope goes) it is more likely to be addressed.

Technology, indeed, can do wonders during natural disasters. Those most affected can use their mobile phones to text both their location and their problems. This information can then be aggregated and visualized on an online map. It may provide no direct relief to those directly affected by the disaster, but it can better inform humanitarian workers about the exact nature of the disaster and thus help in optimizing the allocation of scarce resources. One such tool, Ushahidi, was first designed to report on violence during the postelection crisis in Kenya and since then has been successfully deployed all over the world, including in the devastating earthquakes in Haiti and Chile in early 2010.

But the reason why many projects that rely on crowdsourcing produce trustworthy data in natural disasters is because those are usually
apolitical events. There are no warring sides, and those who report data
do not have any incentives to manipulate it. The problem with using
such crowdsourced tools for other purposes—for example, documenting human rights abuses or monitoring elections, some of the other
uses to which Ushahidi has been put—is that the accuracy of such reports is impossible to verify and easy to manipulate.

After all, anyone can text in deliberately erroneous reports to accuse
their opponents of wrongdoing or, even worse, to sow panic in their
ranks (remember the Nigerian SMS that said that all food was poisoned?). But to be credible, human rights reports and, to a somewhat
lesser degree, reports from election observers need to aim for a 100
percent rate of accuracy. This is because of the peculiar nature of human
rights reporting, especially in conditions where an authoritarian government may dispute the validity of results. One erroneous report—
submitted by mistake or deliberately—is enough to derail the
credibility of the entire database. And once human rights NGOs are
caught producing data of dubious quality, the government gets a good
excuse to shut them down. The *New York Times* praised the power of
Ushahidi when it reported that "as data collects, crisis maps can reveal
underlying patterns of reality: How many miles inland did the hurri-
cane kill? Are the rapes broadly dispersed or concentrated near military
barracks?" But this is quite misleading. At best, such maps could give
us a general idea of the scale and the nature of abuses, but the value of
this as a piece of human rights data is minimal. What's worse, that small
bit of crowdsourced data can easily make all other hard-earned data
about human rights violations easy to dismiss.

Nor do we want to make certain information associated with human
rights abuses publicly accessible on the Internet. In many countries,
there is still a significant social stigma associated with rape. Providing
even the tiniest bits of evidence—say, geographic information about
where rapes have occurred—may reveal the victims, making their lives
even more unbearable. There has always been a certain data protection
mechanism built into human rights reporting, and given the ease with

which information can be collected and disseminated, including by third parties that might be working to impede the work of human rights organizations, it is important to preserve those mechanisms, regardless of the impetus to promote Internet freedom. It's not a given that projects that rely on crowdsourcing won't get this balance right, but we need to resist Internet-centrism and opting for a "more people, more data" kind of approach without considering needs and capabilities.

Ironically, while most of the recent efforts of the digerati have focused on liberating the data from closed databases, the focus of their future efforts may soon shift to squeezing the open data back in or at least finding ways in which to limit the mobility of that data. This is a particularly important problem for various ethnic minorities who suddenly find themselves under threat, as digitized information has publicly identified them in ways they could not anticipate. In Russia a local branch of the Movement Against Illegal Immigration (DPNI), the country's powerful anti-immigration network, created a series of online mash-ups in which they put census data about various ethnic minorities living in the Russian city of Volgograd onto an online map. This was not done to get a better understanding of urban life in Russia but to encourage DPNI's supporters to organize pogroms on those minorities. DPNI is an interesting example of an unabashedly racist organization that has deftly adapted to the Internet era. Not only do they make mash-ups, but their main website runs an online store selling nationalist T-shirts; features the option to translate the site's contents into German, English, and French; and even allows anyone registered on the site to contribute their own news, Wiki-style.

Or take the Burakumin, who are one of Japan's largest social minorities, descending from outcast communities of the feudal era. Since the seventeenth century, members of the one-million-strong Burakumin population have been living outside of Japan's rigid caste system, with other castes pretending that the Burakumin do not exist. When a group of Japanese cartographic enthusiasts overlaid the old maps of the Burakumin communities with Google's satellite images of Tokyo, Osaka, and Kyoto, it first seemed like a good idea. Previously there had been little effort to preserve Burakumin heritage online. Within days, however, a bevy of Japanese nationalists got excited about finally finding the exact locations of the much-hated Burakumin houses, and the Japanese blogosphere was abuzz with discussions of pogroms. After intense pressure from Japanese antidiscrimination NGOs, Google asked the owners of the maps to at least remove the legend that identified the Burakumin ghettos as "scum towns."

Meanwhile, South Korea's xenophobic vigilantes, who formed a group known as Anti-English Spectrum, have been scouring social networking sites searching for personal details of foreigners who come to the country to teach English, in a desperate effort to find any potential misbehavior, so that the foreigners can be thrown out of the country.

Similarly, while China's notorious "human flesh search engines" have mostly been mentioned in the Western media for their courageous online pursuits of corrupt bureaucrats, they also have a darker side. In fact, they have a history of attacking people who voice unpopular political positions (urging respect for ethnic minorities) or simply behaving slightly outside of the accepted norm (being unfaithful to one's spouse). Grace Wang, a student at Duke University, was one of the most famous targets of the angry "human flesh search engines" in 2008, at the height of tensions between China and the West right before the Beijing Olympics. After she urged her fellow Chinese netizens to try to better understand those in Tibet, she had to deal with a wave of personal attacks, with someone even posting directions to her parents' house on a popular Chinese site. (Her family had to go into hiding.)

It may be that what we gain in the ability to network and communicate, we lose in the inevitable empowerment of angry online mobs, who are well-trained to throw "data grenades" at their victims. This may be an acceptable consequence of promoting Internet freedom, but we'd better plan ahead and think of ways in which we can protect the victims. It's irresponsible to put people's lives on the line while hoping we can deal at some later point with the consequences of opening up all the networks and databases.

That the excess of data can pose a danger to freedom and democracy as significant as (if not more significant than) the lack of data has mostly been lost on those cheerleading for Internet freedom. This is hardly surprising, for this may not be such an acute problem in liberal democ-racies, where the dominant pluralist ideology, growing multicultural-ism, and a strong rule of law mitigate the consequences of the data deluge.

But most authoritarian or even transitional states do not have that luxury. Hoping that simply opening up all the networks and uploading all the documents would make a transition to democracy easier or more likely is just an illusion. If the sad experience of the 1990s has taught us anything, it's that successful transitions require a strong state and a relatively orderly public life. The Internet, so far, has posed a major threat to both.

chapter ten

Making History (More Than a Browser Menu)

● • ■ ■

In 1996, when a group of high-profile digerati took to the pages of *Wired* magazine and proclaimed that the "public square of the past" was being replaced by the Internet, a technology that "enables average citizens to participate in national discourse, publish a newspaper, distribute an electronic pamphlet to the world . . . while simultaneously protecting their privacy," many historians must have giggled. From the railways, which Karl Marx believed would dissolve India's caste system, to television, that greatest liberator of the masses, there has hardly appeared a technology that wasn't praised for its ability to raise the level of public debate, introduce more transparency into politics, reduce nationalism, and transport us to the mythical global village. In virtually all cases, such high hopes were crushed by the brutal forces of politics, culture, and economics. Technologies, it seems, tend to overpromise and underdeliver, at least on their initial promises.

This is not to suggest that such inventions didn't have any influence on public life or democracy. On the contrary, they often mattered far more than what their proponents could anticipate. But those effects were often antithetical to the objectives their inventors were originally pursuing. Technologies that were supposed to empower the individual strengthened the dominance of giant corporations, while technologies that were supposed to boost democratic participation produced a population of couch potatoes. Nor is this to suggest that such technologies never had the potential to improve the political culture or make governance more transparent; their potential was immense. Nevertheless, in most cases it was squandered, as the utopian claims invariably attached to those technologies confused policymakers, preventing them from taking the right steps to make good on those early promises of progress.

By touting the uniqueness of the Internet most technology gurus reveal their own historical ignorance, for the rhetoric that accompanied predictions about earlier technologies was usually every bit as sublime as today's quasi-religious discourse about the power of the Internet. Even a cursory look at the history of technology reveals just how quickly public opinion could move from professing an uncritical admiration of certain technologies to eagerly bashing everything they stand for. But acknowledging that criticism of technology is as old as its worship should not lead policymakers to conclude that attempts to minimize the adverse effects of technology on society (and vice versa) are futile. Instead, policymakers need to acquaint themselves with the history of technology so as to judge when the overhyped claims about technology's potential may need some more scrutiny—if only to ensure that at least half of them get realized.

And history does contain plenty of interesting lessons. The telegraph was the first technology predicted to transform the world into a global village. An 1858 editorial in *New Englander* proclaimed: "The telegraph binds together by a vital cord all the nations of the earth. . . . It is impossible that old prejudices and hostilities should longer exist, while such an instrument has been created for an exchange of thought between all the nations of the earth." Speaking in 1868, Edward Thornton, the British ambassador to the United States, hailed the telegraph as "the nerve of international life, transmitting knowledge of events, removing causes of misunderstanding, and promoting peace and harmony throughout the world." The *Bulletin of the American Geographical and*

Statistical Society believed it to be an "extension of knowledge, civilization and truth" that catered to "the highest and dearest interest of the human race." Before long the public saw the telegraph's downside. Those who hailed its power to help find fugitive criminals soon had to concede that it could also be used to spread false alarms and used by the criminals themselves. Perhaps it was a sense of bitter disappointment that prompted the *Charleston Courier* to conclude, just two years after the first American telegraph lines were successfully installed, that "the sooner the [telegraph] posts are taken down the better," while the *New Orleans Commercial Times* expressed its "most fervent wish that the telegraph may never approach us any nearer than it is at present."

The brevity of the telegraph's messages didn't sit well with many literary intellectuals either; it may have opened access to more sources of information, but it also made public discourse much shallower. More than a century before similar charges would be filled against Twitter, the cultural elites of Victorian Britain were getting concerned about the trivialization of public discourse under an avalanche of fast news and "snippets." In 1889, the *Spectator*, one of the empire's finest publications, chided the telegraph for causing "a vast diffusion of what is called 'news,' the recording of every event, and especially of every crime, everywhere without perceptible interval of time. The constant diffusion of statements in snippets . . . must in the end, one would think, deteriorate the intelligence of all to whom the telegraph appeal."

The global village that the telegraph built was not without its flaws and exploitations. At least one contemporary observer of Britain's colonial expansion into India observed that "the unity of feeling and of action which constitutes imperialism would scarcely have been possible without the telegraph." Thomas Misa, a historian of technology at the University of Minnesota, notes that "telegraph lines were so important for imperial communication that in India they were built in advance of railway lines." Many other technological innovations beyond the telegraph contributed to this expansionism. Utopian accounts of technology's liberating role in human history rarely acknowledge the fact that it was the discovery of quinine, which helped to fight malaria, reducing the risk of endemic tropical disease, that eliminated one major barrier to colonialism, or that the invention of printing helped to forge a common Spanish identity and pushed the Spaniards to colonize Latin America.

When the telegraph failed to produce the desired social effects, everyone's attention turned to the airplane. Joseph Corn describes the collective exaltation that surrounded the advent of the airplane in his 2002 book *The Winged Gospel*. According to Corn, in the 1920s and much of the 1930s most people "expected the airplane to foster democracy, equality, and freedom, to improve public taste and spread culture, to purge the world of war and violence; and even to give rise to a new kind of human being." One observer at the time, apparently oblivious to the economic forces of global capitalism, mused that airplanes opened up "the realm of absolute liberty; no tracks, no franchises, no need of thousands of employees to add to the cost," while in 1915 the editor of *Flying* magazine—the *Wired* of its day—enthusiastically proclaimed that the First World War had to be "the last great war in history," because "in less than another decade," the airplane would have eliminated the factors responsible for wars and ushered in a "new period in human relations" (apparently, Adolf Hitler was not a subscriber to *Flying*). As much as one could speak of utopian airplane-centrism of the 1910s, this was it.

But it was the invention of radio that produced the greatest number of unfulfilled expectations. Its pioneers did their share to overhype the democratization potential of their invention. Guglielmo Marconi, one of the fathers of this revolutionary technology, believed that "the coming of the wireless era will make war impossible, because it will make war ridiculous." Gerald Swope, president of General Electric Company, one of the biggest commercial backers of radio at the time, was equally upbeat in 1921, hailing the technology as "a means for general and perpetual peace on earth." Neither Marconi nor Swope could have foreseen that seven decades later two local radio stations would use the airwaves to heighten ethnic tensions, spread messages of hatred, and help fuel the Rwandan genocide.

When Twitter's founders proclaim their site to be a "triumph of humanity," as they did in 2009, the public should save its applause until assessing the possibility of a Twitter-fueled genocide sweeping through some distant foreign land, thousands of miles away from the Bay Area. Then and now, such declarations of technology's benign omnipotence have been nothing more than poorly veiled attempts at creating a favorable regulatory climate—and who would dare to regulate humanity's triumph? But in the earliest stages of its history, radio was also seen as a way to educate the public about politics and raise the level of political discourse; it was widely expected to force politicians to carefully plan their speeches. In the early 1920s, the *New Republic* applauded radio's political effects, for the invention "has found a way to dispense with political middlemen" and even "has restored the demos on which republican government is founded."

Not surprisingly, radio was seen as superior to the previous medium of political communications, the newspaper. As one editorial writer put it in 1924: "Let a legislator commit himself to some policy that is obviously senseless, and the editorial writers must first proclaim his imbecility to the community. But let the radiophone in the legislative halls of the future flash his absurdities into space and a whole state hears them at once." Just the way today's politicians are told to fear their "Macaca moment," politicians of yesteryear were told to fear their "radio moment." Like the Internet today, radio was believed to be changing the nature of political relations between citizens and their governments. In 1928, *Collier's* magazine declared that "the radio properly used will do more for popular government that have most of the wars for freedom and self government," adding that "the radio makes politics personal and interesting and therefore important." But it didn't take long for the public mood to sour again. By 1930 even the initially optimistic *New Republic* reached the verdict that "broadly speaking, the radio in America is going to waste." In 1942 Paul Lazarsfeld, a prominent communications scholar at Columbia University, concluded that "by and large, radio has so far been a conservative force in American life and has produced but few elements of social progress."

The disappointment was caused by a number of factors, not least the dubious uses to which the technology was put by governments. As Asa Briggs and Peter Burke point out in their comprehensive *A Social History of the Media*, "the 'age of radio' was not only the age of Roosevelt and Churchill but also that of Hitler, Mussolini and Stalin." That so many dictators profited so much from radio dampened the nearly universal enthusiasm for the medium, while its commercialization by big business alienated those who hoped it would make the public conversation more serious. It's not hard to guess Lazarsfeld's reaction to the era of Rush Limbaugh.

Radio's fading democratizing potential did not preclude a new generation of pundits, scholars, and entrepreneurs from making equally overblown claims about television. From the 1920s onward, Orrin Dunlap, one of the first television and radio critics for the *New York Times*, was making an argument already familiar to those who studied the history of the telegraph, the airplane, or the radio. "Television," wrote Dunlap, without even a shade of doubt, "will usher in a new era of friendly intercourse between the nations of the earth," while "current conceptions of foreign countries will be changed." David Sarnoff, head of the Radio Corporation of America, believed that another global village was in the making: "When television has fulfilled its ultimate destiny . . . with this may come . . . a new sense of freedom, and . . . a finer and broader understanding between all the peoples of the world."

Lee De Forest, the famed American inventor, held high hopes for the educational potential of television, believing that it could even reduce the number of traffic incidents. "Can we imagine," he asked in 1928, "a more potent means for teaching the public the art of careful driving safety upon our highways than a weekly talk by some earnest police traffic officer, illustrated with diagrams and photographs?" That such programs never really made it to mainstream American television is unfortunate—especially in an era when drivers are texting their way to accidents and even airplane pilots work on their laptops mid-flight— but it is not the limitations of technology that are to blame. Rather, it was the limitations of the political, cultural, and regulatory discourse of the time that soon turned much of American television into, as the chair of the Federal Communications Commission, Newton Minow, put it in 1961, a "vast wasteland."

Like radio before it, television was expected to radically transform the politics of the time. In 1932 Theodore Roosevelt Jr., the son of the late president and then governor-general of the Philippines, predicted that TV would "stir the nation to a lively interest in those who are directing its policies and in the policies themselves," which would result in a "more intelligent, more concerted action from an electorate; the people will think more for themselves and less simply at the direction of local members of the political machines." Thomas Dewey, a prominent Republican who ran against Franklin Delano Roosevelt and Harry Truman in the 1940s, compared television to an X-ray, predicting that "it should make a constructive advance in political campaigning." Anyone watching American television during an election season would be forgiven for disagreeing with Dewey's optimism.

Such enthusiasm about television carried the day until very recently. In 1978, Daniel Boorstin, one of the most famous American historians of the twentieth century, lauded television's power to "disband armies, to cashier presidents, to create a whole new democratic world—democratic in ways never before imagined, even in America." Boorstin wrote these words when many political scientists and policymakers were still awaiting the triumph of "teledemocracy," in which citizens would use television to not only observe but also directly participate in politics. (The hope that new technology could enable more public participation in politics predates television; back in 1940 Buckminster Fuller, the controversial American inventor and architect, was already lauding the virtues of "telephone democracy," which could enable "voting by telephone on all prominent questions before Congress.")

In hindsight, the science-fiction writer Ray Bradbury was closer to the truth in 1953 than Boorstin ever was in 1978. "The television," wrote Bradbury, "is that insidious beast, that Medusa which freezes a billion people to stone every night, staring fixedly, that Siren which called and sang and promised so much and gave, after all, so little."

The advent of the computer set off another utopian craze. A 1950 article in the *Saturday Evening Post* claimed that "thinking machines will bring a healthier, happier civilization than any known heretofore." We are still living in the time of some of its most ridiculous predictions. And while it's easy to be right in hindsight, one needs to remember that there was nothing predetermined about the direction in which radio and television advanced in the last century. The British made a key strategic decision to prioritize public broadcasting and created a behemoth known as the British Broadcasting Corporation; the Americans, for a number of cultural and business reasons, took a more laissez-faire approach. One could debate the merits of either strategy, but it seems undeniable that the American media landscape could have looked very different today, especially if the utopian ideologies promoted by those with a stake in the business were scrutinized a bit more closely.

While it's tempting to forget everything we've learned from history and treat the Internet as an entirely new beast, we should remember that this is how earlier generations must have felt as well. They, too, were tempted to disregard the bitter lessons of previous disappointments and assume a brave new world. Most commonly, it precluded them from making the right regulatory decisions about new technologies. After all, it's hard to regulate divinity. The irony of the Internet is that while it never delivered on the uber-utopian promises of a world without nationalism or extremism, it still delivered more than even the most radical optimists could have ever wished for. The risk here is that given the relative successes of this young technology, some may assume that it would be best to leave it alone rather than subject it to regulation of any kind. This is a misguided view. The recognition of the revolutionary nature of a technology is a poor excuse not to regulate it. Smart regulation, if anything, is a first sign that society is serious about the technology in question and believes that it is here to stay; that it is eager to think through the consequences; and that it wants to find ways to unleash and harvest its revolutionary potential.

No society has ever got such regulatory frameworks right by looking only at technology's bright sides and refusing to investigate how its uses may also produce effects harmful to society. The problem with cyber-optimism is that it simply doesn't provide useful intellectual grounds for regulation of any sort. If everything is so rosy, why even bother with regulation? Such an objection might have been valid in the early 1990s, when access to the Internet was limited to academics, who couldn't possibly foresee why anyone would want to send spam. But as access to the Internet has been democratized, it has become obvious that self-regulation will not always be feasible given such a diverse set of users and uses.

Technology's Double Life

If there is an overarching theme to modern technology it is that it defies the expectations of its creators, taking on functions and roles that were never intended at creation. David Noble, a prolific historian of modern technology, makes this point forcefully in his 1984 book *Forces of Production*. "Technology," writes Noble, "leads a double life, one which conforms to the intentions of designers and interests of power and another which contradicts them—proceeding behind the backs of their architects to yield unintended consequences and unintended possibilities." Even Ithiel de Sola Pool, that naïve believer in the power of information to undermine authoritarianism, was aware that technology alone is not enough to create desired political outcomes, writing that "technology shapes the structure of the battle but not every outcome."

Not surprisingly, futurists often get it wrong. George Wise, a historian associated with General Electric, examined fifteen hundred technology predictions made between 1890 and 1940 by engineers, historians, and other scientists. One-third of those predictions came true, even if somewhat vaguely. The remaining two-thirds either were false or remained in ambiguity.

From a policy perspective, the lesson to be learned from the history of technology and the numerous attempts to foretell it is that few modern technologies are stable enough—in their design, in their applications, in their appeal to the public—to provide for flawless policy planning. This is particularly the case at the early stages of a technology's life cycle. Anyone working on a "radio freedom" policy in the 1920s would have been greatly surprised by the developments—many of them negative—of the 1930s. The problem with today's Internet is that it makes a rather poor companion to a policy planner. Too many stakeholders are involved, from national governments to transnational organizations like ICANN and from the United Nations to users of Internet services; certain technical parts of its architecture may change if it runs out of addresses; malign forces like spammers and cyber-criminals are constantly creating innovations of their own. Predicting the future of the Internet is a process marked by far greater complexity than predicting the future of television because the Web is a technology that can be put to so many different uses at such a cheap price.

It's such essential unpredictably that should make one extremely suspicious of ambitious and yet utterly ambiguous policy initiatives like Internet freedom that demand a degree of stability and maturity that the Internet simply doesn't have, while their advocates are making normative claims of what the Internet *should* look like, as if they already know how to solve all of the problems. But an unruly tool in the hands of overconfident people is a recipe for disaster. It would be far more productive to assume that the Internet is highly unstable; that trying to rebuild one's policies around a tool that is so complex and capricious is not going to work; and that instead of trying to solve what may essentially be unsolvable global problems, one would be well-advised to start on a somewhat smaller scale at which one could still grasp, if not fully master, the connections between the tool and its environment.

But such caution may suit only the intellectuals. Despite the inevitable uncertainty surrounding technology, policymakers need to make decisions, and technology plays a growing role in all of them. Predictions about how technology *might* work are thus inevitable, or paralysis would ensue. The best policymakers can do is to understand why so many people get them wrong so often and then try to create mechanisms and procedures that could effectively weed out excessive hype from the decision-making process.

The biggest problem with most predictions about technology is that they are invariably made based on how the world works today rather than on how it will work tomorrow. But the world, as we know, doesn't stand still: Politics, economics, and culture constantly reshape the environment that technologies were supposed to transform, preferably in accordance with our predictions. Politics, economics, and culture also profoundly reshape technologies themselves. Some, like radio, become cheap and ubiquitous; others, like the airplane, become expensive and available only to a select few. Furthermore, as new technologies come along, some older ones become obsolete (fax machines) or find new uses (TVs as props for playing games on your Wii).

Paradoxically, technologies meant to alleviate a particular problem may actually make it worse. As Ruth Schwartz Cowan, a historian of science at the University of Pennsylvania, shows in her book *More Work for Mother*, after 1870 homemakers ended up working longer hours even though more and more household activities were mechanized. (Cowan notes that in 1950 the American housewife produced single-handedly what her counterpart needed a staff of three or four to produce just a century earlier.) Who could have predicted that the development of "labor-saving devices" had the effect of increasing the burden of housework for most women?

Similarly, the introduction of computers into the workforce failed to produce expected productivity gains (Tetris was, perhaps, part of some secret Soviet plot to halt the capitalist economy). The Nobel Prize–winning economist Robert Solow quipped that "one can see the computer age everywhere but not in the productivity statistics!" Part of the problem in predicting the exact economic and social effects of a technology lies in the uncertainty associated with the scale on which such a technology would be used. The first automobiles were heralded as technologies that could make cities cleaner by liberating them of horse manure. The by-products of the internal combustion engine may be more palatable than manure, but given the ubiquity of automobiles in today's world, they have solved one problem only by making another one—pollution—much worse. In other words, the future uses of a particular technology can often be described by that old adage "It's the economy, stupid."

William Galston, a former adviser to President Clinton and a scholar of public policy at the Brookings Institution, has offered a powerful example of how we tend to underestimate the power of economic forces in conditioning the social impact of technologies. Imagine, he says, a hypothetical academic conference about the social effects of television convened in the early 1950s. The consensus at the conference would almost certainly be that television was poised to strengthen community ties and multiply social capital. Television sets were sparse and expensive, and neighbors had to share and visit each other's houses. Enter today's academic conferences about television, and participants are likely to deplore the pervasive "bedroom culture," whereby the availability of multiple televisions in just one home is perceived as eroding ties within families, not just ties within neighborhoods.

Another reason why the future of a given technology is so hard to predict is that the disappearance of one set of intermediaries is often accompanied by the emergence of other intermediaries. As James Carey, a media scholar at Columbia University, observed, "as one set of borders, one set of social structures is taken down, another set of borders is erected. It is easier for us to see the borders going down." We rarely notice the new ones being created. In 1914 *Popular Mechanics* thought that the age of governments was over, announcing that wireless telegraphy allowed "the private citizen to communicate across great distance without the aid of either the government or a corporation." Only fifteen years later, however, a handful of corporations dominated the field of radio communication, even while the public still maintained some illusions that radio was a free and decentralized media. (The fact that radios were getting cheaper only contributed to those illusions.)

Similarly, just as today's Internet gurus are trying to convince us that the age of "free" is upon us, it almost certainly is not. All those free videos of cats that receive millions of hits on YouTube are stored on powerful server centers that cost millions of dollars to run, usually in electricity bills alone. Those hidden costs will sooner or later produce environmental problems that will make us painfully aware of how expensive such technologies really are. Back in 1990, who could have foreseen that Greenpeace would one day be issuing a lengthy report about the environmental consequences of cloud computing, with some scientists conducting multiyear studies about the impact of email spam on climate change? The fact that we cannot yet calculate all the costs of a given technology—whether financial, moral, or environmental ones—does not mean that it comes free.

No Logic for Old Men

Another recurring feature of modern technology that has been overlooked by many of its boosters is that the emergence of new technologies, no matter how revolutionary their circuitry might be, does not automatically dissolve old practices and traditions. Back in the 1950s, anyone arguing that television would strengthen existing religious institutions was inviting ridicule. And yet, a few decades later, it was television that Pat Robertson and a horde of other televangelists had to thank for their powerful social platform. Who today would bet that the Internet will undermine organized religion?

In fact, as one can currently observe with the revival of nationalism and religion on the Web, new technologies often entrench old practices and make them more widespread. Claude Fischer, who studied how Americans adopted the telephone in the nineteenth century in his book *America Calling*, observes that it was primarily used to "widen and deepen . . . existing social patterns rather than to alter them." Instead of imagining the telephone as a tool that impelled people to embrace modernity, Fischer proposed that we think of it as "a tool modern people have used to various ends, including perhaps the maintenance, even enhancement, of past practices." For the Internet to play a constructive role in ridding the world of prejudice and hatred, it needs to be accompanied by an extremely ambitious set of social and political reforms; in their absence, social ills may only get worse. In other words, whatever the internal logic of the technology at hand, it's usually malleable by the logic of society at large. "While each communication technology does have its own individual properties, especially regarding which of the human senses it privileges and which ones it ignores," writes Susan Douglas, a scholar of communications at the University of Michigan, "the economic and political system in which the device is embedded almost always trumps technological possibilities and imperatives."

And yet this rarely prevents an army of technology experts from claiming that they have cracked that logic and understood what radio, television, or the Internet is all about; the social forces surrounding it are thus deemed mostly irrelevant and can be easily disregarded. Marshall McLuhan, the first pop philosopher, believed that television had a logic: Unlike print, it urges viewers to fill in the gaps in what it is they're seeing, stimulates more senses, and, overall, nudges us closer to the original tribal condition (a new equilibrium that McLuhan clearly favored). The problem is that while McLuhan was chasing the inner logic of television, he might have missed how it could be appropriated by corporate America and produce social effects much more obvious (and uglier) than changes in some obscure sense-ratios that McLuhan so meticulously calculated for each medium.

Things get worse in the international context. The "logic" that the scholars and policymakers supposedly have access to is simply an interpretation of what a particular technology is capable of doing given a particular set of circumstances. Hermann Göring, who put radio to masterful propaganda use in Hitler's Germany, saw its logic in very different terms than, say, Marconi.

Thus, knowing everything about a given technology still tells us little about how exactly it will shape a complex modern society. Economist William Schaniel shares this view, cautioning us that "the analytic focus of a technology transfer should be on the adopting culture and not on the materials being transferred," simply because, while "new technology does create change," this change is not "preordained by the technology adopted." Instead, writes Schaniel, "the adopted technology is adapted by the adopting society to their social processes." When gunpowder was brought to Europe from Asia, Europeans did not concurrently adopt Asian rules and beliefs about it. The adopted gunpowder was adapted by European civilizations according to their own values and traditions.

The Internet is no gunpowder; it's considerably more complex and multidimensional. But this only adds urgency to our quest to understand the societies it is supposed to "reshape" or "democratize." Reshape them it may, but what is of utmost interest to policymakers is the direction in which this reshaping would proceed. The only way for them to understand it is to resist technological determinism and embark on a careful analysis of nontechnological forces that constitute the environments they seek to understand or transform. It may make sense to think about technologies as embodying a certain logic at an early stage of their deployment, but as they mature, their logic usually gives way to more powerful social forces.

The inability to see that the logic of technology, as much as one could say it exists, varies from context to context partly explains the Western failure to grasp the importance of the Internet to authoritarian regimes. Not having a good theory of the internal political and social logic of those regimes, Western observers assume that the dictators and their cronies can't find a regime-strengthening use for the Internet, because under the conditions of Western liberal democracies—and those are the only conditions these observers understand—the Internet has been weakening the state and decentralizing power. Instead of burrowing further into the supposed logic of the Internet, Western do-gooders would be well-advised to get a more refined picture of the political and social logic of authoritarianism under the conditions of globalization. If policymakers lack a good theoretical account of what makes those societies tick, no amount of Internet-theorizing will allow them to formulate effective policies for using the Internet to promote democracy.

Is There History After Twitter?

It's tempting to see technology as some kind of a missing link that can help us make sense of otherwise unrelated events known as human history. Why search for more complex reasons if the establishment of democratic forms of government in Europe could be explained by the invention of the printing press? As the economic historian Robert Heilbroner observed in 1994, "history as contingency is a prospect that is more than the human spirit can bear."

Technological determinism—the belief that certain technologies are bound to produce certain social, cultural, and political effects—is attractive precisely because "it creates powerful scenarios, clear stories, and because it accords with the dominant experience in the West," write Steve Graham and Simon Marvin, two scholars of urban geography. Forcing a link between the role that photocopies and fax machines played in Eastern Europe in 1989 and the role that Twitter played in Iran in 2009 creates a heart-wrenching but also extremely coherent narrative that rests on the widespread belief, rooted in Enlightenment ideals, in the emancipatory power of information, knowledge, and, above all, ideas. It's far easier to explain recent history by assuming that communism dropped dead the moment Soviet citizens understood that there were no queues in Western supermarkets than to search for truth in some lengthy and obscure reports on the USSR's trade balance.

It is for this reason that determinism—whether of the social variety, positing the end of history, or of the political variety, positing the end of authoritarianism—is an intellectually impoverished, lazy way to study the past, understand the present, and predict the future. Bryan Pfaffenberger, an anthropologist at the University of Virginia, believes that the reason why so many of us fall for deterministic scenarios is because it presents the easiest way out. "Assuming technological determinism," writes Pfaffenberger, "is much easier than conducting a fully contextual study in which people are shown to be the active appropriators, rather than the passive victims, of transferred technology."

But it's not only history that suffers from determinism; ethics doesn't fare much better. If technology's march is unstoppable and unidirectional, as a horde of technology gurus keep convincing the public from the pages of technology magazines, it then seems pointless to stand in its way. If radio, television, or the Internet are poised to usher in a new age of democracy and universal human rights, there is little role for us humans to play. However, to argue that a once-widespread practice like lobotomy was simply a result of inevitable technological forces is to let its advocates off the hook. Technological determinism thus obscures the roles and responsibilities of human decision makers, either absolving them of well-deserved blame or minimizing the role of their significant interventions. As Arthur Welzer, a political scientist at Michigan State University, points out, "to the extent that we view ourselves as helpless pawns of an overarching and immovable force, we may renounce the moral and political responsibility that, in fact, is crucial for the good exercise of what power over technology we do possess."

By adopting a deterministic stance, we are less likely to subject technology—and those who make a living from it—to the full bouquet of ethical questions normal for democracy. Should Google be required to encrypt all documents uploaded to its Google Docs service? Should Facebook be allowed to continue making more of their users' data public? Should Twitter be invited to high-profile gatherings of the U.S. government without first signing up with the Global Network Initiative? While many such questions are already being raised, it's not so hard to imagine a future when they would be raised less often, particularly in offices that need to be asking them the most.

Throughout history, new technologies have almost always empowered and disempowered particular political and social groups, sometimes simultaneously—a fact that is too easy to forget under the sway of technological determinism. Needless to say, such ethical amnesia is rarely in the interests of the disempowered. Robert Pippin, a philosopher at the University of Chicago, argues that society's fascination with the technological at the expense of the moral reaches a point where "what ought to be understood as contingent, one option among others, open to political discussion is instead falsely understood as necessary; what serves particular interest is seen without reflection, as of universal interest; what ought to be a part is experienced as a whole." Facebook's executives justifying their assault on privacy by claiming that this is where society is heading anyway is exactly the kind of claim that should be subject to moral and political—not just technological—scrutiny. It's by appealing to such deterministic narratives that Facebook manages to obscure its own role in the process.

Abbe Mowshowitz, professor of computer science at the City College of New York, compares the computer to a seed and concrete historical circumstances to the ground in which the seed is to be planted: "The right combination of seed, ground and cultivation is required to promote the growth of desirable plants and to eliminate weeds. Unfortunately, the seeds of computer applications are contaminated with those of weeds; the ground is often ill-prepared; and our methods of cultivation are highly imperfect." One can't fault Mowshowitz for misreading the history of technology, but there is a more optimistic way to understand what he said: We, the cultivators, can actually intervene in all three stages, and it's up to us to define the terms on which we choose to do so.

The price for not intervening could be quite high. Back in 1974, Raymond Williams, the British cultural critic, was already warning us that technological determinism inevitably produces a certain social and cultural determinism that "ratifies the society and culture we now have, and especially its most powerful internal directions." Williams worried that placing technology at the center of our intellectual analysis is bound to make us view what we have traditionally understood as a problem of politics, with its complex and uneasy questions of ethics and morality, as instead a problem of technology, either eliminating or obfuscating all the unresolved philosophical dilemmas. "If the medium—whether print or television—is the cause," wrote Williams in his best-selling *Television: Technology and Cultural Form*, "all other causes, all that men ordinarily see as history, are at once reduced to effects." For Williams, it was not the end of *history* that technology was ushering in; it was the end of *historical thinking*. And with the end of historical thinking, the questions of justice lose much of their significance as well.

Williams went further in his criticism, arguing that technological determinism also prevents us from acknowledging what is political about technology itself (the kind of practices and outcomes it tends to favor), as its more immediately observable features usually occupy the lion's share of the public's attention, making it difficult to assess its other, more pernicious features. "What are elsewhere seen as effects, and as such subject to social, cultural, psychological and moral questioning," wrote Williams, "are excluded as irrelevant by comparison with the direct physiological and therefore 'psychic' effects of the media as such." In other words, it's far easier to criticize the Internet for making us stupid than it is to provide a coherent moral critique of its impact on democratic citizenship. And under the barrage of ahistorical blurbs about the Internet's liberating potential, even posing such moral questions may seem too contrarian. Considering how the world reacted to Iran's

Twitter Revolution, it's hard not to appreciate the prescience of Williams's words. Instead of talking about religious, demographic, and cultural forces that were creating protest sentiment in the country, all we cared about was Twitter's prominent role in organizing the protests and its resilience in the face of censorship.

Similarly, when many Western observers got carried away discussing the implications of Egypt's Facebook Revolution in April 2008—when thousands of young Egyptians were mobilized via the Internet to express their solidarity with the textile workers who were on strike in the poor industrial city of Mahala—few bothered to ask what it was the workers actually wanted. As it turns out, they were protesting extremely low wages at their factory. It was primarily a protest about labor issues, which was successfully linked to a broader anti-Mubarak constitutional reform campaign. Once, for various reasons, the labor component to the protests fizzled, other attempts at a Facebook revolution—the one with consequences in the physical world—failed to resonate, even though they attracted hundreds of thousands of supporters online. As was to be expected, most reports in the Western media focused on Facebook rather than on labor issues or demands on Mubarak to end the emergency rule imposed on Egypt since 1981. This is yet another powerful reminder that by focusing on technologies, as opposed to the social and political forces that surround them, one may be drawn to wrong conclusions. As long as such protests continue to be seen predominantly through the lens of the technology through which they were organized—rather than, say, through the demands and motivation of the protesters—little good will come of Western policies, no matter how well-intentioned.

What is, therefore, most dangerous about succumbing to technological determinism is that it hinders our awareness of the social and the political, presenting it as the technological instead. Technology as a Kantian category of understanding the world may simply be too expansionist and monopolistic, subsuming anything that has not yet been properly understood and categorized, regardless of whether its roots and nature are technological. (This is what the German philosopher

Martin Heidegger meant when he said that "the essence of technology is by no means anything technological.") Since technology, like gas, will fill in any conceptual space provided, Leo Marx, professor emeritus at the Massachusetts Institute of Technology, describes it as a "hazardous concept" that may "stifle and obfuscate analytic thinking." He notes, "Because of its peculiar susceptibility to reification, to being endowed with the magical power of an autonomous entity, technology is a major contributant to that gathering sense . . . of political impotence. The popularity of the belief that technology is the primary force shaping the postmodern world is a measure of our . . . neglect of moral and political standards, in making decisive choices about the direction of society."

The neglect of moral and political standards that Leo Marx is warning about is on full display in the sudden urge to promote Internet freedom without articulating how exactly it fits the rest of the democracy-promotion agenda. Hoping that the Internet may liberate the Egyptians or the Azeris from authoritarian oppression is no good excuse to continue covertly supporting the very sources of that oppression. To her credit, Hillary Clinton avoided falling for technological determinism in her Internet freedom speech, saying that "while it's clear that the spread of these [information] technologies is transforming our world, it is still unclear how that transformation will affect the human rights and welfare of much of the world's population." On second reading, however, this seems like a very strange statement to make. If it's not clear how such technologies will affect human rights, what is the point of promoting them? Is it just because there is little clarity as to what Internet freedom means and does? Such confusion in the ranks of policymakers is only poised to increase, since they are formulating policies around a highly ambiguous concept.

Leo Marx suggests that the way to address the hazards of the concept of technology is to rethink whether it is still worth putting it at the center of any intellectual inquiry, let alone a theory of action. The more we learn about technology, the less it makes sense to focus on it alone, in isolation from other factors. Or as Marx himself puts it, "the paradoxical result of ever greater knowledge and understanding of technology is to cast doubt on the rationale for making 'technology,' with its unusually obscure boundaries, the focus of a discrete field of specialized historical (or other disciplinary) scholarship." In other words, it's not clear what it is we gain by treating technology as a historical actor in its own right, for it usually hides more about society, politics, and power than it reveals.

As far as the Internet is concerned, scholarship has so far moved in the opposite direction. Academic centers dedicated to the study of the Internet—the intellectual bulwarks of Internet-centrism—keep proliferating on university campuses and, in the process, contribute to its further reification and decontextualization. That virtually any newspaper or magazine today boasts of interviews with "Internet gurus" is a rather troubling sign, for however deep their knowledge of the architecture of the Internet and its diverse and playful culture, it doesn't make up for their inadequate understanding of how societies, let alone non-Western societies, function. It's a sign of how deeply Internet-centrism has corrupted the public discourse that people who have a rather cursory knowledge of modern Iran have become the go-to sources on Iran's Twitter Revolution, as if a close look at all Iran-related tweets could somehow open a larger window on the politics of this extremely complicated country than the careful scholarly study of its history.

Why Technologies Are Never Neutral

If technological determinism is dangerous, so is its opposite: a bland refusal to see that certain technologies, by their very constitution, are more likely to produce certain social and political outcomes than other technologies, once embedded into enabling social environments. In fact, there is no misconception more banal, ubiquitous, and profoundly misleading than "technology is neutral." It all depends, we are often told, on how one decides to use a certain tool: A knife can be used to kill somebody, but it can also be used to carve wood.

The neutrality of technology is a deep-rooted theme in the intellectual history of Western civilization. Boccaccio raised some interesting questions about it in *The Decameron* back in the mid-fourteenth century. "Who doesn't know what a boon wine is to the healthy . . . and how dangerous to the sick? Are we to say, then, that wine is bad simply because it is injurious to the fevered? . . . Weapons safeguard the welfare of those who desire to live in peace; nevertheless; they often shed blood, not through any evil inherent in them, but through the wickedness of the men who use them to unworthy ends."

The neutrality of the Internet is frequently invoked in the context of democratization as well. "Technology is merely a tool, open to both noble and nefarious purposes. Just as radio and TV could be vehicles of information pluralism and rational debate, so they could also be commandeered by totalitarian regimes for fanatical mobilization and total state control," writes Hoover Institution's Larry Diamond. Neutrality-speak crept into Hillary Clinton's Internet freedom speech as well, when she noted that "just as steel can be used to build hospitals or machine guns and nuclear energy can power a city or destroy it, modern information networks and the technologies they support can be harnessed for good or ill." The most interesting thing about Clinton's analogy between the Internet and nuclear energy is that it suggests that there needs to be *more* not *less* oversight and control over the Internet. No one exactly advocates that nuclear plants should be run as their proprietors wish; the notion of "nuclear freedom" as a means of liberating the world sounds rather absurd.

Product designers like to think of tools as having certain perceived qualities. Usually called "affordances," these qualities suggest—rather than dictate—how tools are to be used. A chair may have the affordance for sitting, but it may also have the affordance for breaking a window; it all depends on who is looking and why. The fact that a given technology has multiple affordances and is open to multiple uses, though, does not obviate the need to closely examine its ethical constitution, compare the effects of its socially beneficial uses with those of its socially harmful uses, estimate which uses are most likely to prevail, and, finally, decide whether any mitigating laws and policies should be established to amplify or dampen some of the ensuing effects. On paper, nuclear technology is beautiful, complex, safe, and brilliantly designed;

in reality, it has one peculiar "affordance" that most societies cannot afford, or at least they cannot afford it without significant safeguards.

Similarly, the reason why most schools ban their students from carrying knives is because this behavior could lead to bloodshed. That we do not know how exactly knives will be used in the hands of young people in every particular situation is not a strong enough reason to allow them; knowing how they can be misused, on the other hand, even if the chance of misuse is small, provides us with enough information to craft a restricting policy. Thus, most societies want to avoid some of the affordances of knives (such as their ability to hurt people) in certain contexts (such as schools).

The main problem with the "technology is neutral" thesis, therefore, is its complete uselessness for the purposes of policymaking. It may offer a useful starting point for some academic work in design, but it simply doesn't provide any foundation for sensible policymaking, which is often all about finding the right balance between competing goods in particular contexts. If technology is neutral and its social effects are unknowable—it all depends on who uses it and when—it appears that policymakers and citizens can do painfully little about controlling it. The misuses of some simple technologies, however, are so widespread and easy to grasp that their undesirability in certain contexts is nothing short of obvious; it's hard to imagine anyone making the case that knives are merely tools, open to both noble and nefarious contexts, at a PTA meeting. But when it comes to more complex technologies—and especially the Internet, with its plethora of applications—their conditional undesirability becomes far less obvious, save, perhaps, for highly sensitive issues (e.g., children gaining access to online pornography).

The view that technology is neutral leaves policymakers with little to do but scrutinize the social forces around technologies, not technologies themselves. Some might say that when it comes to the co-optation of the Internet by repressive regimes, one shouldn't blame the Internet but only the dictators. This is not a responsible view either. Even those who argue that the logic of technology is malleable by the logic of society that adopts it don't propose to stop paying attention to the former. Iran's police may continue monitoring social networking sites forever, but it's easy to imagine a world where Facebook offers better data protection to its users, thus making it harder for the police to learn more about Iranians on Facebook. Likewise, it's easy to imagine a world where Facebook doesn't change how much user data it discloses to the public without first soliciting explicit permission from the user.

Thus, one can believe that authoritarian regimes will continue being avid users of the Internet, but one can make it hard for them to do so. The way forward is to clearly scrutinize both the logic of technology *and* the logic of society that adopts it; under no circumstances should we be giving technologies—whether it's the Internet or mobile phones—a free pass on ethics. All too often the design of technologies simply conceals the ideologies and political agendas of their creators. This alone is a good enough reason to pay closer attention to whom they are most likely to benefit and hurt. That technologies may fail to achieve the objectives their proponents intended should not distract us from analyzing the desirability of those original agendas. The Internet is no exception. The mash-up ethos of Web 2.0, whereby new applications can be easily built out of old ones, is just more proof that the Internet excels at generating affordances. There is nothing about it suggesting that all such affordances would be conducive to democratization. Each of them has to be evaluated on its own terms, not lumped under some mythical "tool neutrality." Instead, we should be closely examining which of the newly created affordances are likely to have democracy-enhancing qualities and which are likely to have democracy-suppressing qualities. Only then will we be able to know which affordances we need to support and which ones we need to counter.

It's inevitable that in many contexts, some of the affordances of the Web, like the ability to remain anonymous while posting sensitive information, could be interpreted both ways, for example, positively as a means of avoiding government censorship but also negatively as a means of producing effective propaganda or launching cyber-attacks. There will never be an easy solution to such predicaments. But then this is also the kind of complex issue that, instead of being glossed over or assumed to be immutable, should be addressed by democratic deliberation. Democracies run into such issues all the time. What seems undeniable, however, is that refusing to even think in terms of affordances and positing "tool neutrality" instead is not a particularly effective way to rein in some of technology's excesses.

chapter eleven

The Wicked Fix

● ● ■ ■

In 1966 the *University of Chicago Magazine* published a brief but extremely provocative essay by Alvin Weinberg, a prominent physicist and head of Oak Ridge National Laboratory, once an important part of the Manhattan Project. Titled "Can Technology Replace Social Engineering?" the essay, best described as an engineer's cri de coeur, argued that "profound and infinitely complicated social problems" can be circumvented and reduced to simpler technological problems. The latter, in turn, can be solved by applying "quick technological fixes" to them, fixes that are "within the grasp of modern technology, and which would either eliminate the original social problem without requiring a change in the individual's social attitudes, or would so alter the problem as to make its resolution more feasible."

One of the reasons why the essay received so much attention was because Weinberg's ultimate technological fix—the one that could end all wars—was the hydrogen bomb. As it "greatly increases the provocation that would precipitate large-scale war," he argued, the Soviets would recognize its destructive power and hold considerably less militarist attitudes as a result. This was an interesting argument to make in 1966, and the essay still has relevance today. Weinberg's fascination with "technological fixes" was largely the product of an engineer's frustration with the other, invariably less tractable, and more controversial alternative of the day: social engineering. Social engineers, as opposed to technologists, tried to influence popular attitudes and social behavior of citizens through what nontechnologists refer to as "policy" but what Weinberg described as "social devices": education, regulation, and a complicated mix of behavioral incentives.

Given that technology could help accomplish the same objectives more effectively, Weinberg believed that social engineering was too expensive and risky. It also helped that "technological fixes" required no profound changes in human behavior and were thus more reliable. If people are given to bouts of excessive drinking, Weinberg's preferred response would be not to organize a public campaign to caution them to drink responsibly or impose heavier fines for drunk driving but to design a pill that would help to dampen the influence of the alcohol. Human nature was corrupt, and Weinberg's solution was to simply accept this and work around it. Weinberg was under no illusion that he was eliminating the root causes of the problem; he knew that technological fixes can't do that. All technology could do was to mitigate the social consequences of that problem, "to provide the social engineer broader options, to make intractable social problems less intractable . . . and [to] buy time—that precious commodity that converts social revolution into acceptable social evolution." It was a pragmatic approach of a pragmatic man.

Upon publication, Weinberg's essay launched a heated debate between technologists and social engineers. This debate is still raging today, in part because Google, founded by a duo of extremely ambitious engineers on a crusade to "organize the world's information and make it universally accessible and useful," has put the production of technological fixes on something of an industrial scale. Make the world's knowledge available to everyone? Take photos of all streets in the world? How about feeding the world's books into a scanner and dealing with the consequences later? Name a problem that has to deal with information, and Google is already on top of it.

Why the Ultimate Technological Fix Is Online

It's not all Google's fault. There is something about the Internet and its do-it-yourself ethos that invites an endless production of quick fixes, bringing to mind the mathematician John von Neumann's insightful observation that "technological possibilities are irresistible to man. If man can go to the moon, he will. If he can control the climate, he will" (even though on that last point, von Neumann may have been a bit off). With the Internet, it seems, everything is irresistible, if only because everything is within easy grasp. It's the Internet, not nuclear power, that is widely seen as the ultimate technological fix to all of humanity's problems. It won't solve them, but it could make them less visible or less painful.

As the Internet makes technological fixes cheaper, the temptation to apply them even more aggressively and indiscriminately also grows. And the easier it is to implement them, the harder it is for internal critics to argue that such fixes should not be tried at all. In most organizations, low cost—and especially in times of profound technological change— is usually a strong enough reason to try something, even if it makes little strategic sense at the time. When technology promises so much and demands so little, the urge to find a quick fix is, indeed, irresistible. Policymakers are not immune to such temptations either. When it's so easy and cheap to start a social networking site for activists in some authoritarian country, a common gut reaction is usually "It should be done." That cramming personal details of all dissidents on one website and revealing connections among them may outweigh the benefits of providing activists with a cheaper mode of communication only becomes a concern retroactively. In most cases, if it can be done, it will be done. URLs will be bought, sites will be set up, activists will be imprisoned, and damning press releases will be issued. Likewise, given the undeniable mobilization advantages of the mobile phone, one may start singing its praises before realizing that it has also provided the secret police with a unique way to track and even predict where the protests may break out.

The problem with most technological fixes is that they come with costs unknown even to their fiercest advocates. Historian of science

Lisa Rosner argues that "technological fixes, because they attack symptoms but don't root out causes, have unforeseen and deleterious side effects that may be worse than the social problem they were intended to solve." It's hard to disagree, even more so in the case of the Internet. When digital activism is presented as the new platform for campaigning and organizing, one begins to wonder whether its side effects—further disengagement between traditional oppositional forces who practice real politics, no matter how risky and boring, and the younger generation, passionate about campaigning on Facebook and Twitter—would outweigh the benefits of cheaper and leaner communications. If the hidden costs of digital activism include the loss of coherence, morality, or even sustainability of the opposition movement, it may not be a solution worth pursuing.

Another problem with technological fixes is that they usually rely on extremely sophisticated solutions that cannot be easily understood by laypeople. The claims of their advocates are, thus, almost impenetrable to external scrutiny, while their ambitious promise—the elimination of some deeply entrenched social ill—makes such scrutiny, even if it is possible, hard to mount. Not surprisingly, the dangerous fascination with solving previously intractable social problems with the help of technology allows vested interests to disguise what essentially amounts to advertising for their commercial products in the language of freedom and liberation. It's not by coincidence that those who are most vocal in proclaiming that the most burning problems of Internet freedom can be solved by breaking a number of firewalls happen to be the same people who develop and sell the technologies needed to break them. Obviously they have no incentive to point out that one needs to be fighting other, nontechnological problems or to disclose problems with their own technologies. The founders of Haystack rarely bothered to highlight the flaws in their own software—let alone disclose that it was still in the testing stage—and the media never bothered to ask. As the Haystack fiasco so clearly illustrates, even being able to ask the right technological questions requires a good grasp of the sociopolitical context in which a given technology is supposed to be used.

This points to another commonly overlooked problem: Our growing commitment to the instruments we use to implement "technological fixes" for what may be important global problems greatly restrains our ability to criticize those who own the rights to those fixes. Every new article or book about a Twitter Revolution is not a triumph of humanity; it is a triumph of Twitter's marketing department. In fact, Silicon Valley's marketing geniuses may have a strong interest in misleading the public about the similarity between the Cold War and today: The Voice of America and Radio Free Europe still enjoy a lot of goodwill with policymakers, and having Twitter and Facebook be seen as their digital equivalents doesn't hurt their publicity.

What We Talk About When We Talk About Code

Perhaps most disturbingly, reframing social problems as a series of technological problems distracts policymakers from tackling problems that are nontechnological in nature and cannot be reframed. As the media keep trumping the role that mobile phones have played in fueling economic growth in Africa, policymakers cannot afford to forget that innovation by itself will not rid African nations of the culture of pervasive corruption. Such an achievement will require a great deal of political will. In its absence, even the fanciest technology would go to waste. The funds for the computerization of Sudan would remain unspent, and computers would remain untouched, as long as many of the region's politicians are "more used to carrying AK-47s and staging ambushes than typing on laptops," as a writer for the *Financial Times* so aptly put it.

On the contrary, when we introduce a multipurpose technology like a mobile phone into such settings, it can often have side effects that only aggravate existing social problems. Who could have predicted that, learning of the multiple money transfer opportunities offered by mobile banking, corrupt Kenyan police officers would demand that drivers now pay their bribes with much-easier-to-conceal transfers of air time rather than cash? In the absence of strong political and social institutions, technology may only precipitate the collapse of state power, but it is easy to lose sight of real-world dynamics when one is so enthralled by the supposed brilliance of a technological fix. Otherwise policymakers risk falling into unthinking admiration of technology as panacea, which the British architect Cedric Price once ridiculed by pondering, "Technology is the answer, but what was the question?"

When technological fixes fail, their proponents are usually quick to suggest another, more effective technological fix as a remedy—and fight fire with fire. That is, they want to fight technology's problems with even more technology. This explains why we fight climate change by driving cars that are more fuel-efficient and protect ourselves from Internet surveillance by relying on tools that encrypt our messages and conceal our identity. Often this only aggravates the situation, as it precludes a more rational and comprehensive discussion about the root causes of a problem, pushing us to deal with highly visible and inconsequential symptoms that can be cured on the cheap instead. This creates a never-ending and extremely expensive cat-and-mouse game in which, as the problem gets worse, the public is forced to fund even newer, more powerful tools to address it. Thus we avoid the search for a more effective nontechnological solution that, while being more expensive (politically or financially) in the short-term, could end the problem once and for all. We should resist this temptation to fix technology's excesses by applying even more technology to them.

How, for example, do most Western governments and foundations choose to fight Internet censorship by authoritarian governments? Usually by funding and promoting technology that helps circumvent it. This may be an appropriate solution for some countries—think, for example, of North Korea, where Western governments have very little diplomatic and political leverage—but this is not necessarily the best approach to handle countries that are nominally Western allies.

In such cases, a nearly exclusive focus on fighting censorship with anticensorship tools distracts policymakers from addressing the root causes of censorship, which most often have to do with excessive restrictions that oppressive governments place on free speech. The easy availability of circumvention technology should not preclude policymakers from more ambitious—and ultimately more effective—ways of engagement. Otherwise, both Western and authoritarian governments get a free pass. Democratic leaders pretend that they are once again heroically destroying the Berlin Wall, while their authoritarian counterparts are happy to play along, for they have found other effective ways to control the Internet.

In an ideal world, the Western campaign to end Internet censorship in Tunisia or Kazakhstan would primarily revolve around exerting political pressure on their West-friendly authoritarian rulers and would deal with the offline world of newspapers and magazines as well. In many of these countries, muzzling journalists would continue to be the dominant tactic of suppressing dissent until, at least, more of their citizens get online and start using it for more activities than just using email or chatting with their relatives abroad. Allowing a handful of bloggers in Tajikistan to circumvent the government's system of Internet controls means little when the vast majority of the population get their news from radio and television.

Except for his ruminations about hydrogen bombs and war, Weinberg did not discuss how technological fixes might affect foreign policy. Nevertheless, one can still trace how a tendency to frame foreign policy problems in terms of technological fixes has affected Western thinking about authoritarian rule and the role that the Internet can play in undermining it. One of the most peculiar features of Weinberg's argument was his belief that the easy availability of clear-cut technological solutions can help policymakers better grasp and identify the problems they face. "The [social] problems are, in a way, harder to identify just because their solutions are never clear-cut," wrote Weinberg. "By contrast, the availability of a crisp and beautiful technological solution often helps focus on the problem to which the new technology is the solution."

In other words, just because policymakers have "a crisp and beautiful technological solution" to break through firewalls, they tend to believe that the problem they need to solve is, indeed, that of breaking firewalls, while often this is not the case at all. Similarly, just because the Internet—that ultimate technological fix—can help mobilize people around certain causes, it is tempting to conceptualize the problem in terms of mobilization as well. This is one of those situations in which the unique features of technological fixes prevent policymakers from discovering the multiple hidden dimensions of the challenge, leading them to identify and solve problems that are easily solvable rather than those that require immediate attention.

Many calls to apply technological fixes to complex social problems smack of the promotion of technology for technology's own sake—a technological fetishism of an extreme variety—which policymakers should resist. Otherwise, they run the risk of prescribing their favorite medicine based only on a few common symptoms, without even bothering to offer a diagnosis. But as it is irresponsible to prescribe cough medicine for someone who has cancer, so it is to apply more technology to social and political problems that are not technological in nature.

Taming the Wicked Authoritarianism

The growing supply of technological and even social fixes presupposes that the problem of authoritarianism can be fixed. But what if it is simply an unsolvable problem to begin with? To ask this question is not to suggest that there will always be evil and dictators in the world; rather, it is to question whether, from a policy-planning perspective, one can ever find the right mix of policies and incentives that could be described as a "solution" and could then be applied in completely different environments.

In 1972, Horst Rittel and Melvin Webber, two influential design theorists at the University of California at Berkeley, published an essay with the unpromising title of "Dilemmas in a General Theory of Planning." The essay, which quickly became a seminal text in the theory of planning, argued that, with the passing of the industrial era, the modern planner's traditional focus on *efficiency*—performing specific tasks with low inputs of resources—has been replaced by a focus on *outputs*, entrapping the planner in an almost never-ending ethical investigation of whether the produced outputs were socially desirable. But the growing complexity of modern societies made such investigations difficult to conduct. As planners began to "see social processes as the links tying open systems into large and interconnected networks of systems, such that outputs from one become inputs to others," they were no longer certain of "*where* and *how* [to] intervene even if [they] do happen to know what aims [they] seek." In a sense, the sheer complexity of the modern world has led to planning paralysis, as the very solutions to older problems inevitably create problems of their own. This was a depressing thought.

Nevertheless, Rittel and Webber proposed that instead of glossing over the growing inefficiency of both technological and social fixes, planners—and policymakers more generally—should confront this gloomy reality and acknowledge that no amount of careful planning would resolve many of the problems they were seeking to tackle. To better understand the odds of success, they proposed to distinguish between "wicked" and "tame" problems. Tame or benign problems can be precisely defined, and one can easily tell when such problems have been solved. The solutions may be expensive but are not impossible and, given the right mix of resources, can usually be found. Designing a car that burns less fuel and attempting to accomplish checkmate in five moves in chess are good examples of typical tame problems.

Wicked problems, on the other hand, are more intellectually challenging. They are hard to define—in fact, they cannot be defined until a solution has been found. But they also have no stopping rule, so it's hard to know when that has happened. Furthermore, every wicked problem can be considered a symptom of another, "higher-level" problem and thus should be tackled on the highest possible level, for "if . . . the problem is attacked on too low a level, then success of resolution may result in making things worse, because it may become more difficult to deal with the higher problems."

Solutions to such problems are never true or false, like they are in chess, but rather good or bad. As such, there could never be a single "best" solution to a wicked problem, as "goodness" is too contentious of a term to satisfy everyone. Worse, there is no immediate or ultimate test for the effectiveness of such solutions, as their side effects may take time to surface. In addition, any such solution is also a one-shot operation. Since there is no opportunity to learn by trial and error, every trial counts. Unlike a lost chess game, which is seldom consequential for other games or non–chess-players, a failed solution to a wicked problem has long-term and largely unpredictable implications far beyond its original context. Every solution, as the authors put it, "leaves traces that cannot be undone."

The essay contained more than a taxonomy of various planning problems. It also contained a valuable moral prescription: Rittel and Webber thought that the task of the planner was not to abandon the fight in disillusionment but to acknowledge its challenges and find ways to distinguish between tame and wicked problems, not least because it was "morally objectionable for the planner to treat a wicked problem as though it were a tame one." They argued that the planner, unlike the scientist, has no right to be wrong: "In the world of planning . . . the aim is not to find the truth, but to improve some characteristic of the world where people live. Planners are liable for the consequences of the actions they generate." It's a formidable moral imperative.

Even though Rittel and Webber wrote the essay with highly technical domestic policies in mind, anyone concerned with the future of democracy promotion and foreign policy in general would do well to heed their advice. Modern authoritarianism, by its very constitution, is a wicked, not a tame, problem. It cannot be "solved" or "engineered away" by a few lines of genius computer code or a stunning iPhone app. The greatest obstacle that Internet-centric initiatives like Internet freedom pose to this fight is that they misrepresent uber-wicked problems as tame ones. They thus allow policymakers to forget that the very act of choosing one solution over another is pregnant with political repercussions; it is not a mere chess game they are playing. But while it is hard to deny that wicked problems defy easy solutions, it doesn't mean that some solutions wouldn't be more effective (or at least less destructive) than others.

From this perspective, a "war on authoritarianism"—or its younger digital sibling, a "war for Internet freedom"—is as misguided as a "war on terror." Not only does such terminology mask the wicked nature of many problems associated with authoritarianism, concealing a myriad of complex connections between them, it suggests—falsely—that such a war can be won if only enough resources are mobilized. Such aggrandizement is of little help to a policy planner, who instead should be trying to grasp how exactly particular wicked problems relate to their context and what may be done to isolate and tackle them while controlling for side effects. The overall push, thus, is away from the grandiose and the rhetorical—qualities inherent in highly ambiguous terms like "Internet freedom"—and toward the miniscule and the concrete.

Assuming that wicked problems lumped under the banner of Internet freedom could be reduced to tame ones won't help either. Western policymakers can certainly work to undermine the information trinity of authoritarianism—propaganda, censorship, and surveillance—but they should not lose sight of the fact that all of them are so tightly interrelated that by fighting one pillar, they may end up strengthening the other two. And even their perception of this trinity may simply be a product of their own cognitive limitations, with their minds portraying the pillars they *can* fight rather than the pillars they *should* fight.

Furthermore, it's highly doubtful that wicked problems can ever be resolved on a global scale; some local accomplishments—preferably not only of the rhetorical variety—is all a policymaker can hope for. To build on the famous distinction drawn by the Austrian philosopher Karl Popper, policymakers should not, as a general rule, preoccupy themselves with utopian social engineering—ambitious, ambiguous, and often highly abstract attempts to remake the world according to some grand plan—but rather settle for piecemeal social engineering instead. This approach might be less ambitious but often more effective; by operating on a smaller scale, policymakers can still stay aware of the complexity of the real world and can better anticipate and mitigate the unintended consequences.

Prophecies Versus Profits

Technological fetishism and a constant demand for technological fixes inevitably breed demand for technological expertise. Technological experts, as clever as they may be on matters concerning technology, are rarely familiar with the complex social and political context in which the solutions they propose are to be implemented.

Nevertheless, whenever nontechnological problems are viewed through the lens of technology, it's technological experts who get the last word. They design solutions that are often more complex than the problems they were trying to solve, while their effectiveness is often impossible to evaluate, as multiple solutions are being tried at once and their individual contributions are often hard to verify. Even the experts themselves have no full control over those technologies, for they trigger effects that could not have been anticipated. Still, this doesn't prevent the inventors from claiming their technologies behave according to a plan. It is hard to disagree with John Searle, an American philosopher at the University of California at Berkeley, when he writes that "the two worst things that experts can do when explaining . . . technology to the general public are first to give the readers the impression that they understand something they do not understand, and second to give the impression that a theory has been established as true when it has not."

Chances are that the technological visionaries we count on to guide us into a brighter digital future may excel at solving the wrong kind of problems. Their proposed solutions are technological by definition, for it's only by touting the benefits of technology that these visionaries have become publicly essential (or as the writer Chuck Klosterman poignantly remarked, "the degree to which anyone values the Internet is proportional to how valuable the Internet makes that person"). Since the only hammer such visionaries have is the Internet, it's not surprising that every possible social and political problem is presented as an online nail.

Thus, most digital visionaries see the Web as a Swiss army knife ready for any job at hand. They rarely alert us to the information black holes created by the Internet, from the sprawling surveillance apparatus facilitated by the public nature of social networking to the persistence of myth making and propaganda, which is much easier to produce and distribute in a world where every fringe movement blogs, tweets, and Facebooks. The very existence of such black holes suggests that we may not always be able to shape the effects of the Internet as we would like.

The political philosopher Langdon Winner was right when he observed in 1986 that "the sheer dynamism of technical and economic activity in the computer industry evidently leaves its members little time to ponder the historical significance of their own activity." Winner could not foresee that the situation would only get worse in the era of the Internet, now that the perpetual revolution it has unleashed has shortened the time and space left for analytical thinking. Nevertheless, Winner's conclusion—that "don't ask; don't tell" is "the unspoken motto for today's technological visionaries"—still rings true today. Their technological fetishism combined with a strong penchant for populism—perhaps just a way of making the "little guys" in their fan base, now armed with iPhones and iPads, feel important—prevents most Internet gurus from asking uncomfortable questions about the social and political effects of the Internet. And why would they ask those questions if they might reveal that they, too, have little control over the situation? It's for this reason that the kind of future predicted by such gurus—and they do need to predict some plausible future to argue that their "fix" would actually work—is rarely reflective of the past.

The technologists, especially technology visionaries who invariably pop up to explain technology to the wider public, "largely extrapolate from today or tomorrow while showing painfully limited interest in the past," as Howard Segal, another historian of technology, once mused. This, perhaps, explains the inevitable barrage of utopian claims every time a new invention comes along. After all, it's not historians of technology but futurists—those who prefer to fantasize about the bright but unknowable future rather than confront the dark but knowable past—that make the most outrageous claims about the fundamental, world-transforming significance of any new technology, especially if it is already on its way to making the cover of *Time* magazine.

As a result, excessive optimism about what technology has to offer, bordering at times on irrational exuberance, overwhelms even those with superior knowledge of history, society, and politics. For better or worse, many such people don't have the resources (and time) for studying how every new iPhone app contributes to the progress of civilization and are thus in desperate need of expert judgment on how technology really transforms the world. It's thanks to their overblown claims about yet another digital revolution that so many Internet gurus end up advising those in positions of power, compromising their own intellectual integrity and ensuring the presence of Internet-centrism in policy planning for decades to come.

Hannah Arendt, one of America's most treasured public intellectuals, was aware of this problem back in the 1960s, when the "scientifically minded brain trusters"—Alvin Weinberg was just one of many; another whiz kid with a penchant for computer modeling, Robert McNamara, was put in charge of the Vietnam War—were beginning to penetrate the corridors of power and influence government policy. "The trouble [with such advisers] is not that they are cold-blooded enough to 'think the unthinkable,'" cautioned Arendt in "On Violence," "but that they do not 'think.'" "Instead of indulging in such an old-fashioned, uncomputerizable activity," she wrote, "they reckon with the consequences of certain hypothetically assumed constellations without, however, being able to test their hypothesis against actual occurrences." A cursory glimpse at the overblown and completely unsubstantiated rhetoric that followed Iran's Twitter Revolution is enough to assure us that not much has changed.

It was more than just the constant glorification of technical, largely quantitative expertise at the expense of erudition that bothered Arendt. She feared that increased reliance on half-baked predictions uttered by self-interested technological visionaries and the futuristic theories they churn out on an hourly basis would prevent policymakers from facing the highly political nature of the choices in front of them. Arendt worried that "because of their inner consistency . . . [such theories] have a hypnotic effect; they put to sleep our common sense." The ultimate irony of the modern world, which is more dependent on technology than ever, is that, as technology becomes ever more integrated into political and social life, less and less attention is paid to the social and political dimensions of technology itself. Policymakers should resist any effort to take politics out of technology; they simply cannot afford to surrender to the kind of apolitical hypnosis that Arendt feared. The Internet is too important a force to be treated lightly or to be outsourced to know-all consultants. One may not be able to predict its impact on a particular country or social situation, but it would be foolish to deny that some impact is inevitable. Understanding how exactly various stakeholders—citizens, policymakers, foundations, journalists—can influence the way in which technology's political future unfolds is a quintessential question facing any democracy.

More than just politics lies beyond the scope of technological analysis; human nature is also outside its grasp. Proclaiming that societies have entered a new age and embraced a new economy does not automatically make human nature any more malleable, nor does it necessarily lead to universal respect for humanist values. People still lust for power and recognition, regardless of whether they accumulate it by running for office or collecting Facebook friends. As James Carey, the Columbia University media scholar, put it: "The 'new' man and woman of the 'new age' strikes one as the same mixture of greed, pride, arrogance and hostility that we encounter in both history and experience." Technology changes all the time; human nature hardly ever.

The fact that do-gooders usually mean well does not mitigate the disastrous consequences that follow from their inability (or just sheer lack of ambition) to engage with broader social and political dimensions of technology. As the German psychologist Dietrich Dörner observed in *The Logic of Failure*, his masterful account of how decision-makers' ingrained psychological biases could aggravate existing problems and blind them to the far more detrimental consequences of proposed solutions, "it's far from clear whether 'good intentions plus stupidity' or 'evil intentions plus intelligence' have wrought more harm in the world." In reality, the fact that we mean well should only give us extra reasons for scrupulous self-retrospection, for, according to Dörner, "incompetent people with good intentions rarely suffer the qualms of conscience that sometimes inhibit the doings of competent people with bad intentions."

After Utopia: The Cyber-Realist Manifesto

A few months after Hillary Clinton's speech on Internet freedom, Ethan Zuckerman, a senior researcher at Harvard University's Berkman Center for Internet and Society and a widely respected expert on Internet censorship, penned a poignant essay titled "Internet Freedom: Beyond Circumvention," one of the first serious attempts to grapple with the policy implications of Washington's new favorite buzzword. In it, Zuckerman made an important argument that building tools to break through authoritarian firewalls wouldn't be enough, because there are too many Internet users in China to make it affordable and too many nontechnological barriers to freedom of expression on the Web. "We can't circumvent our way around censorship. . . . The danger in heeding Secretary Clinton's call is that we increase our speed, marching in the wrong direction," he wrote.

His own contribution to the debate was to elucidate several theories that may help policymakers better understand how the Internet can nudge authoritarian societies toward democratization. "To figure out how to promote internet freedom, I believe we need to start addressing the question: 'How do we think the Internet changes closed societies?'" wrote Zuckerman. He listed three good potential answers. One such theory states that providing access to suppressed information may eventually push people to change opinion of their governments, precipitating a revolution. Another one posits that if citizens have access to various social networking sites and communication tools like Skype, they are able to better plan and organize their antigovernment activity. A third theory predicts that by providing a rhetorical space where different ideas can be debated, the Internet will gradually empower a new generation of leaders with a more modern set of demands.

As Zuckerman correctly points out, all of these theories have some intellectual merit. The additional assumptions that he makes, either explicitly or implicitly, is that the American government has a separate pot of money to spend on Internet freedom issues; that most of this money would invariably go to fund technological rather than political solutions; and that the best thing to do is to prioritize which tools are needed the most. Zuckerman's suggestion, then, is that policymakers first need to figure out which theory is to guide their efforts in online space and then rely on it to allocate their resources. Thus, if they expect to enact change by mobilizing citizens to rise up against their governments, they need to ensure that tools like Twitter and Facebook are widely available and resistant to both attempts to block access to them and DDoS attacks. In contrast, if they stick to the "liberated by facts" theory, they would need to prioritize access to blogs of the opposition as well as websites like Wikipedia, BBC News, and so forth.

Instead of formulating a better theory to complement Zuckerman's, one needs to ponder what breeds demand for such theories in the first place. While it is hard to disagree with his warning that, in their pursuit of Internet freedom nirvana, policymakers may be speeding up in the wrong direction, Zuckerman's neo-Weinbergian philosophy of action seems much more ambiguous. It is founded on a belief that once policymakers understand the "logic" of the Internet, which in Zuckerman's interpretation, inherently favors those challenging autocracy and power but in ways that we may not yet understand, they will be able to formulate smarter Internet policies and can then pursue a host of technological solutions to accomplish the objectives of those policies. Thus, from Zuckerman's perspective, it's important to articulate numerous theories by which the Internet may be transforming autocracies and then act on those that best match the empirical reality.

In the meantime, the mental gymnastics of proposing and evaluating theories may also add meaning to the term "Internet freedom," which even Zuckerman acknowledges to be currently empty. It's this last point that is most troubling: Even though Zuckerman agrees that Internet freedom offers a poor foundation for effective foreign policy, he is nevertheless eager to propose—somewhat cynically—all sorts of fixes to make this foundation last for a year or two longer than it might otherwise. Unfortunately, those rare intellectuals who do know a great deal about both the Internet and the rest of the world—Zuckerman is also an Africa expert—prefer to spend their time seeking marginal improvements to wrong-headed policies, unable or unwilling to see through the pernicious Internet-centrism that permeates them and to reject their very foundation. (The situation is certainly not helped by the fact that the State Department funds some of Zuckerman's projects at Harvard, as he himself acknowledged in the essay.)

But an even greater problem with Zuckerman's approach is that, should the "logic" of the Internet defy his expectations and prove elusive, nonexistent, or inherently antidemocratic, the rest of the proposed course of action also falls apart and is at best irrelevant and at worst deceptive. That the Internet may also be strengthening rather than undermining authoritarian regimes; that placing it at the cornerstone of foreign policy helps Internet companies deflect the criticism they so justly deserve; that a dedication to the highly abstract goal of promoting Internet freedom complicates a thorough assessment of other parts of foreign and domestic policies—these are not the kind of insights one is likely to gain while groping for a theory to justify one's own penchant for cyber-utopianism or Internet-centrism. As a result, many of these concerns barely register when future policies are being crafted.

The way forward is not to keep coming up with new theories until they match one's existing biases about what the logic of the Internet is or should be like. Instead, one should seek to come up with a philosophy of action to help design policies that have no need for such logic as their inputs. But while it's becoming apparent that policymakers need to abandon both cyber-utopianism and Internet-centrism, if only for the lack of accomplishment, it is not yet clear what can take their place. What would an alternative, more down-to-earth approach to policymaking in the digital age—let's call it cyber-realism—look like? Here are some preliminary notes that future theorists may find useful.

Instead of trying to build a new shiny pillar to foreign policy, cyberrealists would struggle to find space for the Internet in existing pillars, not least on the desks of regional officers who are already highly sensitive to the political context in which they operate. Instead of centralizing decision making about the Internet in the hands of a select few digerati who know the world of Web 2.0 start-ups but are completely lost in the world of Chinese or Iranian politics, cyber-realists would defy any such attempts at centralization, placing as much responsibility for Internet policy on the shoulders of those who are tasked with crafting and executing regional policy.

Instead of asking the highly general, abstract, and timeless question of "How do we think the Internet changes closed societies?" they would ask "How do we think the Internet is affecting our existing policies on country X?" Instead of operating in the realm of the utopian and the ahistorical, impervious to the ways in which developments in domestic and foreign policies intersect, cyber-realists would be constantly searching for highly sensitive points of interaction between the two. They would be able to articulate in concrete rather than abstract terms how specific domestic policies might impede objectives on the foreign policy front. Nor would they have much tolerance for a black-and-white color scheme. As such, while they would understand the limitations of doing politics online, they wouldn't label all Internet activism as either useful or harmful based solely on its outputs, its inputs, or its objectives. Instead, they would evaluate the desirability of promoting such activism in accordance with their existing policy objectives.

Cyber-realists wouldn't search for technological solutions to problems that are political in nature, and they wouldn't pretend that such solutions are even possible. Nor would they give the false impression that on the Internet concerns over freedom of expression trump those over energy supplies, when this is clearly not the case. Such acknowledgments would only be factual rather than normative statements—it may well be that concerns over freedom of expression *should* be more important than concerns over energy supplies—but cyber-realists simply would not accept that any such radical shifts in the value system of the entire policy apparatus could or should happen under the pressure of the Internet alone.

Now would cyber-realists search for a silver bullet that could destroy authoritarianism—or even the next-to-silver-bullet, for the utopian dreams that such a bullet can even exist would have no place in their conception of politics. Instead, cyber-realists would focus on optimizing their own decision-making and learning processes, hoping that the right mix of bureaucratic checks and balances, combined with the appropriate incentive structure, would identify wicked problems before they are misdiagnosed as tame ones, as well as reveal how a particular solution to an Internet problem might disrupt solutions to other, non-Internet problems.

Most important, cyber-realists wouldn't allow themselves to get dragged into the highly abstract and high-pitched debates about whether the Internet undermines or strengthens democracy. Instead, they would accept that the Internet is poised to produce different policy outcomes in different environments and that a policymaker's chief objective is not to produce a thorough philosophical account of the Internet's impact on society at large but, rather, to make the Internet an ally in achieving specific policy objectives.

Cyber-realists would acknowledge that by continuing to flirt with Internet-centrism and cyber-utopianism, policymakers are playing a risky game. Not only do they squander plenty of small-scale opportunities for democratization that the Internet has to offer because they look from too distant a perspective, but they also inadvertently embolden dictators and turn everyone who uses the Internet in authoritarian states into unwilling prisoners. Cyber-realists would argue that this is a terribly expensive and ineffective way to promote democracy; worse, it threatens to corrupt or crowd out cheaper and more effective alternatives. For them, the promotion of democracy would be too important an activity to run it out of a Silicon Valley lab with a reputation for exotic experiments. Above all, cyber-realists would believe that a world made of bytes may defy the law of gravity but absolutely nothing dictates that it should also defy the law of reason.

ACKNOWLEDGMENTS

I won't be overstating the case if I say that this book would not have been possible without the generous support—moral, intellectual, financial—of the Open Society Foundations. I was lucky to get an OSF scholarship very early on (I was still in a Belarusian high school) that allowed me to pursue my undergraduate studies abroad. Were it not for this scholarship, I may have well ended up on the wrong side of the digital barricades.

Later, some of my work at Transitions Online was also funded by grants from OSF's Information Program. Even if some of our projects at Transitions were somewhat cyber-utopian in spirit, OSF was always the funder eager to take the most risks and take the most unconventional approaches. That's more than an NGO can usually wish for!

After leaving Transitions, I was lucky to be invited to join the board of OSF's Information Program, which gave me a great opportunity to think about the numerous political and social implications of the Internet from a philanthropic perspective. There is hardly a better gig in the world to see just how important and political technology is—and how getting it right matters.

Most important, being one of the first recipients of an Open Society fellowship, I tremendously benefited from the necessary support and flexibility to conduct much of the research for this book. A list of all my friends at the extended OSF family is too long to include here, but I'd like to thank Darius Cuplinskas, Janet Haven, Stephen Hubbell, Sasha Post, Bipasha Ray, Istvan Rev, Anthony Richter, Laura Silber, Ethan Zuckerman, and especially Leonard Benardo for all their help as well as their willingness to tolerate my contrarian streak.

Between September 2009 and May 2010 I spent a wonderful academic year at Georgetown University's Institute for the Study for Diplomacy. This was a charming experience, not least because Charles Dolgas, Paula Newberg, and Jim Seevers helped to make it so by providing me with a superb intellectual environment to work in. I'd also like to thank Tony Arendt for letting me test some of my ideas on Georgetown graduate students in my class about the Internet and democracy.

My students deserve a dedicated thanks of their own. They were a very open-minded bunch who helped to challenge my arguments in every possible way. In addition, Chanan Weissman and Tracy Huang provided me with excellent research assistance. I'd also like to thank the folks at Yahoo's Business and Human Rights Program, who made my stay at Georgetown possible. I've been pleasantly surprised by their genuine dedication to intellectual inquiry into the politics of the Internet, even if that inquiry often clashed with their own interests.

I'd also like to thank Joshua Cohen for first inviting me to write for *Boston Review*, where I published several essays about the Internet, and then for extending an invitation to spend the 2010–2011 year at Stanford. Andres Martinez and Steve Coll at the New American Foundation were also kind enough not only to award me a fellowship for the same period but to let me spend the bulk of my fellowship time at Stanford, an extremely flexible arrangement that I greatly appreciate.

Dozens of people have helped me to hone and package my ideas on stage and in the media. Of those, I'd particularly like to thank the *Foreign Policy* team—Susan Glasser, Blake Hounshell, Joshua Keating, and Moises Naim—who gave me a chance to test my ideas both in their pages and in a dedicated technology blog. I have many other editors to thank: Ryan Sager at the *Wall Street Journal*, David Goodhart and James Crabtree at *Prospect*, and Michael Walzer at *Dissent*. Without their guidance and encouragement, it would have taken much longer to get my argument out into the open. The TED Conference, where I had the privilege to speak in 2009, helped to spread my ideas even further. I'd like to thank Chris Anderson, June Cohen, Bruno Giussani, Logan McClure, and Tom Rielly for all their assistance on this front.

My editors at PublicAffairs, Niki Papadopoulos and Lindsay Jones, have been a true delight to work with. I'd also like to thank Peter Osnos, Clive Priddle, and Susan Weinberg for taking on this project and giving me plenty of autonomy to shape its outcome. Max and John Brockman have done as superb a job as literary agents could do.

Despite the fact that they don't fully grasp what it is that I do, my family back in Belarus have all been very supportive of my intellectual quest. Even though my current vocation hardly meets their definition of "a real job," my parents and my sister have shown tremendous flexibility in letting me pursue the life of the mind unburdened by the grim Belarusian reality. This book would have been impossible without their support and encouragement.

Above all, I'd like to thank Aernout van Lynden, who inspired me to write, taught me how to ask challenging questions, and who, as a distinguished war correspondent, showed me what courage and decency look like. I've had the privilege to study under Aernout as well as to discuss politics over a glass of wine with him; his influence on my own intellectual development has been immense, and it is with utmost pleasure that I dedicate this book to him.

SAN FRANCISCO,
AUGUST 30, 2010

BIBLIOGRAPHY

CHAPTER 1

Abadi, Cameron. "Iran, Facebook, and the Limits of Online Activism." *Foreign Policy*, February 12, 2010.

Alexiou, Philip. "A 'Twitter Moment' in Politics?" *Voice of America*, July 29, 2010.

"A Look at Twitter in Iran." *Sysomos Blog*, June 21, 2009. blog.sysomos.com/2009/06/21/a-look-at-twitter-in-iran/.

Ambinder, Marc. "The Revolution Will Be Twittered." *Marc Ambinder's Politics Blog*, September 15, 2009. www.theatlantic.com/politics/archive/2009/06/the-revolution-will-be-twittered/19376/.

Anderson, Chris. "Q&A with Clay Shirky on Twitter and Iran." *TED Blog*. blog.ted.com/2009/06/qa_with_clay_sh.php.

Athanasiadis, Iason. "Iran Uses Internet as Tool Against Protesters." *Christian Science Monitor*, January 4, 2010.

Bajkowski, Julian. "Al Jazeera Offers Reality Check for the Twitterverse." *MIS Australia.com*, February 22, 2010. www.misaustralia.com/viewer.aspx?EDP://1266801386432.

Berkeley, B. "Bloggers vs. Mullahs: How the Internet Roils Iran." *World Policy Journal* 23, no. 1 (2006): 71.

"Besporjadki V Irane Shli Po Moldavskomu Scenariju—SShA Propalilis.'" *Evrazia.org*, June 18, 2009. evrazia.org/news/8648.

Best, M. L., and K. W. Wade. "The Internet and Democracy: Global Catalyst or Democratic Dud?" *Bulletin of Science, Technology & Society* 30, no. 3 (June 2009).

Bozorgmehr, Nahmeh. "Tehran Calls for Informants on Protesters." *Financial Times*, January 5, 2010.

"Chat Rooms and Chadors." *Newsweek*, August 21, 1995.

"China Group: Facebook Used to Sow Unrest." Associated Press, July 9, 2010.

"China Military Paper Issues Warning over Twitter, YouTube 'Subversion.'" *BBC Monitoring Media*, August 7, 2009.

"China News Agency Views Twitter's Role in Iran Political Crisis." *BBC Monitoring Media*, June 30, 2009.

"The Clinton Internet Doctrine." *Wall Street Journal*, January 23, 2010.

"CNN Responds to Iranian Hacking Accusation." *CNN.com*, June 22, 2009. edition.cnn.com/2009/WORLD/meast/06/22/cnn.iran.claim/.

"Communicators with Alec Ross." *Communicators* (video). C-SPAN, April 14, 2010. www.c-spanvideo.org/program/293002-1.

"Communicators with Tim Sparapani." *Communicators* (video). C-SPAN, March 8, 2010. www.c-spanvideo.org/program/292422-1.

Dabashi, Hamid. "A Tale of Two Cities." *Al-Ahram Weekly*, August 20, 2009. weekly.ahram.org.eg/2009/961/op51.htm.

Dickie, Mure. "China Traps Online Dissent." *Financial Times*, November 12, 2007.

"Don't Be Evil." *New Republic*, April 21, 2010.

Eltahawy, Mona. "Facebook, YouTube and Twitter Are the New Tools of Protest in the Arab World." *Washington Post*, August 7, 2010.

Esfandiari, Golnaz. "Authorities Warn Iranians Not to Protest—By SMS." *Transmission Blog (RFE/RL)*, November 20, 2009. www.rferl.org/content/Authorities_Warn_Iranians_Not_To_Protest_By_SMS/1883679.html.

———. "Iranian Social Networking, Hard-Line Style." Radio Free Europe / Radio Liberty, July 28, 2010.

———. "The Twitter Devolution." *Foreign Policy*, June 7, 2010.

———. "Why Did Iran Unblock Facebook?" Radio Free Europe / Radio Liberty, March 14, 2009.

Fassihi, Farnaz. "Iranian Crackdown Goes Global." *Wall Street Journal*, December 3, 2009.

"Fax Against Fictions." *Time*, June 19, 1989.

"Fresh Iran Hearing Focuses on 'Internet Plot.'" *Press TV*, September 14, 2009.

Fukuyama, Francis. *The End of History and the Last Man*. New York: Free Press, 2006.

Gapper, John. "Technology Is a Tool for Revolution (and Repression)." *Financial Times*, June 20, 2009.

Gheytanchi, Elham, and Babak Rahimi. "The Politics of Facebook in Iran." *openDemocracy*, June 1, 2009. www.opendemocracy.net/article/email/the -politics-of-facebook-in-iran.

Giroux, Henry A. "The Iranian Uprisings and the Challenge of the New Media: Rethinking the Politics of Representation." *Fast Capitalism* 5, no. 2 (2009).

"Google Calls an End to Business as Usual in China." *Age* (Melbourne), January 18, 2010.

"Gosdep SShA Pozabotilsja Ob Iranskih Studentah." *Novosti NTV*, June 17, 2009. www.ntv.ru/novosti/165016/.

Gross, Doug. "Awards Honor Top 10 Internet Moments of the Decade." *CNN.com*, November 19, 2009. cnn.com/2009/TECH/11/18/top.internet .moments/.

Hornby, Lucy. "China Paper Slams U.S. for Cyber Role in Iran Unrest." Reuters, January 24, 2010.

Houghton, D. P. "The Role of Self-Fulfilling and Self-Negating Prophecies in International Relations." *International Studies Review* 11, no. 3 (2009): 552–584.

"Internet 'in Running' for Nobel Peace Prize." *BBC News*, March 10, 2010.

"Iran Hacked Twitter's Database and Arrested Its Users." *BBC Monitoring Media*, January 17, 2010.

"Iran Police Say Public Help in Arresting Rioters." Reuters, January 19, 2010.

"Iran's Police Vow No Tolerance Towards Protesters." Reuters, February 6, 2010.

Johnson, A. Ross. "A Brief History of RFE/RL." Radio Free Europe / Radio Liberty, December 2008. www.rferl.org/section/history/133.html.

Johnston, Nancy. "Twitter Gets Serious." *Baltimore Sun*, June 20, 2009.

Kalathil, Shanthi, and Taylor C. Boas. *Open Networks, Closed Regimes: The Impact of the Internet on Authoritarian Rule*. Washington, DC: Carnegie Endowment for International Peace, 2003.

Kaminsky, Ross. "Iran's Twitter Revolution." *Human Events*, June 18, 2009.

Keohane, R. O., and J. S. Nye Jr. "Power and Interdependence in the Information Age." *Foreign Affairs* 77, no. 5 (1998): 81–94.

Kerry, John. "Standing up for Internet Freedom." TPMCafe, *Talking Points Memo*, January 21, 2010. tpmcafe.talkingpointsmemo.com/2010/01/21/ standing_up_for_internet_freedom/index.php.

Khiabany, G., and A. Sreberny. "The Politics of/in Blogging in Iran." *Comparative Studies of South Asia, Africa and the Middle East* 27, no. 3 (2007): 563.

Kimmage, Daniel. "Fight Terror with YouTube." *New York Times*, June 26, 2008.

Kirkpatrick, David. *The Facebook Effect: The Inside Story of the Company That Is Connecting the World*. New York: Simon & Schuster, 2010.

Kristof, Nicholas D. "Tear Down This Cyberwall!" *New York Times*, June 17, 2009.

Krugman, Paul. "Understanding Globalization." *Washington Monthly*, June 1999.

Lakshmanan, Indira A. R. "Supporting Dissent with Technology." *New York Times*, February 23, 2010.

Landler, Mark. "Google Searches for a Foreign Policy." *New York Times*, March 28, 2010.

Landler, Mark, and Brian Stelter. "Washington Taps into a Potent New Force in Diplomacy." *New York Times*, June 16, 2009.

Last, Jonathan V. "Tweeting While Tehran Burns." *Weekly Standard*, August 17, 2009.

Lee, Tom. "The Cost of Hashtag Revolution." *American Prospect*, June 29, 2009.

Marian, Boris. "Sorry Moldova!" *Nezavisimaya Moldova*, July 3, 2009. www.nm.md/daily/article/2009/07/03/0905.html.

McMillan, Robert. "In Iran, Cyber-Activism Without the Middle-Man." *Computer World*, June 18, 2009.

Meet the Press. Transcript. MSNBC, May 23, 2010.

Mungiu-Pippidi, A., and I. Munteanu. "Moldova's 'Twitter Revolution.'" *Journal of Democracy* 20, no. 3 (2009): 136–142.

Musgrove, Mike. "Twitter Is a Player in Iran's Drama." *Washington Post*, June 17, 2009.

"News Roundup: Hour 2." *The Diane Rehm Show*, WAMU, June 19, 2009.

Pfeifle, Mark. "A Nobel Peace Prize for Twitter?" *Christian Science Monitor*, July 6, 2009.

Quinn, James. "Iran Shuts Down Google Mail." *Daily Telegraph*, February 10, 2010.

"Revolutionbook." *Financial Times*, May 27, 2009.

Rhoads, Christopher. "Activists Skirt Web Crackdown to Reach the Outside World." *Wall Street Journal*, December 8, 2009.

Rutten, Tim. "Tyranny's New Nightmare: Twitter." *Los Angeles Times*, June 24, 2009.

Schleifer, Yigal. "Why Iran's Twitter Revolution Is Unique." *Christian Science Monitor*, June 19, 2009.

Schorr, Daniel. "In Iran, a Struggle over Cyberspace." *All Things Considered*. National Public Radio, June 17, 2009.

Sheridan, Barrett. "The Internet Helps Build Democracies." *Newsweek*, April 30, 2010.

Shirk, S.L. "Changing Media, Changing Foreign Policy in China." *Japanese Journal of Political Science* 8, no. 1 (2007): 43–70.

Sohrabi-Haghighat, M. Hadi, and Shohre Mansouri. "'Where's My Vote?' ICT Politics in the Aftermath of Iran's Presidential Election." *International Journal of Emerging Technologies and Society* 8, no. 1 (2010): 24–41.

Sreberny, A., and G. Khiabany. "Becoming Intellectual: The Blogestan and Public Political Space in the Islamic Republic." *British Journal of Middle Eastern Studies* 34, no. 3 (2007): 267–286.

"State Department Is Taking Right Steps to Foster Internet Freedom." *Washington Post*, July 21, 2010.

Stone, Brad, and Noam Cohen. "Tweeting Their Way to Freedom?" *New York Times*, October 5, 2009.

Sullivan, Andrew. "The Revolution Will Be Twittered." The Daily Dish, *Atlantic*, June 13, 2009. andrewsullivan.theatlantic.com/the_daily_dish/2009/06/the-revolution-will-be-twittered-1.html.

———. "Twitter Maintenance?" The Daily Dish, *Atlantic*, June 15, 2009. andrewsullivan.theatlantic.com/the_daily_dish/2009/06/twitter-maintenance.html.

Tait, Robert. "Iran Moves to Silence Opposition with Internet Crime Unit." *Guardian*, November 15, 2009.

"Tehran Clashes Reported on Iran Vote Anniversary." *BBC News*, June 12, 2010.

Tehrani, Hamid. "Iranian Officials 'Crowd-Source' Protester Identities." *Global Voices*, June 27, 2009. globalvoicesonline.org/2009/06/27/iranian-officials-crowd-source-protester-identities-online/.

Viner, Katharine. "Internet Has Changed Foreign Policy for Ever, Says Gordon Brown." *Guardian*, June 19, 2009.

Weaver, Matthew. "Iran's 'Twitter Revolution' Was Exaggerated, Says Editor." *Guardian*, June 9, 2010.

———. "Oxfordgirl vs Ahmadinejad: the Twitter User Taking on the Iranian Regime." *Guardian*, February 10, 2010.

Webster, George. "Street in Palestinian Refugee Camp Named After Twitter Account." *CNN.com*, October 5, 2009. edition.cnn.com/2009/WORLD/meast/10/01/twitter.street/index.html.

Weisberg, Jacob. "Publishers Should Beware Apple's iPad." *Newsweek*, May 15, 2010.

Winner, Langdon. *Autonomous Technology: Technics-Out-of-Control as a Theme in Political Thought.* Cambridge, MA: MIT Press, 1978.

"*Wired* Nominates the Internet for the Nobel Peace Prize." Press release. Internet for Peace, November 17, 2009. www.internetforpeace.org/media detail.cfm?pressid=1.

Zia-Ebrahimi, Reza. "Bombard Iran . . . with Broadband." *Guardian*, February 24, 2010.

CHAPTER 2

Agenda. Conference on Cyber Dissidents: Global Successes and Challenges. George W. Bush Institute, April 19, 2010.

Albrecht, H., and O. Schlumberger. "'Waiting for Godot': Regime Change Without Democratization in the Middle East." *International Political Science Review / Revue internationale de science politique* 25, no. 4 (2004): 371.

Anderson, Perry. "A Ripple of the Polonaise." *London Review of Books*, November 25, 1999.

Arias-King, F. "Orange People: A Brief History of Transnational Liberation Networks in East Central Europe." *Demokratizatsiya: The Journal of Post-Soviet Democratization* 15, no. 1 (2007): 29–72.

Ascherson, Neal. "The Media Did It." *London Review of Books*, June 21, 2007.
———. "They're Just Not Ready." *London Review of Books*, January 7, 2010.

Barme, Geremie R., and Sang Ye. "The Great Firewall of China." *Wired*, February 1, 1996.

Bennett, A. "The Guns That Didn't Smoke: Ideas and the Soviet Non-Use of Force in 1989." *Journal of Cold War Studies* 7, no. 2 (2005): 81–109.

Bilalic, M., P. McLeod, and F. Gobet. "Why Good Thoughts Block Better Ones: The Mechanism of the Pernicious Einstellung (Set) Effect." *Cognition* 108, no. 3 (2008): 652–661.

Bildt, Carl. "Tear Down These Walls Against Internet Freedom." *Washington Post*, January 25, 2010.

Bollinger, Lee C. "A Free Press for a Global Society." *Chronicle of Higher Education*, February 21, 2010.

Brooks, S. G., and W. C. Wohlforth. "Power, Globalization, and the End of the Cold War: Reevaluating a Landmark Case for Ideas." *International Security* 25, no. 3 (2001): 5–53.

Brown, C. "History Ends, Worlds Collide." *Review of International Studies* 25 (1999): 41–57.

Brownback, Sam. "Twitter Against Tyrants: New Media in Authoritarian Regimes." Commission on Security and Cooperation in Europe, October 22, 2009.

Bunce, V. J., and S. L. Wolchik. "Defeating Dictators: Electoral Change and Stability in Competitive Authoritarian Regimes." *World Politics* 62, no. 1 (2009): 43–86.

Burnell, P. "From Evaluating Democracy Assistance to Appraising Democracy Promotion." *Political Studies* 56, no. 2 (2008): 414–434.

Campbell, J. L. "Institutional Analysis and the Role of Ideas in Political Economy." *Theory and Society* 27, no. 3 (1998): 377–409.

Carothers, T. "The Backlash Against Democracy Promotion." *Foreign Affairs* 85, no. 2 (2006): 55–68.

Centeno, M. A. "Between Rocky Democracies and Hard Markets: Dilemmas of the Double Transition." *Annual Review of Sociology* 20, no. 1 (1994): 125–147.

Chen, C. "Institutional Legitimacy of an Authoritarian State: China in the Mirror of Eastern Europe." *Problems of Post-Communism* 52, no. 4 (2003): 3–13.

Clinton, Hillary. "Remarks on Internet Freedom." The Newseum, Washington, DC, January 21, 2010.

Cohen, B. J. "A Grave Case of Myopia." *International Interactions* 35, no. 4 (2009): 436–444.

Cohen, Roger. "Target Iran's Censors." *International Herald Tribune*, February 18, 2010.

Cox, M. "Why Did We Get the End of the Cold War Wrong?" *British Journal of Politics & International Relations* 11, no. 2 (2009): 161–176.

Critchlow, J. "Public Diplomacy During the Cold War: The Record and Its Implications." *Journal of Cold War Studies* 6, no. 1 (2004): 75–89.

———. "Western Cold War Broadcasting." *Journal of Cold War Studies* 1, no. 3 (1999): 168–175.

Crovitz, Gordon L. "The Internet and Political Freedom." *Wall Street Journal*, March 15, 2010.

Danyi, E. "Xerox Project: Photocopy Machines as a Metaphor for an 'Open Society.'" *Information Society* 22, no. 2 (2006): 111–115.

Ding, X. L. "Institutional Amphibiousness and the Transition from Communism: The Case of China." *British Journal of Political Science* (1994): 293–318.

"Excerpts: Bush to Remain 'Committed' to War on Terror." *Washington Times*, January 11, 2005.

Falk, B. J. "1989 and Post-Cold War Policymaking: Were the "Wrong" Lessons Learned from the Fall of Communism?" *International Journal of Politics, Culture, and Society* 22, no. 3 (2009): 291–313.

Feith, David. "Senate to Hillary: Support Cyber Dissident." *Wall Street Journal*, July 23, 2009.

Fowler, Geoffrey A., and Loretta Chao. "U.S. Urged to Act on Internet Freedoms." *Wall Street Journal*, March 25, 2010.

"Freedom vs. the Firewall; The Senate Can Help Fend Off Authoritarian Censorship." *Washington Post*, July 7, 2009.

Fukuyama, Francis. *Our Posthuman Future: Consequences of the Biotechnology Revolution*. New York: Farrar, Straus and Giroux, 2002.

Fukuyama, F., and M. McFaul. "Should Democracy Be Promoted or Demoted?" *Washington Quarterly* 31, no. 1 (2008): 23–45.

Garton Ash, Timothy. "1989!" *New York Review of Books*, November 5, 2009.

Gedmin, Jeffrey. "Democracy Isn't Just a Tweet Away." *USA Today*, April 22, 2010.

Glassman, James K. "Statement by James K. Glassman on the Conference on Cyber Dissidents." George W. Bush Institute, April 19, 2010.

Goldstone, Jack. "Towards a Fourth Generation of Revolutionary Theory." *Annual Review of Political Science* 4 (2001): 139–187.

Granville, Johanna C. "Radio Free Europe's Impact on the Kremlin in the Hungarian Crisis of 1956: Three Hypotheses." *Canadian Journal of History* 39, no. 3 (2004): 515–546.

Grodsky, B. "Lessons (Not) Learned: A New Look at Bureaucratic Politics and US Foreign Policy-Making in the Post-Soviet Space." *Problems of Post-Communism* 56, no. 2 (2009): 43–57.

Hachten, W. "The Triumph of Western News Communication." *Fletcher Forum of World Affairs* 17 (1993): 17.

Henderson, Scott. "Patriotic Chinese Hackers Attack Website of Melamine Poisoned Children." *Dark Visitor*, January 23, 2009. www.thedarkvisitor.com/2009/01/patriotic-chinese-hackers-attack-website-of-melamine-poisoned-children/.

Hokenos, Paul. "Past Forward." *Boston Review,* March 2010.

Jacoby, Jeff. "Despite Forecasts, Freedom Takes More Than Technology." *Boston Globe*, April 25, 2010.

———. "Medium Isn't the Message." *Boston Globe*, April 28, 2010.

Jervis, R. "Bridges, Barriers, and Gaps: Research and Policy." *Political Psychology* 29, no. 4 (2008): 571–592.

———. "Understanding Beliefs." *Political Psychology* 27, no. 5 (2006): 641–663.

Judt, Tony. "A Story Still to Be Told." *New York Review of Books*, March 23, 2006.

Kahneman, D., and G. Klein. "Conditions for Intuitive Expertise: A Failure to Disagree." *American Psychologist* 64, no. 6 (2009): 515–526.

Kalandadze, K., and M. A. Orenstein. "Electoral Protests and Democratization: Beyond the Color Revolutions." *Comparative Political Studies* 42, no. 11 (2009): 1403.

Kaldor, M. H. "The Ideas of 1989: The Origins of the Concept of Global Civil Society." *Transnational Law & Contemporary Problems* 9 (1999): 475.

Kaminski, M. M. "How Communism Could Have Been Saved: Formal Analysis of Electoral Bargaining in Poland in 1989." *Public Choice* 98, no. 1 (1999): 83–109.

Kegley, C. W., Jr. "How Did the Cold War Die? Principles for an Autopsy." *Mershon International Studies Review* 38, no. 1 (1994): 11–41.

Kopstein, J. "1989 as a Lens for the Communist Past and Post-Communist Future." *Contemporary European History* 18, no. 3 (2009): 289–302.

———. "The Transatlantic Divide over Democracy Promotion." *Washington Quarterly* 29, no. 2 (2006): 85–98.

Kotkin, Stephen, with Jan T. Gross. *Uncivil Society: 1989 and the Implosion of the Communist Establishment.* New York: Modern Library, 2009.

Kramer, Mark. "The Collapse of East European Communism and the Repercussions Within the Soviet Union (Part 1)." *Journal of Cold War Studies* 5, no. 4 (Fall 2003): 178–256.

———. "The Collapse of East European Communism and the Repercussions Within the Soviet Union (Part 3)." *Journal of Cold War Studies* 7, no. 1 (Winter 2005): 3–96.

———. "Special Issue: The Collapse of the Soviet Union (Part 2): Introduction." *Journal of Cold War Studies* 5, no. 4 (Fall 2003): 3–42.

Kuran, T. "Now Out of Never: The Element of Surprise in the East European Revolution of 1989." *World Politics: A Quarterly Journal of International Relations* 44, no. 1 (1991): 7–48.

Kurki, M. "Critical Realism and Causal Analysis in International Relations." *Millennium: Journal of International Studies* 35, no. 2 (2007): 361.

Lake, D. A., R. Powell, S. Choice, et al. "Adapting International Relations Theory to the End of the Cold War." *Journal of Cold War Studies* 5, no. 3 (2003): 96–101.

Lake, Eli. "Hacking the Regime." *New Republic* 240, no. 16 (2009).

Lakoff, George, and Mark Johnson. *Metaphors We Live By.* Chicago: University of Chicago Press, 1980.

Lane, D. "'Coloured Revolution' as a Political Phenomenon." *Journal of Communist Studies and Transition Politics* 25, no. 2 (2009): 113–135.

Lawson, G. "Historical Sociology in International Relations: Open Society, Research Programme and Vocation." *International Politics* 44, no. 4 (2007): 343–368.

Leedom-Ackerman, Joanne. "The Intensifying Battle over Internet Freedom." *Christian Science Monitor*, February 24, 2009.

Levy, J. S. "Learning and Foreign Policy: Sweeping a Conceptual Minefield." *International Organization* 48, no. 2 (1994): 279–312.

Lohmann, S. "Collective Action Cascades: An Informational Rationale for the Power in Numbers." *Journal of Economic Surveys* 14, no. 5 (2000): 655–684.

———. "The Dynamics of Informational Cascades: The Monday Demonstrations in Leipzig, East Germany, 1989–91." *World Politics* 47, no. 1 (1994): 42–101.

Mahoney, J., E. Kimball, and K. L. Koivu. "The Logic of Historical Explanation in the Social Sciences." *Comparative Political Studies* 42, no. 1 (2009): 114.

Mahoney, J., and R. Snyder. "Rethinking Agency and Structure in the Study of Regime Change." *Studies in Comparative International Development* 34, no. 2 (1999): 3–32.

McConnell, Mike. "Mike McConnell on How to Win the Cyber-War We're Losing." *Washington Post*, February 28, 2010.

McFaul, M. "The Fourth Wave of Democracy and Dictatorship: Noncooperative Transitions in the Postcommunist World." *World Politics* 54, no. 2 (2002): 212–244.

Moe, H. "Everyone a Pamphleteer? Reconsidering Comparisons of Mediated Public Participation in the Print Age and the Digital Era." *Media, Culture & Society* 32, no. 4 (2010): 691.

Nairn, Tom. "Where's the Omelette?" *London Review of Books*, November 23, 2008.

Nelson, Michael. *War of the Black Heavens: The Battles of Western Broadcasting in the Cold War*. Syracuse, NY: Syracuse University Press, 1997.

Osgood, K. A. "Hearts and Minds: The Unconventional Cold War." *Journal of Cold War Studies* 4, no. 2 (2002): 85–107.

Oushakine, S. A. "The Terrifying Mimicry of Samizdat." *Public Culture* 13, no. 2 (2001): 191.

Palmer, Mark. *Breaking the Real Axis of Evil: How to Oust the World's Last Dictators by 2025*. Lanham, MD: Rowman & Littlefield, 2003.

Patterson, E., and J. Amaral. "Presidential Leadership and Democracy Promotion." *Public Integrity* 11, no. 4 (2009): 327–346.

Pierskalla, J H. "Protest, Deterrence, and Escalation: The Strategic Calculus of Government Repression." *Journal of Conflict Resolution* 54, no. 1 (2010): 117.

Posner, Michael H., and Alec Ross. "Briefing on Internet Freedom and 21st Century Statecraft." U.S. Department of State, Washington, DC, January 22, 2010.

Puddington, Arch. *Broadcasting Freedom: The Cold War Triumph of Radio Free Europe and Radio Free Liberty*. Lexington: University Press of Kentucky, 2000.

Ray, J. L., and B. Russett. "The Future as Arbiter of Theoretical Controversies: Predictions, Explanations and the End of the Cold War." *British Journal of Political Science* 26, no. 4 (1996): 441–470.

"Reagan Urges 'Risk' on Gorbachev: Soviet Leader May Be Only Hope for Change, He Says." *Los Angeles Times*, June 13, 1989.

Rosati, J. A. "A Cognitive Approach to the Study of Foreign Policy." *Foreign Policy Analysis: Continuity and Change in Its Second Generation* (1995): 49–70.

Rose, R., and D. C. Shin. "Democratization Backwards: The Problem of Third-Wave Democracies." *British Journal of Political Science* 31, no. 02 (2001): 331–354.

Said, Edward. "Hey, Mister, You Want Dirty Book?" *London Review of Books*, September 20, 1999.

Saunders, Doug. "In Czechoslovakia, Human Network Made the Message Go Viral." *Globe and Mail* (Toronto), October 29, 2009.

Saxonberg, S. "The 'Velvet Revolution' and the Limits of Rational Choice Models." *Czech Sociological Review* 7, no. 1 (1999): 23–36.

Schipani-Adúriz, A. "Through an Orange-Colored Lens: Western Media, Constructed Imagery, and Color Revolutions." *Demokratizatsiya: The Journal of Post-Soviet Democratization* 15, no. 1 (2007): 87–115.

Schmitter, P. C., and T. L. Karl. "What Democracy Is . . . and Is Not." *Journal of Democracy* 2, no. 3 (1991): 75–88.

Schmitter, P. C., and J. Santiso. "Three Temporal Dimensions to the Consolidation of Democracy." *International Political Science Review/Revue internationale de science politique* 19, no. 1 (1998): 69–92.

Senor, Dan, and Christian Whiton. "Five Ways Obama Could Promote Freedom in Iran." *Wall Street Journal,* June 17, 2009.

Shah, N. "From Global Village to Global Marketplace: Metaphorical Descriptions of the Global Internet." *International Journal of Media and Cultural Politics* 4, no. 1 (2008): 9–26.

Shane, Scott. *Dismantling Utopia: How Information Ended the Soviet Union.* Chicago: I. R. Dee, 1995.

Sharman, J. C. "Culture, Strategy, and State-Centered Explanations of Revolution, 1789 and 1989." *Social Science History* 27, no. 1 (2003).

Shimko, K. L. "Metaphors and Foreign Policy Decision Making." *Political Psychology* 15, no. 4 (1994): 655–671.

———. "Psychology and Cold War History: A Review Essay." *Political Psychology* 15, no. 4 (1994): 801–806.

———. "Reagan on the Soviet Union and the Nature of International Conflict." *Political Psychology* 13, no. 3 (1992): 353–377.

Shirk, S. L. "Changing Media, Changing Foreign Policy in China." *Japanese Journal of Political Science* 8, no. 1 (2007): 43–70.

Silitski, V. "What Are We Trying to Explain?" *Journal of Democracy* 20, no. 1 (2009): 86–89.

Snyder, R. S. "The End of Revolution?" *Review of Politics* 61, no. 1 (1999): 5–28.

Sontag, Susan. *Illness as Metaphor; and AIDS and Its Metaphors.* New York: Picador, 2001.

Specter, Arlen. "Attack the Cyberwalls!" *Pittsburgh Post-Gazette,* July 7, 2009.

Sterling, B. "Triumph of the Plastic People." *Wired,* January 22, 1995.

Stier, Ken. "U.S. Girds for a Fight for Internet Freedom." *Time,* February 6, 2010.

Sunstein, C. R. "Hazardous Heuristics." *University of Chicago Law Review* 70, no. 2 (2003): 751–782.

Suri, J. "Explaining the End of the Cold War: A New Historical Consensus?" *Journal of Cold War Studies* 4, no. 4 (2002): 60–92.

Sweller, J., R. F. Mawer, and W. Howe. "Consequences of History-Cued and Means-End Strategies in Problem Solving." *American Journal of Psychology* 95, no. 3 (1982): 455–483.

Tait, Robert. "Hardliners Turn on Ahmadinejad for Watching Women Dancers." *Guardian*, December 5, 2006.

Tetlock, P. E., and C. McGuire. "Cognitive Perspectives on Foreign Policy." In *Political Behavior Annual*, edited by S. Long, 255–273. Boulder: Westview, 1986.

Tilly, C. "Trust and Rule." *Theory and Society* 33, no. 1 (2004): 1–30.

Tsui, Lokman. "The Great Firewall as Iron Curtain 2.0: The Implications of China's Internet Most Dominant Metaphor for US Foreign Policy." Paper presented at 6th annual Chinese Internet Research Conference, University of Hong Kong, June 13–14, 2008. jmsc.hku.hk/blogs/circ/files/2008/06/tsui_lokman.pdf.

———. "An Inadequate Metaphor: The Great Firewall and Chinese Internet Censorship." *Global Dialogue* 9, no. 1-2 (2007).

Ungar, S. "Misplaced Metaphor: A Critical Analysis of the 'Knowledge Society.'" *Canadian Review of Sociology/Revue canadienne de sociologie* 40, no. 3 (2003): 331–347.

Vasina, John I. "Transformation of Eastern Europe in 1989: Impacts of Information, Communications Technology and Globalization Process on Change." Master's thesis, Excelsior College, March 7, 2007.

Vertzberger, Y. Y. I. "Foreign Policy Decisionmakers as Practical-Intuitive Historians: Applied History and Its Shortcomings." *International Studies Quarterly* 30, no. 2 (1986): 223–247.

Way, L. "The Real Causes of the Color Revolutions." *Journal of Democracy* 19, no. 3 (2008): 55–69.

Way, L. A., and S. Levitsky. "Linkage, Leverage, and the Post-Communist Divide." *East European Politics and Societies* 21, no. 1 (2007): 48.

Wells, W. G., Jr. "Politicians and Social Scientists: An Uneasy Relationship." *American Behavioral Scientist* 26, no. 2 (1982): 235.

Wyatt, S. "Danger! Metaphors at Work in Economics, Geophysiology, and the Internet." *Science, Technology & Human Values* 29, no. 2 (2004): 242.

Youngs, R. "European Approaches to Democracy Assistance: Learning the Right Lessons?" *Third World Quarterly* 24, no. 1 (2003): 127–138.

CHAPTER 3

"America's Emobyte Deficit." *Economist*, November 27, 2007.

Ballard, J. G. "Aldous Huxley: An English Intellectual by Nicholas Murray." *Guardian*, April 13, 2002.

Barboza, David. "Internet Boom in China Is Built on Virtual Fun." *New York Times*, February 5, 2007.

———. "The People's Republic of Sex Kittens and Metrosexuals." *New York Times*, March 4, 2007.

Berdahl, Daphne. *Where the World Ended: Re-unification and Identity in the German Borderland*. Berkeley: University of California Press, 1999.

Betts, P. "The Twilight of the Idols: East German Memory and Material Culture." *Journal of Modern History* 72, no. 3 (2000): 731–765.

Breitenborn, U. "'Memphis Tennessee' in Borstendorf: Boundaries Set and Transcended in East German Television Entertainment." *Historical Journal of Film, Radio and Television* 24, no. 3 (2004): 391–402.

Chang, Anita. "Some Internet Porn Sites in China Now Accessible." Associated Press, July 22, 2010.

Coleman, Peter. "Thinking About Life: A Brave New World." United Press International, July 22, 2002.

Cooper, Robert. "Freedom for Sale: How We Made Money and Lost Our Liberty by John Kampfner / Democracy Kills by Humphrey Hawksley." *Sunday Times*, September 6, 2009.

Daly, Peter M., et al., eds. *Germany Reunified: A Five- and Fifty-Year Retrospective*. New York: P. Lang, 1997.

Darnton, R. "Censorship, a Comparative View: France, 1789–East Germany, 1989." *Representations* (1995): 40–60.

Daves, William. "Is It Aldous Huxley or George Orwell?" *New Statesman*, August 1, 2005.

Demich, Barbara. "For Chinese, Getting into Harvard Is a Class Act." *Los Angeles Times*, June 4, 2010.

"Despite Authoritarian Rule, Myanmar Art Grows." *New York Times*, March 25, 2010.

DeYoung, Karen. "U.S. Media Campaign Has Failed to Influence Cuba, Senators Say." *Washington Post*, May 4, 2010.

Dittmar, C. "GDR Television in Competition with West German Programming." *Historical Journal of Film, Radio and Television* 24, no. 3 (2004): 327–343.

Eickelman, D. F. "Islam and the Languages of Modernity." *Daedalus* 129, no. 1 (2000): 119–135.

Farrer, J. "A Chinese-Led Global Sexual Revolution." *Contexts* 7, no. 3 (2008): 58–60.

Forney, Matthew. "China's Loyal Youth." *New York Times*, April 13, 2008.

"For the Young Hillary Rodham, Which Individuals and Books Were Most Influential?" *Pittsburgh Post-Gazette*, April 11, 1993.

Fukuyama, Francis. *Our Posthuman Future: Consequences of the Biotechnology Revolution.* New York: Farrar, Straus and Giroux, 2002.

Fulbrook, Mary. *History of Germany, 1918–2000: The Divided Nation.* 2nd ed. Malden, MA: Wiley-Blackwell, 2002.

Fung, A. "'Think Globally, Act Locally': China's Rendezvous with MTV." *Global Media and Communication* 2, no. 1 (2006): 71.

Gleye, Paul. *Behind the Wall: An American in East Germany, 1988–89.* Carbondale: Southern Illinois University Press, 1991.

Gordon, Wendy J. "Listen Up, Ready or Not." *New York Times,* November 21, 1993.

Hargittai, E. "Digital Na (t) ives? Variation in Internet Skills and Uses Among Members of the 'Net Generation.'" *Sociological Inquiry* 80, no. 1 (2010): 92–113.

Havel, Václav, et al. *The Power of the Powerless: Citizens Against the State in Central-Eastern Europe.* Edited by John Keane. Armonk, NY: M. E. Sharpe, 1985.

Hendelman-Baavur, L. "Promises and Perils of Weblogistan: Online Personal Journals and the Islamic Republic of Iran." *Middle East Review of International Affairs* 11, no. 2 (2007): 77.

Hesse, K. R. "Cross-Border Mass Communication from West to East Germany." *European Journal of Communication* 5, no. 2 (1990): 355.

Hille, Kathrin, and Robin Kwong. "Chinese Consumers Opt for Bigger TVs than American Counterparts." *Financial Times,* October 24, 2009.

Hirschman, A. O. "Exit, Voice, and the Fate of the German Democratic Republic: An Essay in Conceptual History." *World Politics* (1993): 173–202.

Hoff, P., and W. Mühll-Benninghaus. "Depictions of America in GDR Television Films and Plays, 1955–1965." *Historical Journal of Film, Radio and Television* 24, no. 3 (2004): 403–410.

Hoff, P., and L. Willmot. "'Continuity and Change': Television in the GDR from Autumn 1989 to Summer 1990." *German History* 9, no. 2 (1991): 184.

Huxley, Aldous. *Brave New World and Brave New World Revisited.* New York: HarperCollins, 2004.

"Iran Pours Oil Fund Billions into Wooing Disaffected Youth." *Independent,* September 1, 2005.

Junker, Detlef, and Philipp Gassert. *The United States and Germany in the Era of the Cold War, 1945–1990: A Handbook.* Vol. 1. New York: Cambridge University Press, 2004.

Kampfner, John. *Freedom for Sale: How We Made Money and Lost Our Liberty.* Kindle edition. New York: Simon & Schuster, 2009.

Kellner, D. "From 1984 to One-Dimensional Man: Critical Reflections on Orwell and Marcuse." *Current Perspectives in Social Theory* 10 (1990): 223–252.

Kern, H. L. "Foreign Media and Protest Diffusion in Authoritarian Regimes: The Case of the 1989 East German Revolution." *Comparative Political Studies*, March 16, 2010.

Kern, H. L., and J. Hainmueller. "Opium for the Masses: How Foreign Media Can Stabilize Authoritarian Regimes." *Political Analysis* 17, no. 4 (2009): 377–399.

Khiabany, G. "Iranian Media: The Paradox of Modernity." *Social Semiotics* 17, no. 4 (2007): 479–501.

Klein, Naomi. "China's All-Seeing Eye." *Rolling Stone*, May 14, 2008.

Kohák, E. "Ashes, Ashes . . . Central Europe After Forty Years." *Daedalus* 121, no. 2 (1992): 197–215.

Korosteleva, Elena. "Was There a Quiet Revolution? Belarus After the 2006 Presidential Election." *Journal of Communist Studies and Transition Politics* 25, no. 2 (June 2009): 324–346.

Kraidy, M. M. "Reality TV and Multiple Arab Modernities: A Theoretical Exploration." *Middle East Journal of Culture and Communication* 1, no. 1 (2008): 49–59.

Lagerkvist, J. "Internet Ideotainment in the PRC: National Responses to Cultural Globalization." *Journal of Contemporary China* 17, no. 54 (2008): 121–140.

Manaev, Oleg. "The Influence of Western Radio on the Democratization of Soviet Youth." *Journal of Communication* 41, no. 2 (June 1991): 72–91.

Marcuse, Herbert. *One-Dimensional Man: Studies in the Ideology of Advanced Industrial Society.* New York: Routledge, 2002.

Maza, Erik. "Cuban Commies Give Avatar the Thumbs-Down." *Miami New Times*, February 18, 2010.

Meyen, M., and W. Hillman. "Communication Needs and Media Change: The Introduction of Television in East and West Germany." *European Journal of Communication* 18, no. 4 (2003): 455.

Meyen, M., and U. Nawratil. "The Viewers: Television and Everyday Life in East Germany." *Historical Journal of Film, Radio and Television* 24, no. 3 (2004): 355–364.

Meyen, M., and K. Schwer. "Credibility of Media Offerings in Centrally Controlled Media Systems: A Qualitative Study Based on the Example of East Germany." *Media, Culture & Society* 29, no. 2 (2007): 284.

Miller Llana, Sara. "Cuba's Youth: Restless but Not Often Political." *Christian Science Monitor*, July 26, 2008.

Mirsky, Jonathan. "Vietnam Now." *New York Review of Books*, June 24, 2010.

"Mobile Phones Change Dating Habits as Saudis Search for New Identity." *Irish Times*, March 3, 2009.

Moore, Matthew. "China to Lead World Scientific Research by 2020." *Daily Telegraph*, January 25, 2010.

Mostaghim, Ramin, and Borzou Daragahi. "Iran's Other Youth Movement." *Los Angeles Times*, June 10, 2007.

Murphy, Caryle. "Saudi Women Revel in Online Lives." *Global Post*, February 4, 2010. www.globalpost.com/dispatch/saudi-arabia/100203/internet-women.

Nelson, Michael. *War of the Black Heavens: The Battles of Western Broadcasting in the Cold War*. Syracuse, NY: Syracuse University Press, 1997.

Orwell, George. *1984*. New York: Signet Classic, 1977.

Pfaff, S., and H. Kim. "Exit-Voice Dynamics in Collective Action: An Analysis of Emigration and Protest in the East German Revolution." *American Journal of Sociology* 109, no. 2 (2003): 401–444.

"Porn Dominates Saudi Mobile Use." *BBC News*, April 25, 2007.

Posner, R. A. "Orwell Versus Huxley: Economics, Technology, Privacy, and Satire." *Philosophy and Literature* 24, no. 1 (2000): 1–33.

Postman, Neil. *Amusing Ourselves to Death: Public Discourse in the Age of Show Business*. Twentieth anniversary edition. New York: Penguin Books, 2006.

Prior, Markus. *Post-Broadcast Democracy: How Media Choice Increases Inequality in Political Involvement and Polarizes Elections*. New York: Cambridge University Press, 2007.

Problems of Communism. Vol. 12. Washington, DC: Documentary Studies Section, International Information Administration, U.S. Department of State, 1963.

Quester, George H. *Before and After the Cold War: Using Past Forecasts to Predict the Future*. Portland, OR: Frank Cass, 2002.

Rantanen, T. "The Old and the New: Communications Technology and Globalization in Russia." *New Media & Society* 3, no. 1 (2001): 85.

Rosen, S. "The Victory of Materialism: Aspirations to Join China's Urban Moneyed Classes and the Commercialization of Education." *China Journal* (2004): 27–51.

Rosenstiel, Thomas B. "TV, VCRs Fan Fire of Revolution." *Los Angeles Times*, January 18, 1990.

Roth, Philip. "A Conversation in Prague." *New York Review of Books*, April 12, 1990.

Rueschemeyer, Marilyn, and Christiane Lemke. *The Quality of Life in the German Democratic Republic: Changes and Developments in a State Socialist Society*. Armonk, NY: M. E. Sharpe, 1989.

"Russian Internet Users Care Most About Love, Weight Losing." Associated Press, March 22, 2010.

Schielke, S. "Boredom and Despair in Rural Egypt." *Contemporary Islam* 2, no. 3 (2008): 251–270.

Schwoch, J. "Cold War Telecommunications Strategy and the Question of German Television." *Historical Journal of Film, Radio and Television* 21, no. 2 (2001): 109–121.

Semuels, Alana. "Unlikely Forum for Iran's Youth." *Los Angeles Times*, January 2, 2008.

"Sex, Social Mores, and Keyword Filtering: Microsoft Bing in the 'Arabian Countries.'" OpenNet Initiative, March 4, 2010. opennet.net/sex-social-mores-and-keyword-filtering-microsoft-bing-arabian-countries.

Snyder, Alvin A. *Warriors of Disinformation: American Propaganda, Soviet Lies, and the Winning of the Cold War*. New York: Arcade, 1997.

Sola Pool, Ithiel De. "Communication in Totalitarian Societies." In *Handbook of Communication*, edited by Ithiel De Sola Pool, W. Schramm, N. Maccoby, and E. Parker, 463–474. Chicago: Rand McNally, 1973.

Sreberny, A. "The Analytic Challenges of Studying the Middle East and its Evolving Media Environment." *Middle East Journal of Culture and Communication* 1, no. 1 (2008): 8–23.

Sreberny-Mohammadi, A. "Small Media for a Big Revolution: Iran." *International Journal of Politics, Culture, and Society* 3, no. 3 (1990): 341–371.

Sullivan, Kevin. "Saudi Youth Use Cellphone Savvy to Outwit the Sentries of Romance." *Washington Post*, August 6, 2006.

Sydell, Laura. "Chinese Fans Follow American TV Online—for Free." National Public Radio, June 24, 2008.

Tait, Robert. "Iran Launches First Online Supermarket." *Guardian*, February 4, 2010.

Talbot, David. "Bing Dinged on Arab Sex Censorship." *Technology Review Editors' Blog*, March 4, 2010. www.technologyreview.com/blog/editors/24891/?utm_source=twitterfeed&utm_medium=twitter.

Thussu, Daya Kishan. *News as Entertainment: The Rise of Global Infotainment*. Thousand Oaks, CA: Sage, 2007.

Trachtenberg, Jeffrey A. "Philip Roth on 'The Humbling.'" *Wall Street Journal*, October 30, 2009.

"Vietnam to Offer Movies about Sex on Internet." Reuters, July 20, 2006.

Wang, X. "The Post-Communist Personality: The Spectre of China's Capitalist Market Reforms." *China Journal* (2002): 1–17.

"Web Link to China Teen Pregnancy." *BBC News*, July 10, 2007.

Weitz, Eric D. *Creating German Communism, 1890–1990: From Popular Protests to Socialist State*. Princeton, NJ: Princeton University Press, 1997.

Wheary, Jennifer. "The Global Middle Class Is Here: Now What?" *World Policy Journal* 26, no. 4 (January 2010): 75–83.

Wilson, Ben. *What Price Liberty?* London: Faber and Faber, 2009.

Zhang, L. L. "Are They Still Listening? Reconceptualizing the Chinese Audience of the Voice of America in the Cyber Era." *Journal of Radio Studies* 9 (2002): 317.

CHAPTER 4

Adams, P. C. "Protest and the Scale Politics of Telecommunications." *Political Geography* 15, no. 5 (1996): 419–441.

Agre, P. E. "Real-Time Politics: The Internet and the Political Process." *Information Society* 18, no. 5 (2002): 311–331.

Albrecht, H. "How Can Opposition Support Authoritarianism? Lessons From Egypt." *Democratization* 12, no. 3 (2005): 378–397.

Al Hussaini, Amira. "Arabeyes: No to 'Offensive' Blogs." *Global Voices*, February 14, 2008. globalvoicesonline.org/2008/02/14/arabeyes-no-to-offensive-blogs/.

Ameripour, Aghil, Brian Nicholson, and Michael Newman. "Internet Usage Under Authoritarian Regimes: Conviviality, Community, Blogging and Online Campaigning in Iran." *IDPM Working Paper 43* (2009). www.sed.manchester.ac.uk/idpm/research/publications/wp/di/di_Wp43.htm.

"Anger in China over Web Censorship." *BBC News*, June 30, 2009.

Arkhipov, Ilya, and Lyubov Pronina. "Putin Fire Webcams Show Power Pyramid Makes Local Leaders Reluctant to Act." *Bloomberg News*, August 25, 2010.

Baber, Z. "Engendering or Endangering Democracy? The Internet, Civil Society and the Public Sphere." *Asian Journal of Social Science* 30, no. 2 (2002): 287–303.

Bambauer, Derek E. "Cybersieves." *Duke Law Journal* 59 (2009).

Becker, J. "Lessons From Russia: A Neo-Authoritarian Media System." *European Journal of Communication* 19, no. 2 (2004): 139.

Bhagwati, Jagdish. "Made in China." *New York Times Book Review*, February 18, 2007.

Blaydes, Lisa. "Authoritarian Elections and Elite Management: Theory and Evidence from Egypt." Presented at the Conference on Dictatorships: Their Governance and Social Consequences, Princeton University, April 2008. www.princeton.edu/~piirs/Dictatorships042508/Blaydes.pdf.

Boix, Carles, and Milan Svolik. "The Foundations of Limited Authoritarian Government: Institutions and Power-Sharing in Dictatorships." Social Science Research Network Working Paper Series, June 1, 2010.

Bueno De Mesquita, Ethan. "Regime Change and Revolutionary Entrepreneurs." *American Political Science Review* (forthcoming).

Builder, Carl H., and Steven C. Bankes. *The Etiology of European Change.* Santa Monica, CA: RAND, 1990.

Burrows, Peter. "Internet Censorship, Saudi Style." *Business Week*, November 13, 2008.

Cheng, C. T. "New Media and Event: A Case Study on the Power of the Internet." *Knowledge, Technology & Policy* 22, no. 2 (2009): 145–153.

Chiou, Jing-Yuan, Mark Dincecco, and David Rahman. "Private Information and Institutional Change: The Case of Foreign Threats." Social Science Research Network Working Paper Series, December 22, 2009.

Collier, D., and S. Levitsky. "Democracy with Adjectives: Conceptual Innovation in Comparative Research." *World Politics* 49, no. 3 (1997): 430–451.

Converse, P. E. "Power and the Monopoly of Information." *The American Political Science Review* 79, no. 1 (1985): 1–9.

Cooper, B. "The Western Connection: Western Support for the East German Opposition." *German Politics and Society* 21, no. 4 (2003): 74–93.

Corrales, J., and F. Westhoff. "Information Technology Adoption and Political Regimes." *International Studies Quarterly* 50, no. 4 (2006): 911–933.

Deibert, Ronald, John G. Palfrey, Rafal Rohozinski, and Jonathan Zittrain. *Access Denied: The Practice and Policy of Global Internet Filtering.* Cambridge, MA: MIT Press, 2008.

———. *Access Controlled: The Shaping of Power, Rights, and Rule in Cyberspace.* Cambridge, MA: MIT Press, 2010.

Drezner, Daniel. "Weighing the Scales: The Internet's Effect on State-Society Relations." *Brown Journal of World Affairs* 16, no. 2 (2010).

Egorov, G., S. Guriev, and K. Sonin. "Why Resource-Poor Dictators Allow Freer Media: A Theory and Evidence from Panel Data." *American Political Science Review* 103, no. 04 (2009): 645–668.

Elhadj, Elie. *The Islamic Shield: Arab Resistance to Democratic and Religious Reforms.* Boca Raton, FL: BrownWalker Press, 2007.

Fletcher, Owen. "China Pays Web Users to Find Porn Amid Crackdown." IDG News Service, January 19, 2010.

Friedman, Thomas L. *The Lexus and the Olive Tree.* New York: Farrar, Straus, Giroux, 2000.

Friedrich, Carl Joachim, and Zbigniew Brzezinski. *Totalitarian Dictatorship and Autocracy.* Cambridge, MA: Harvard University Press, 1965.

Gandhi, J., and A. Przeworski. "Authoritarian Institutions and the Survival of Autocrats." *Comparative Political Studies* 40, no. 11 (2007): 1279.

Gans-Morse, J. "Searching for Transitologists: Contemporary Theories of Post-Communist Transitions and the Myth of a Dominant Paradigm." *Post-Soviet Affairs* 20, no. 4 (2004): 320–349.

Geddes, B. "Why Parties and Elections in Authoritarian Regimes?" Presented to the Annual Meeting of the American Political Science Association, 2005.

Glikin, M., and Kostenko, N. "Chudo Vozmozhno." *Vedomosti*, February 15, 2010. www.vedomosti.ru/newspaper/article/2010/02/15/225543.

Habermas, J. "Political Communication in Media Society: Does Democracy Still Enjoy an Epistemic Dimension? The Impact of Normative Theory on Empirical Research." *Communication Theory* 16, no. 4 (2006): 411–426.

He, B., and M. E. Warren. "Authoritarian Deliberation: The Deliberative Turn in Chinese Political Development." In *Proceedings of the American Political Science Association Annual Meeting*, vol. 28. Boston: American Political Science Association, 2008.

Hearn, K. "The Management of China's Blogosphere Boke (Blog)." *Continuum* 23, no. 6 (2009): 887–901.

Hearn, K., and B. Shoesmith. "Exploring the Roles of Elites in Managing the Chinese Internet." *Javnost—The Public*, Vol 11 (2004) 101–114.

Hille, Kathrin. "The Net Closes." *Financial Times*, July 18, 2009.

Hoover, D., and D. Kowalewski. "Dynamic Models of Dissent and Repression." *Journal of Conflict Resolution* 36, no. 1 (1992): 150–182.

Huang, Haifeng. "Media Freedom, Governance, and Regime Stability in Authoritarian States." Unpublished Paper, 2008.

Huntington, Samuel P. *The Third Wave: Democratization In the Late Twentieth Century.* Norman: University of Oklahoma Press, 1993.

Jiang, M., and H. Xu. "Exploring Online Structures on Chinese Government Portals: Citizen Political Participation and Government Legitimation." *Social Science Computer Review* 27, no. 2 (2009): 174–195.

Johnson, Erica, and Beth Kolko. "E-Government and Transparency in Authoritarian Regimes: Comparison of National- and City-Level E-Government

Websites in Central Asia." Presented at the Annual Meeting of the International Studies Association, 2010.

Johnston, H., and C. Mueller. "Unobtrusive Practices of Contention in Leninist Regimes." *Sociological Perspectives* 44, no. 3 (2001): 351–375.

Kaplan, Jeremy A. "China Expanding Censorship to Text Messages." *FOXNews .com*, January 20, 2010. www.foxnews.com/scitech/2010/01/20/china -expanding-censorship-text-messages/.

Kapstein, E. B., and N. Converse. "Why Democracies Fail." *Journal of Democracy* 19, no. 4 (2008): 57–68.

Kaufman, Stephen. "Bloggers in Mauritania Form a Union." *America.gov*, August 8, 2008. www.america.gov/st/democracy-english/2008/April/2008 0408172637liameruoy0.9660608.html.

Kennedy, J. J. "Maintaining Popular Support for the Chinese Communist Party: The Influence of Education and the State-Controlled Media." *Political Studies* 57, no. 3 (2009): 517–536.

Kluver, R. "The Architecture of Control: A Chinese Strategy for E-Governance." *Journal of Public Policy* 25, no. 1 (2005): 75–97.

———. "US and Chinese Policy Expectations of the Internet." *China Information* 19, no. 2 (2005): 299.

Kluver, R., and C. Yang. "The Internet in China: A Meta-Review of Research." *Information Society* 21, no. 4 (2005): 301–308.

Kramer, Andrew E., and Jenna Wortham. "Professor Main Target of Assault on Twitter." *New York Times*, August 7, 2009.

Kristof, Nicholas D. "Death by a Thousand Blogs." *New York Times*, May 24, 2005.

Lacharite, J. "Electronic Decentralisation in China: A Critical Analysis of Internet Filtering Policies in the People's Republic of China." *Australian Journal of Political Science* 37, no. 2 (2002): 333–346.

Lagerkvist, J. "The Techno-Cadre's Dream: Administrative Reform by Electronic Governance in China Today?" *China Information* 19, no. 2 (2005): 189.

Latham, K. "SMS, Communication, and Citizenship in China's Information Society." *Critical Asian Studies* 39, no. 2 (2007): 295–314.

Legezo, Denis. "Medvedev chitaet svoj ZhZh i reshaet problemy." *CNews*, October 8, 2009. www.cnews.ru/news/top/index.shtml?2009/10/08/364944.

Levitsky, S., and L. A. Way. "The Rise of Competitive Authoritarianism." *Journal of Democracy* 13, no. 2 (2002): 51–65.

Levy, Clifford J. "Videos Rouse Russian Anger Toward Police." *New York Times*, July 27, 2010.

Li, S. "The Online Public Space and Popular Ethos in China." *Media, Culture & Society* 32, no. 1 (2010): 63.

MacKinnon, Rebecca. "Flatter World and Thicker Walls? Blogs, Censorship and Civic Discourse in China." *Public Choice* 134, no. 1 (2008): 31–46.

———. "Liberty or Safety? Both—or Neither." *IEEE Spectrum*, May 2010.

Magaloni, B., and J. Wallace. "Citizen Loyalty, Mass Protest and Authoritarian Survival." Presented at the Conference on Dictatorships: Their Governance and Social Consequences, Princeton University, April 2008.

Markoff, John. "Iranians and Others Outwit Net Censors." *New York Times*, April 30, 2009.

"Medvedev Looks to Singapore for Electronic Efficiency." Reuters, November 11, 2009.

Meng, B. "Moving Beyond Democratization: A Thought Piece on the China Internet Research Agenda." *International Journal of Communication* 4 (2010): 501–508.

De Mesquita, B. B., and G. W. Downs. "Development and Democracy." *Foreign Affairs* 84, no. 5 (2005): 77–86.

Miradova, Mina. "Azerbaijan: Webcams Used for 'Transparent' Municipal Elections." *EurasiaNet.org*, December 22, 2009. www.eurasianet.org/departments/insight/articles/122309.shtml.

Moore, Malcolm. "China's Internet Porn Reward Drives Rise in Online Erotica Searches." *Daily Telegraph*, December 7, 2009.

O'Brien, K. J. "How Authoritarian Rule Works." *Modern China* 36, no. 1 (2010): 79–86.

"Our Chip Has Come In." *New Republic* 200, no. 24 (1989): 7–8.

Pickel, A. "Authoritarianism or Democracy? Marketization as a Political Problem." *Policy Sciences* 26, no. 3 (1993): 139–163.

"Press Conference with Secretary Gates and Adm. Mullen." Transcript. U.S. Department of Defense, June 18, 2009.

Ramstad, Evan. "Gulags, Nukes and a Water Slide: Citizen Spies Lift North Korea's Veil." *Wall Street Journal*, May 22, 2009.

Roberts, Hal. "China Bans the Letter 'F.'" *Watching Technology*, June 12, 2009. blogs.law.harvard.edu/hroberts/2009/06/12/china-bans-the-letter-f/.

Rodan, G. "The Internet and Political Control in Singapore." *Political Science Quarterly* 113, no. 1 (1998): 63–89.

Rosen, Stanley. "Is the Internet a Positive Force in the Development of Civil Society, a Public Sphere, and Democratization in China?" *International Journal of Communication* 4 (2010): 509–516.

Rothstein, B. "Creating Political Legitimacy: Electoral Democracy Versus Quality of Government." *American Behavioral Scientist* 53, no. 3 (2009): 311.

Rubin, M. "The Telegraph, Espionage, and Cryptology in Nineteenth Century Iran." *Cryptologia* 25, no. 1 (2001): 18–36.

"Russian Opposition Newspaper Comes Under Hacker Attack." Agence France-Presse, January 26, 2010.

Satarov, Georgy. "Don't Expect Miracles From Russia's 'Authoritarian Modernization.'" Radio Free Europe / Radio Liberty, February 21, 2010.

"Saudi Religious Police Launch Facebook Group." *Al Arabiya*, November 8, 2009.

Saunders, R. "Wiring the Second World: The Geopolitics of Information and Communications Technology in Post-Totalitarian Eurasia." *Russian Cyberspace Journal* 1 (2009).

Schucher, Günter. "Liberalization in Times of Instability: Margins of Unconventional Participation in Chinese Authoritarianism. Social Science Research Network Working Paper Series, December 5, 2009.

Schuppan, T. "E-Government in Developing Countries: Experiences from Sub-Saharan Africa." *Government Information Quarterly* 26, no. 1 (2009): 118–127.

Seligson, A. L., and J. A. Tucker. "Feeding the Hand That Bit You: Voting for Ex-Authoritarian Rulers in Russia and Bolivia." *Demokratizatsiya: The Journal of Post-Soviet Democratization* 13, no. 1 (2005): 11–44.

Shepherd, T. "Twittering in the OECD's 'Participative Web': Microblogging and New Media Policy." *Global Media Journal* 2, no. 1 (2009): 149–165.

Shilton, K. "Four Billion Little Brothers? Privacy, Mobile Phones, and Ubiquitous Data Collection." *Communications of the ACM* 52, no. 11 (2009): 48–53.

Shultz, G. P. "New Realities and New Ways of Thinking." *Foreign Affairs* 63, no. 4 (1985): 705–721.

Siegel, D. A. "Social Networks and Collective Action." *American Journal of Political Science* 53, no. 1 (2009): 122–138.

Sola Pool, Ithiel De. "Communication in Totalitarian Societies." In *Handbook of Communication*, edited by Ithiel De Sola Pool, W. Schramm, N. Maccoby, and E. Parker, 463–474. Chicago: Rand McNally, 1973.

"S. Mironov predlagaet veduwim pol'zovateljam Runeta i blogeram vyrabotat' pravila setevoj cenzury." IA "Prajm-TASS," September 30, 2009. en.rian.ru/russia/20100126/157686584.html.

"Tehnologii: U Dmitrija Medvedeva Pojavilsja iPad." *Lenta.ru*, May 18, 2010. www.lenta.ru/news/2010/05/18/ipad/.

"Thai Website to Protect the King." *BBC News*, February 5, 2009.

Thornton, P. M. "Censorship and Surveillance in Chinese Cyberspace: Beyond the Great Firewall." In *Chinese Politics: State, Society and the Market*, edited by P. H. Gries and S. Rosen, 179–198. New York: Routledge, 2010.

Thussu, Daya Kishan. *News as Entertainment: The Rise of Global Infotainment*. Thousand Oaks, CA: Sage, 2007.

Toffler, Alvin. *The Third Wave*. New York: Bantam Books, 1981.

Union, S., and H. Steve. "Deliberative Institutions as Mechanisms for Managing Social Unrest: The Case of the 2008 Chongqing Taxi Strike." *China: An International Journal* 7 (2009): 336–352.

Wang, X. "Mutual Empowerment of State and Society: Its Nature, Conditions, Mechanisms, and Limits." *Comparative Politics* 31, no. 2 (1999): 231–249.

Warschauer, M. "Singapore's Dilemma: Control Versus Autonomy in IT-Led Development." *Information Society* 17, no. 4 (2001): 305–311.

Webster, Graham. "Writing 'Bass Ackwards' to Defeat Censorship in China." *Sinobyte: China and Technology, CNET News*, July 2, 2008. news.cnet.com/8301-13908_3-9982672-59.html.

Weiss, C. "Science, Technology and International Relations." *Technology in Society* 27, no. 3 (2005): 295–313.

Wohlstetter, Albert. "The Fax Shall Make You Free." Address to President Havel's Peaceful Road to Democracy Conference, Prague, July 4–6, 1990. profiles.nlm.nih.gov/BB/A/R/X/K/_/bbarxk.ocr.

Wong, Alicia. "More Government Bodies Listen to Views Online." *Today* (Singapore), November 4, 2009.

Wriston, W. B. "Bits, Bytes, and Diplomacy." *Foreign Affairs* 76, no. 5 (1997): 172–182.

CHAPTER 5

Akhvlediani, M. "The Fatal Flaw: The Media and the Russian Invasion of Georgia." *Small Wars & Insurgencies* 20, no. 2 (2009): 363–390.

Amir-Ebrahimi, M. "Blogging from Qom, Behind Walls and Veils." *Comparative Studies of South Asia, Africa and the Middle East* 28, no. 2 (2008): 235.

Andreev, Leha. "Konstantin Rykov: 'Net nikakoj politicheskoj angazhirovannosti.'" *Webplaneta*, December 2, 2008. www.webplanet.ru/interview/business/2008/12/02/rykov.html.

Bandurski, David. "China's Guerrilla War for the Web." *Far Eastern Economic Review*, July 2008.

Barry, Ellen. "War on a Cultural Battlefield: On Anniversary of Fight over South Ossetia, Rage Is Still Raw in the Region." *International Herald Tribune*, August 8, 2009.

Beloborodova, Olesja. "Vmesto Srednego Klassa Voznikli 'Novye Serditye.'" *Vzgljad*, July 1, 2010. www.vz.ru/politics/2010/7/1/415114.html.

"Bloggery ZhZh Nachali Prodavat' i Pokupat' Populjarnost'." *Webplaneta*, November 9, 2006. webplanet.ru/news/life/2006/11/09/rykov.html.

Boudreaux, Richard. "Bucks Populi: Making Democracy a Going Concern in Kiev." *Wall Street Journal*, February 5, 2010.

Brady, Anne-Marie. "The Beijing Olympics as a Campaign of Mass Distraction." *China Quarterly* 197 (2009): 1–24.

———. *Marketing Dictatorship: Propaganda and Thought Work in Contemporary China*. Lanham, MD: Rowman & Littlefield, 2008.

———. "Mass Persuasion as a Means of Legitimation and China's Popular Authoritarianism." *American Behavioral Scientist* 53, no. 3 (2009): 434.

———. "Regimenting the Public Mind: The Modernization of Propaganda in the PRC." *International Journal* 57 (2001): 563.

———. "'Treat Insiders and Outsiders Differently': The Use and Control of Foreigners in the PRC." *China Quarterly* 164 (2009): 943–964.

Budaragin, Mihail. "Medvedevu i Obame ne Nuzhny Pervyj Kanal i CNN." *Vzgljad*, January 12, 2010. www.vz.ru/politics/2010/1/12/364498.html.

Cairncross, Frances. *The Death of Distance: How the Communications Revolution Will Change Our Lives*. Cambridge, MA: Harvard Business Press, 1997.

Cancel, Daniel. "Chavez Adds iPod to Portfolio After Embracing BlackBerry, Twitter Account." *Bloomberg News*, July 21, 2010.

———. "Chavez Says Twitter, Blackberry Are 'Secret Weapon.'" *Business Week*, April 29, 2010.

———. "Chavez to Join Obama, Castro in Adding Twitter to Media Arsenal." *Business Week*, April 26, 2010.

"Chavez Beefs Up Twitter Moves with State Funds, Staff." Agence France-Presse, November 5, 2010.

"China's Internet 'Spin Doctors.'" *BBC News*, December 16, 2008.

Chinea, Eyanir. "Venezuela's Chatty Leader Chavez Joins Twitter." Reuters, April 27, 2010.

Crovitz, Gordon L. "China's Web Crackdown Continues." *Wall Street Journal*, January 10, 2010.

Diaz, Marianne. "President Chávez and His 'Communicational Guerrilla.'" *Global Voices Advocacy*, April 16, 2010. advocacy.globalvoicesonline.org/ 2010/04/16/president-chavez-and-his-communicational-guerrilla/.

Ding, S. "Informing the Masses and Heeding Public Opinion: China's New Internet-Related Policy Initiatives to Deal with Its Governance Crisis." *Journal of Information Technology & Politics* 6, no. 1 (2009): 31–42.

Eadie, William F. *Twenty-First Century Communication: A Reference Handbook*. Vol. 1. Thousand Oaks, CA: Sage, 2009.

"Edinoross Rykov Zapustil Onlajn-TV." *Webplaneta*, October 10, 2007. webplanet .ru/news/life/2007/10/10/rykov_edrussia.html.

Ernkvist, M., and P. Strom. "Enmeshed in Games with the Government: Governmental Policies and the Development of the Chinese Online Game Industry." *Games and Culture* 3, no. 1 (2008): 98.

Esfandiari, Golnaz. "Iranian Social Networking, Hard-Line Style." Radio Free Europe / Radio Liberty, July 29, 2010.

Fareed, Malik. "China Joins a Turf War." *Guardian*, September 22, 2008.

Fossato, F. "Web Captives." *Index on Censorship* 38, no. 3 (2009): 132.

Fowler, Geoffrey A., and Juliet Ye. "Chinese Bloggers Scale the 'Great Firewall' in Riot's Aftermath." *Wall Street Journal*, July 2, 2008.

Franchetti, Mark. "Maria Sergeyeva: Putin's Rising Star." *Sunday Times*, March 8, 2009.

French, Howard W. "As Chinese Students Go Online, Little Sister Is Watching." *New York Times*, May 9, 2006.

Fu, Rocky. "China's Anti-Blogging Strategy: Tell the Truth, and Fast." *China Internet Watch*, July 30, 2009. www.chinainternetwatch.com/193/china %E2%80%99s-anti-blogging-strategy-tell-the-truth-and-fast/.

Geddes, B., and J. Zaller. "Sources of Popular Support for Authoritarian Regimes." *American Journal of Political Science* 33, no. 2 (1989): 319–347.

Goodin, Dan. "Security Boss Calls for End to Net Anonymity." Register, October 16, 2009. www.theregister.co.uk/2009/10/16/kaspersky_rebukes _net_anonymity/.

"Gosduma Podderzhala Komp'juternyj Patriotizm." *BFM.ru*, February 15, 2010. www.bfm.ru/articles/2010/02/15/gosduma-predlagaet-igrat-s-patriot izmom.html.

Gribanov, A., and M. Kowell. "Samizdat According to Andropov." *Poetics Today* 30, no. 1 (2009): 89.

Hassan, Amro. "Egypt: Gamal Mubarak Turns to the Web." *Babylon & Beyond, Los Angeles Times*, August 13, 2009. latimesblogs.latimes.com/babylonbe yond/2009/08/egypt-gamal-mubarak-reaches-out-for-internet-users .html.

He, Z. "SMS in China: A Major Carrier of the Nonofficial Discourse Universe." *Information Society* 24, no. 3 (2008): 182–190.

Hearn, K., and A. Willis. "Lei Feng Lives On in Cyberspace." In *Proceedings of the 3rd International Conference on Digital Interactive Media in Entertainment and Arts*, 248–255. ACM International Conference Proceeding Series, vol. 349. 2008.

Hille, Kathrin. "Internet Anger Forces Jail Death U-Turn." *Financial Times*, February 28, 2009.

———. "The Net Closes." *Financial Times*, July 18, 2009.

Hodge, Nathan. "Kremlin Launches 'School of Bloggers.'" *Danger Room, Wired*, May 27, 2009. www.wired.com/dangerroom/2009/05/kremlin -launches-school-of-bloggers/.

"Internet Razdvinul Granicy Vremeni." *Dni.ru*, April 21, 2009. www.dni.ru/ tech/2009/4/21/164475.html.

"Interview: Maria Sergeyeva." *Moscow News*, March 5, 2009.

"Iran's Bloggers Thrive Despite Blocks." *BBC News*, December 15, 2008.

Jacobs, Andrew. "China's Answer to a Crime Includes Amateur Sleuths." *New York Times*, February 24, 2009.

Jakemenko, Boris. "Cerkov' Dolzhna Borot'sja za Molodezh' v Internete." *Regions.ru*, February 5, 2009. www.regions.ru/news/location01824/2194524/.

Jung, Sung-ki. "Russian Experts Arrive to Review Cheonan Findings." *Korea Times*, May 31, 2010.

Keohane, Joe. "How Facts Backfire." *Boston Globe*, July 11, 2010.

Kim, Tae-gyu. "'Sleeper Effect' Causes People to Be Swayed by Rumors." *Korea Times*, June 2, 2010.

Kshetri, N. "The Evolution of the Chinese Online Gaming Industry." *Journal of Technology Management in China* 4, no. 2 (2009): 158–179.

LaFraniere, Sharon. "China Adds a Feature to Phones: Patriotism." *New York Times*, September 30, 2009.

Landsberger, S. R. "Learning by What Example? Educational Propaganda in Twenty-First-Century China." *Critical Asian Studies* 33, no. 4 (2001): 541–571.

Latynina, Yulia. "Why Putin Isn't Afraid of a Free Internet." *Moscow Times*, March 24, 2010.

Lee, C. C., Z. He, and Y. Huang. "'Chinese Party Publicity Inc.' Conglomerated: The Case of the Shenzhen Press Group." *Media, Culture & Society* 28, no. 4 (2006): 581.

Levine, Yasha. "Blogger Gorshkov Will Dish Dirt on Russian Politicians." *Wired*, August 16, 2010.

Li, Datong. "China's Leaders, the Media and the Internet." *OpenDemocracy*, July 8, 2008. www.opendemocracy.net/article/china-s-leaders-and-the -internet.

Li Xiaoming, "深圳警方认定林嘉祥行为不构成猥亵," Sina, news.sina.com.cn/c/ 2008-11-06/025914685329s.shtml (Accessed April 8, 2010).

Liñán, Miguel Vázquez. "History as a Propaganda Tool in Putin's Russia." *Communist and Post-Communist Studies* 43, no. 2 (June 2010): 167–178.

———. "Putin's Propaganda Legacy." *Post-Soviet Affairs* 25, no. 2 (2009): 137– 159.

Lollar, X. L. "Assessing China's E-Government: Information, Service, Transparency and Citizen Outreach of Government Websites." *Journal of Contemporary China* 15, no. 46 (2006): 31–41.

Lu, J., and I. Weber. "State, Power and Mobile Communication: A Case Study of China." *New Media & Society* 9, no. 6 (2007): 925.

Mallpas, Anna. "The Bombastic Blonde of the Blogosphere." *Moscow Times*, February 13, 2009.

Millan, Mark. "USocial CEO: 'We're Gaming Digg.'" *Technology Blog, Los Angeles Times*, March 5, 2009. latimesblogs.latimes.com/technology/2009/ 03/usocial-digg.html.

Molinski, Dan. "Venezuelan President's First Tweets Lure 204,000 Followers." *Speakeasy Blog, Wall Street Journal*, May 5, 2010. blogs.wsj.com/speakeasy/ 2010/05/05/hugo-chavez-joins-twitter-asks-whats-up/.

Nagornyh, Irina, and Natal'ja Bespalova. "'Edinaja Rossija' Predlozhit Videoservis." *Kommersant*, August 21, 2008. www.kommersant.ru/doc .aspx?DocsID=1013903.

Nathan, Andrew. "Medals and Rights." *New Republic*, July 9, 2008.

Noelle-Neumann, Elisabeth. *The Spiral of Silence: Public Opinion, Our Social Skin*. Chicago: University of Chicago Press, 1993.

Novruzov, Ali S. "Facebook and Plans of the Party." *In Mutatione Fortitudo Blog*, June 5, 2010. blog.novruzov.az/2010/06/facebook-and-plans-of -party.html.

Olshansky, Elliot. "A 'Russian Sarah Palin'? Meet Pro-Putin Activist Maria Sergeyeva, Russia's Rising Political Star." *New York Daily News*, March 9, 2009.

Osipovich, Alexander. "NoizeMC, aka Ivan Alexeyev, and Russian Rap Inspire a Movement." *Wall Street Journal*, July 24, 2010.

Page, Lewis. "NSA Offering 'Billions' for Skype Eavesdrop Solution." *Register*, February 12, 2009. www.theregister.co.uk/2009/02/12/nsa_offers_billions_for_skype_pwnage/.

Podger, Corinne. "China Marshalls Army of Bloggers." *Connect Asia*. Radio Australia, August 21, 2008.

Pretel, Enrique Andres. "Twitter's Heady Rise Has Venezuela's Hugo Chavez in Spin." Reuters, March 30, 2010.

"Prjamaja Rech." Prilozhenie Telekom, *Kommersant*, November 16, 2006. www.kommersant.ru/doc.aspx?DocsID=720472.

Ramzy, Austin. "In China, Suspicious Jail Deaths on the Rise." *Time*, March 7, 2008.

Rogers, Paul. "Streisand's Home Becomes Hit on Web." *Mercury News* (San Jose, California), June 24, 2003.

Romero, Simon. "Building a New History By Exhuming Bolívar." *New York Times*, August 3, 2010.

Shaer, Matthew. "From the Tumult in Iran, Twitter Emerges as a Powerful Social Tool." *Christian Science Monitor*, June 17, 2009.

Shambaugh, D. "China's Propaganda System: Institutions, Processes and Efficacy." *China Journal* 57 (January 2007): 25.

Si-Soo, Park. "Police Hunt for Cheonan Rumors." *Korea Times*, June 1, 2010.

Soldatov, A. "Kremlin.com." *Index on Censorship* 39, no. 1 (2010): 71.

Solove, Daniel J. *Understanding Privacy*. Cambridge, MA: Harvard University Press, 2008.

Stewart, Will. "Putin's Poster Girl: Pin-up Politician Who Hates the West . . . but Loves Thatcher." *Daily Mail* (London), March 15, 2009.

Sunstein, C. R., and A. Vermeule. "Conspiracy Theories: Causes and Cures." *Journal of Political Philosophy* 17, no. 2 (2009): 202–227.

Tamayo, Juan. "Cuba Fighting Blogs with Blogs." *Miami Herald*, December 13, 2009.

Taylor, D. G. "Pluralistic Ignorance and the Spiral of Silence: A Formal Analysis." *Public Opinion Quarterly* 46, no. 3 (1982): 311.

"Texting in China: Well-red." *Economist*, February 18, 2010.

"Torgovlja 'Ideologicheskoj' Kartoshkoj." *OPRF*, May 26, 2009. www.oprf.ru/newsblock/news/2468/chamber_news?returnto=0&n=1.

Troianovski, Anton, and Peter Finn. "Kremlin Seeks to Extend Its Reach in Cyberspace." *Washington Post*, October 28, 2007.

"Umaru Yar'adua Regime Launches $5 Million Online War." *Sahara Reporters*, June 16, 2009. saharareporters.com.

"Vietnam to Tighten Control over Online Games." *Vietnam News Briefs*, April 15, 2010.

"V Lenoblasti Startoval Konkurs Blogerov 'Mestnyj blog-2009.'" *Regions.ru*, August 6, 2009. www.regions.ru/news/2231453/.

"V Mahachkale Otkroetsja Shkola Bloggerov." *RIA Dagestan*, February 19, 2010. www.riadagestan.ru/news/2010/02/19/92692/.

Volkova, Marina. "Church to Enhance Its Influence Through Blogging." *Voice of Russia*, July 19, 2010. english.ruvr.ru/2010/07/19/12756896.html.

Von Twickel, Nikolaus. "Putin Calls Hip-Hop a Cure for Booze, Drugs." *Moscow Times*, November 16, 2009.

Walker, Christopher. "Repressing the Media." *Miami Herald*, April 30, 2010.

Wang, Z. "Explaining Regime Strength in China." *China: An International Journal* 4, no. 2 (2006): 217–237.

Watts, Jonathan. "Old Suspicions Magnified Mistrust into Ethnic Riots in Urumqi." *Guardian*, July 10, 2009.

Wong, Edward. "Vice President Turns to Speech Editing." *New York Times*, May 13, 2010.

Wu, Zhong. "China's Internet Awash with State Spies." *Asia Times Online*, August 14, 2008. www.atimes.com/atimes/China/JH14Ad01.html.

Ye, Juliet. "'Hidden Cat': A Prisoner's Death Gives New Meaning to Children's Game." *China Real Time Report Blog, Wall Street Journal*, February 18, 2009. blogs.wsj.com/chinarealtime/2009/02/18/hidden-cat-a-prisoners-death-gives-new-meaning-to-childrens-game/.

Yiannopoulos, Milo. "What is 'The Streisand Effect'?" *Milo Yiannopoulos Blog, Daily Telegraph* (London), January 31, 2009. blogs.telegraph.co.uk/technology/miloyiannopoulos/8248311/What_is_The_Streisand_Effect/.

Zhang, J. "Will the Government 'Serve the People'? The Development of Chinese E-Government." *New Media & Society* 4, no. 2 (2002): 163.

Zhou, Yu. *The Inside Story of China's High-Tech Industry: Making Silicon Valley in Beijing*. Lanham, MD: Rowman & Littlefield, 2008.

CHAPTER 6

Andrianov, Konstantin, and Andrej Kozenko. "'Memorial' Povernul Obysk Vspjat." *Kommersant*, March 21, 2009. www.kommersant.ru/doc.aspx?DocsID=1142280.

Armstrong Moore, Elizabeth. "Google Flu Trends: Take with Grain of Salt." *CNET News*, May 17, 2010. news.cnet.com/8301-27083_3-20005150-247.html.

"Azerbaijani Authorities Interrogate Music Fan over Eurovision Vote for Armenia." Radio Free Europe / Radio Liberty, August 14, 2009.

Balduzzi, M., et al. "Abusing Social Networks for Automated User Profiling." *International Secure Systems Lab* (March 2010).

Bannon, L. J. "Forgetting as a Feature, not a Bug: The Duality of Memory and Implications for Ubiquitous Computing." *CoDesign* 2, no. 1 (2006): 3–15.

Bennett, C. J., and C. D. Raab. *The Governance of Privacy: Policy Instruments in Global Perspective*. Farnham, UK: Ashgate, 2003.

Bilton, Nick. "An Average American Consumes 34 Gigabytes a Day, Study Says." *New York Times*, December 9, 2009.

Blanchette, J. F., and D. G. Johnson. "Data Retention and the Panoptic Society: The Social Benefits of Forgetfulness." *Information Society* 18, no. 1 (2002): 33–45.

"Bloggery Soobwajut, Chto FSB Prosit Udaljat' Posty na Temu Akcij Protesta." *Rambler-Novosti*, December 24, 2008. news.rambler.ru/Russia/head/1634066/?abstroff=0.

Bonneau, J., J. Anderson, R. Anderson, and F. Stajano. "Eight Friends Are Enough: Social Graph Approximation via Public Listings." In *Proceedings of the Second ACM EuroSys Workshop on Social Network Systems*, 13–18. 2009.

Bunyan, T. "Just over the Horizon: The Surveillance Society and the State in the EU." *Race & Class* 51, no. 3 (2010): 1.

"Cambodia Shuts Off SMS Ahead of Elections." Associated Press, April 2, 2007.

Carver, G. A., Jr. "Intelligence in the Age of Glasnost." *Foreign Affairs* 69 (1989): 147.

Clover, Charles. "Stalin-Era Files Raided in 'War over Memory.'" *Financial Times*, December 7, 2008.

Cohen, Jared. *Children of Jihad: A Young American's Travels Among the Youth of the Middle East*. New York: Gotham, 2007.

Dautin, Aleksandr. "Chem Bol'she Oppozicionerov, Tem Huzhe Svjaz'?" *Belorusskie Novosti*, March 29, 2006. naviny.by/rubrics/mobile/2006/03/29/ic_articles_127_133799/.

Dementis, G., and G. Sousa. "A Legal Reasoning Component of a Network Security Command and Control System." Master's thesis, Naval Postgraduate School, Monterey, CA, 2010.

Dodge, M., and R. Kitchin. "The Ethics of Forgetting in an Age of Pervasive Computing." *CASA Working Papers* 92 (2005).

Elliott, Christopher. "Hotels Connecting Dots to Online Reviewers." *Tribune Media Services*, June 4, 2010.

Faris, Robert, Hal Roberts, and Stephanie Wang. "China's Green Dam: The Implications of Government Control Encroaching on the Home PC." OpenNet Initiative, June 2009. opennet.net/chinas-green-dam-the-implica tions-government-control-encroaching-home-pc.

Fassihi, Farnaz. "Iranian Crackdown Goes Global." *Wall Street Journal*, December 3, 2009.

"FBI Backs Record-Keeping on Prepaid Cell Phones." Associated Press, July 31, 2010.

George-Cosh, David. "Blackberry Maker Silent on UAE Security Talks." *National* (Abu Dhabi), July 29, 2010.

"Google: Critics of Vietnam Mine Face Online Attack." Associated Press, March 31, 2010.

"Google Says Vietnam Mine Opponents Under Cyber Attack." *BBC News*, March 31, 2010.

"Government to Banish Unbranded Mobile Phones." *SiliconIndia*, July 14, 2010. www.siliconindia.com/shownews/Government_to_banish_unbrand ed_mobile_phones_-nid-69662.html.

Graham, S., and D. Wood. "Digitizing Surveillance: Categorization, Space, Inequality." *Critical Social Policy* 23, no. 2 (2003): 227.

Heintz, Jim. "Text Messages Warn of Violence in Belarus." Associated Press, March 18, 2006.

Helft, Miguel. "Google Uses Web Searches to Track Flu's Spread." *New York Times*, November 11, 2008.

Henckel von Donnersmarck, Florian. *The Lives of Others*. Sony Pictures, 2006.

Hille, Kathrin. "Censorship Fears Grow as IBM Tool Bolsters Chinese Spam Curb." *Financial Times*, March 25, 2010.

———. "Censorship Fears over China Spam Curb." *Financial Times*, March 24, 2010.

———. "China Bolsters Internet Censors' Scrutiny." *Financial Times*, January 5, 2009.

Holland, H. Brian. "Privacy Paradox 2.0." *Widener Law Journal* (forthcoming).

"Home Office Targets Terror-Related Websites." *BBC News*, February 1, 2010.

Huber, M., S. Kowalski, M. Nohlberg, and S. Tjoa. "Towards Automating Social Engineering Using Social Networking Sites." In *2009 International Conference on Computational Science and Engineering*, 117–124. 2009.

"India Bans Chinese Telecom Equipment." Associated Press, April 30, 2010.

"Iran's Police Vow No Tolerance Towards Protesters." Reuters, February 6, 2010.

Johnson, Carolyn Y. "Project 'Gaydar': An MIT Experiment Raises New Questions About Online Privacy." *Boston Globe*, September 20, 2009.

Jonietz, Erika. "Augmented Identity." *Technology Review*, February 23, 2010. www.technologyreview.com/computing/24639/?a=f.

Kazmin, Amy. "Police Threaten Kashmiri Facebook Users." *Financial Times*, July 19, 2010.

Kirk, Jeremy. "Vietnam Rebuffs Hacking Claims from Google." IDG News Service, April 6, 2010.

Kondrat'ev, Aleksandr, and Taras Podrez. "Minkomsvjazi Vydelit 10 mln Rub. na E-mail Adresa na Kirillice." *Gazeta Marker*, June 4, 2010. www.marker.ru/news/891.

Krishna, Jai. "India Minister: Security Agencies Have Concerns over Blackberry Services." *Wall Street Journal*, July 27, 2010.

Lemos, Robert. "Your Groups Tell Hackers Who You Are." *Technology Review*, July 23, 2010. www.technologyreview.com/printer_friendly_article.aspx?id=25852&channel=web&section=.

Lukyanenko, Peter, and Cathy Young. "Tricks of the KGB trade." *Harper's Magazine*, January 1992.

Mayer-Schonberger, Viktor. *Delete: The Virtue of Forgetting in the Digital Age*. Princeton, NJ: Princeton University Press, 2009.

McMillan, Robert. "Activists Worry About a New 'Green Dam' in Vietnam." IDG News Service, June 4, 2010.

Morar, Natalia. "Blog for Democracy, from the Streets of Moldova." *OpenDemocracy*, April 8, 2009. www.opendemocracy.net/article/email/blog-for-democracy-from-the-streets-of-moldova.

"Moskovskaja Milicija Monitorit Twitter." *Vebplaneta*, November 10, 2009. webplanet.ru/news/life/2009/11/10/iopasnaitrudna.html.

Mumford, Lewis. *The Pentagon of Power*. New York: Harcourt Brace Jovanovich, 1974.

"Nigeria: New Policy on Mobile Phone SIM Cards." *Daily Trust*, January 11, 2010. allafrica.com/stories/201001110286.html.

"Note to Readers." *New York Times*, September 13, 2009. www.nytimes.com/2009/09/13/business/media/13note.html?_r=1.

Owad, Tom. "Data Mining 101: Finding Subversives with Amazon Wishlists." *Applefritter*, January 4, 2006. www.applefritter.com/bannedbooks.

"P2P Comes to the Aid of Audiovisual Search." *PhysOrg.com*. November 18, 2009. www.physorg.com/news177780052.html.

Page, Lewis. "NSA Offering 'Billions' for Skype Eavesdrop Solution." *Register*, February 12, 2009. www.theregister.co.uk/2009/02/12/nsa_offers_bil lions_for_skype_pwnage/.

Palmer, Maija. "Face Recognition Software Gaining a Broader Canvas." *Financial Times*, May 22, 2010.

———. "Google Debates Face Recognition Technology After Privacy Blunders." *Financial Times*, May 20, 2010.

Pankavec, Zmicer. "KDB Verbue Praz vkontakte.ru." *Nasha Niva*, December 19, 2009.

Peterson, Kristina. "Intelligence Agents Borrow Wall Street Trading Technology." *Wall Street Journal*, May 28, 2010.

Pljuwev, Aleksandr. "Poiskovye Sistemy Kak Strategicheskij i Politicheskij Resurs." Tochka, *Jeho Moskvy*, January 24, 2010. echo.msk.ru/programs/tochka/651123-echo/.

Scheck, Justin. "Stalkers Exploit Cellphone GPS." What They Know series, *Wall Street Journal*, August 3, 2010.

Sharma, A. "Cyber Wars: A Paradigm Shift from Means to Ends." *Strategic Analysis* 34, no. 1 (2010): 62–73.

Simonite, Tom. "Surveillance Software Knows What a Camera Sees." *Technology Review*, June 1, 2010. www.technologyreview.com/computing/25439/?a=f.

Soar, Daniel. "Short Cuts." *London Review of Books*, August 14, 2008.

Soghoian, Christopher. "Caught in the Cloud: Privacy, Encryption, and Government Back Doors in the Web 2.0 Era." *Journal of Telecommunications and High Technology Law*, August 17, 2009.

———. "Exclusive: Widespread Cell Phone Location Snooping by NSA?" *Surveillance State, CNET News*, September 8, 2008. news.cnet.com/8301-13739_3-10030134-46.html?tag=mncol;title.

Solove, Daniel J. *The Digital Person: Technology and Privacy in the Information Age*. Ex Machina. New York: New York University Press, 2004.

———. "Do Social Networks Bring the End of Privacy?" *Scientific American*, September 2008.

———. "I've Got Nothing to Hide and Other Misunderstandings of Privacy." *San Diego Law Review* 44 (2007): 745.

———. *Understanding Privacy*. Cambridge, MA: Harvard University Press, 2008.

Sternstein, Aliya. "Lawmaker Questions White House Official's Use of Gmail." *Nextgov*, April 12, 2010. www.nextgov.com/nextgov/ng_20100412_6003.php.

Stone, Brad. "T.M.I.? Not for Sites Focused on Sharing." *New York Times*, April 22, 2010.

Timmons, Heather. "India Wary of Chinese Telecom Equipment." *New York Times*, April 30, 2010.

"Vietnam: Farmers to Get Free Computers in Plan to Boost Production." *Thai Press Reports*, April 14, 2010.

"Vietnam Launches Own Social Network Site." Agence France-Presse, May 22, 2010.

"Vietnam Politics: A State-Run Social Networking Website Is Launched." *EIU ViewsWire*, June 7, 2010.

"Vietnam Steps Up China-Style Internet Control." Agence France-Presse, July 1, 2010.

Villeneuve, Nart, and Greg Walton. "'0day': Civil Society and Cyber Security." Malware Lab, October 28, 2009. malwarelab.org/2009/10/0day-civil-society-and-cyber-security/.

Wells-Dang, A. "Political Space in Vietnam: A View from the 'Rice-Roots.'" *Pacific Review* 23, no. 1 (2010): 93–112.

Wigglesworth, Robin. "UAE Comments Raise Fears of Crackdown on Black-Berry." *Financial Times*, July 26, 2010.

Wines, Michael. "In Restive Chinese Area, Cameras Keep Watch." *New York Times*, August 2, 2010.

Wondracek, G., T. Holz, E. Kirda, S. Antipolis, and C. Kruegel. "A Practical Attack to De-Anonymize Social Network Users." In *2010 IEEE Symposium on Security and Privacy*, 223–238. 2010.

Young, Martin J. "Vietnam Strengthens Firewall." *Asia Times Online*, June 19, 2010. www.atimes.com/atimes/Global_Economy/LF19Dj03.html.

Zetter, Kim. "Tor Researcher Who Exposed Embassy E-mail Passwords Gets Raided by Swedish FBI and CIA." *Threat Level, Wired*, November 14, 2007. www.wired.com/threatlevel/2007/11/swedish-researc/.

CHAPTER 7

Agre, P. E. "The Practical Republic: Social Skills and the Progress of Citizenship." *Community in the Digital Age: Philosophy and Practice* (2004): 201–223.

Alexander, M. G., and S. Levin. "Theoretical, Empirical, and Practical Approaches to Intergroup Conflict." *Journal of Social Issues* 54, no. 4 (1998): 629–639.

Alinsky, Saul David. *Rules for Radicals: A Practical Primer for Realistic Radicals.* New York: Vintage Books, 1989.

Arguello, J., B. S. Butler, E. Joyce, R. Kraut, K. S. Ling, C. Rosé, and X. Wang. "Talk to Me: Foundations for Successful Individual-Group Interactions in Online Communities." In *Proceedings of the SIGCHI Conference on Human Factors in Computing Systems*, 968. 2006.

Bakardjieva, M. "Virtual Togetherness: An Everyday-Life Perspective." *Media, Culture & Society* 25, no. 3 (2003): 291.

Bargh, J. A., and K. Y. McKenna. "The Internet and Social Life." *Annual Review of Psychology* 55 (2004): 573.

Beenen, G., K. Ling, X. Wang, K. Chang, D. Frankowski, P. Resnick, and R. E. Kraut. "Using Social Psychology to Motivate Contributions to Online Communities." In *Proceedings of the 2004 ACM Conference on Computer Supported Cooperative Work*, 221. 2004.

Bennett, Drake. "Protesters' Secret: They're Out There Because It Makes Them Happier." *Boston Globe*, October 11, 2009.

Bennett, W. L. *Changing Citizenship in the Digital Age.* Cambridge, MA: MIT Press, 2008.

Bennett, W. "Communicating Global Activism." *Information, Communication & Society* 6, no. 2 (2003): 143–168.

Billig, M., and H. Tajfel. "Social Categorization and Similarity in Intergroup Behaviour." *European Journal of Social Psychology* 3, no. 1 (1973): 27–52.

Bimber, B., A. J. Flanagin, and C. Stohl. "Reconceptualizing Collective Action in the Contemporary Media Environment." *Communication Theory* 15, no. 4 (2005): 365–388.

Blears, James. "Information Revolution Breathes Life into Alliance of Youth Movements." *Fox News*, October 18, 2009.

Borgmann, Albert. *Technology and the Character of Contemporary Life: A Philosophical Inquiry.* Chicago: University of Chicago Press, 1987.

Bornstein, G., L. Crum, J. Wittenbraker, K. Harring, C. A. Insko, and J. Thibaut. "On the Measurement of Social Orientations in the Minimal Group Paradigm." *European Journal of Social Psychology* 13, no. 4 (1983): 321–350.

Boudreaux, Richard. "Rent-a-Crowd Entrepreneurs in Ukraine Find People Fast to Cheer or Jeer for Any Cause." *Wall Street Journal*, February 5, 2010.

Bratich, Jack Z. "The Fog Machine." *CounterPunch*, June 22, 2009. www.counter punch.org/bratich06222009.html.

Chidambaram, L., and L. L. Tung. "Is Out of Sight, Out of Mind? An Empirical Study of Social Loafing in Technology-Supported Groups." *Information Systems Research* 16, no. 2 (2005): 149.

Chin, M. G., and C. G. McClintock. "The Effects of Intergroup Discrimination and Social Values on Level of Self-Esteem in the Minimal Group Paradigm." *European Journal of Social Psychology* 23, no. 1 (1993): 63–75.

Christopherson, K. M. "The Positive and Negative Implications of Anonymity in Internet Social Interactions." *Computers in Human Behavior* 23, no. 6 (2007): 3038–3056.

Dahlgren, P. "Doing Citizenship: The Cultural Origins of Civic Agency in the Public Sphere." *European Journal of Cultural Studies* 9, no. 3 (2006): 267.

Davis, A. "New Media and Fat Democracy: The Paradox of Online Participation." *New Media & Society* (2009).

Davis, Angela Y. *Abolition Democracy: Beyond Empire, Prisons, and Torture.* New York: Seven Stories Press, 2005.

Daum, Megham. "Not Sold on RentAFriend." *Los Angeles Times*, July 8, 2010.

Della Porta, D., and L. Mosca. "Global-Net for Global Movements? A Network of Networks for a Movement of Movements." *Journal of Public Policy* 25, no. 1 (2005): 165–190.

Diehl, M. "The Minimal Group Paradigm: Theoretical Explanations and Empirical Findings." *European Review of Social Psychology* 1, no. 1 (1990): 263–292.

"Digital Revolution." *Statesman*, June 9, 2009.

Dreyfus, Hubert L. "Anonymity Versus Commitment: The Dangers of Education on the Internet." *Ethics and Information Technology* 1, no. 1 (1999): 15–20.

———. "Kierkegaard on the Internet: Anonymity vs. Commitment in the Present Age." *Kierkegaard Studies: Yearbook* (1999): 96–109.

———. *On the Internet.* Thinking in Action. New York: Routledge, 2001.

Duncan, W. J. "Why Some People Loaf in Groups While Others Loaf Alone." *Academy of Management Executive (1993–2005)* 8, no. 1 (1994): 79–80.

Faris, D. "Revolutions Without Revolutionaries? Network, Theory, Facebook, and the Egyptian Blogosphere." *Arab Media and Society* (2008).

Flanagin, A. J., C. Stohl, and B. Bimber. "Modeling the Structure of Collective Action." *Communication Monographs* 73, no. 1 (2006): 29–54.

"Funding Opportunity Title: New Empowerment Communication Technologies: Opportunities in the Middle East and North Africa." U.S. Department of State, September 15, 2009. mepi.state.gov/opportunities/129624.htm.

Garff, Joakim. *Søren Kierkegaard: A Biography*. Princeton, NJ: Princeton University Press, 2005.

Garrett, R. K. "Protest in an Information Society: A Review of Literature on Social Movements and New ICTs." *Information, Communication & Society* 9, no. 2 (2006): 202–224.

Geen, R. G. "Social Motivation." *Annual Review of Psychology* 42, no. 1 (1991): 377–399.

Giridharadas, Anand. "'Buycotting': Boycotts Minus the Pain." *New York Times*, October 10, 2009.

Greer, C., and E. McLaughlin. "We Predict a Riot? Public Order Policing, New Media Environments and the Rise of the Citizen Journalist." *British Journal of Criminology* (2010).

Harkins, S. G., and R. E. Petty. "Effects of Task Difficulty and Task Uniqueness on Social Loafing." *Journal of Personality and Social Psychology* 43, no. 6 (1982): 1214–1229.

Heil, Alan L. *Voice of America: A History*. New York: Columbia University Press, 2003.

Hesse, Monica. "Facebook Activism: Lots of Clicks, but Little Sticks." *Washington Post*, July 2, 2009.

Higgs, Eric, Andrew Light, and David Strong, eds. *Technology and the Good Life?* Chicago: University of Chicago Press, 2000.

"Homosexuality in Nigeria: Go Online If You're Glad to Be Gay." *Economist*, February 11, 2010.

Hutchins, B., and L. Lester. "Environmental Protest and Tap-Dancing with the Media in the Information Age." *Media, Culture & Society* 28, no. 3 (2006): 433.

Huyke, H. J. "Technologies and the Devaluation of What Is Near." *Techné: The Journal of the Society for Philosophy and Technology* 6, no. 3 (2003): 1–17.

Ingham, A. G., J. Graves, and V. Peckham. "The Ringelmann Effect: Studies of Group Size and Group Performance." *Journal of Experimental Social Psychology* 10, no. 4 (1974): 371–384.

Jayson, Sharon. "Are Social Networks Making Students More Narcissistic?" *USA Today*, August 25, 2009.

Kierkegaard, Søren. *Either/Or: A Fragment of Life*. Edited by Victor Eremita. Translated by Alastair Hannay. New York: Penguin Books, 1992.

———. *Fear and Trembling*. Edited by C. Stephen Evans and Sylvia Walsh. Translated by Sylvia Wash. New York: Cambridge University Press, 2006.

Klar, M., and T. Kasser. "Some Benefits of Being an Activist: Measuring Activism and Its Role in Psychological Well-Being." *Political Psychology* 30, no. 5 (2009): 755–777.

Levine, John M., and Richard L. Moreland, eds. *Small Groups: Key Readings*. New York: Psychology Press, 2006.

Lombaard, C. "Fleetingness and Media-ted Existence: From Kierkegaard on the Newspaper to Broderick on the Internet." *Communication* 35, no. 1 (2009): 17–29.

Lupia, A., and G. Sin. "Which Public Goods Are Endangered? How Evolving Communication Technologies Affect the Logic of Collective Action." *Public Choice* 117, no. 3 (2003): 315–331.

Lysenko, V. V., and K. C. Desouza. "Cyberprotest in Contemporary Russia: The Cases of Ingushetiya.ru and Bakhmina.ru." *Technological Forecasting and Social Change* 77, no. 7 (September 2010).

Martin, B., and W. Varney. "Nonviolence and Communication." *Journal of Peace Research* 40, no. 2 (2003): 213.

Mazzoleni, G., and W. Schulz. "'Mediatization' of Politics: A Challenge for Democracy?" *Political Communication* 16, no. 3 (1999): 247–261.

McCarthy, Caroline. "Facebook, Google, Others Sponsor Youth Activism Summit." *CNET News*, November 18, 2008. news.cnet.com/8301-13577 _3-10101653-36.html.

Moqadam, Afsaneh. *Death to the Dictator! A Young Man Casts a Vote in Iran's 2009 Election and Pays a Devastating Price*. New York: Farrar, Straus and Giroux, 2010.

Morozov, Evgeny. "From Slacktivism to Activism." *Net Effect, Foreign Policy*, September 5, 2009. neteffect.foreignpolicy.com/posts/2009/09/05/from _slacktivism_to_activism.

———. "It Feels Like Activism." *Newsweek International*, June 29, 2009.

Nielsen, Rasmus Kleis. "The Labors of Internet-Assisted Activism: Overcommunication, Miscommunication, and Communicative Overload." *Journal of Information Technology & Politics* 6, no. 3 (July 2009): 267–280.

Oliver, P. E. "Formal Models of Collective Action." *Annual Review of Sociology* 19, no. 1 (1993): 271–300.

Olson, Mancur. *The Logic of Collective Action: Public Goods and the Theory of Groups*. Harvard Economic Studies, vol. 124. Cambridge, MA: Harvard University Press, 1971.

Piezon, S. L., and R. L. Donaldson. "Online Groups and Social Loafing: Understanding Student-Group Interactions." *Online Journal of Distance Learning Administration* 8, no. 4 (2005).

Rev, Istvan. "Just Noise?" Paper presented at the Conference on Cold War Broadcasting Impact, Hoover Institution, Stanford University, Stanford, California, October 13–15, 2004.

Rutkowski, A. F., D. Vogel, M. van Genuchten, and C. Saunders. "Communication in Virtual Teams: Ten Years of Experience in Education." *IEEE Transactions on Professional Communication* 3 (2008): 302–312.

Ryan, Alan. "Exaggerated Hopes and Baseless Fears." *Social Research* 64, no. 3 (Fall 1997).

Scola, Nancy. "The Next Diplomatic Cable." *American Prospect,* July 27, 2009.

"Secretary Rice Remarks with U.K. Foreign Secretary David Miliband and Google Senior Vice President David Drummond." U.S. Department of State, May 22, 2008.

Shah, D. V., J. Cho, W. P. Eveland, et al. "Information and Expression in a Digital Age: Modeling Internet Effects on Civic Participation." *Communication Research* 32, no. 5 (2005): 531.

Shapiro, Samantha M. "Revolution, Facebook-Style: Can Social Networking Turn Young Egyptians into a Force for Democratic Change?" *New York Times Magazine,* January 22, 2009.

Shaw, Gillian. "Social Media Don't Promote Significant Social Change, Author Says." *Vancouver Sun,* April 9, 2010.

Shirky, Clay. *Here Comes Everybody: The Power of Organizing Without Organizations.* New York: Penguin Books, 2009.

Shiue, Y. C., C. M. Chiu, and C. C. Chang. "Exploring and Mitigating Social Loafing in Online Communities." *Computers in Human Behavior* (2010).

Skitka, L. J., and E. G. Sargis. "The Internet as Psychological Laboratory." *Annual Review of Psychology* 57 (2006): 529.

Slee, Tom. "Digital Activism: If Information Is Not the Problem, Information Is Not the Solution." *Whimsley,* January 10, 2010. whimsley.typepad.com/whimsley/2010/01/digital-activism-if-information-is-not-the-problem-information-is-not-the-solution.html.

Starobin, Paul. "In New Media, Image Is Still Everything." *National Journal Magazine,* September 12, 2009.

Stolle, D., M. Hooghe, and M. Micheletti. "Politics in the Supermarket: Political Consumerism as a Form of Political Participation." *International Political Science Review/ Revue internationale de science politique* 26, no. 3 (2005): 245.

Stone, Brad. "Users of Social Networks Use Posts to Support Charity." *New York Times,* November 11, 2009.

Suleiman, J., and R. T. Watson. "Social Loafing in Technology-Supported Teams." *Computer Supported Cooperative Work* 17, no. 4 (2008): 291–309.

Tajfel, H., M. G. Billig, R. P. Bundy, and C. Flament. "Social Categorization and Intergroup Behaviour." *European Journal of Social Psychology* 1, no. 2 (1971): 149–178.

Tripathi, A. K. "On the Internet: Thinking in Action." *Information Technology & People* 15, no. 4 (2002): 136.

Twenge, Jean M., and W. Keith Campbell. *The Narcissism Epidemic: Living in the Age of Entitlement*. New York: Simon & Schuster, 2009.

"U.S. Official Discusses Alliance of Youth Movements Summit." U.S. Department of State, December 1, 2008.

"Undersecretary Glassman and Jared Cohen Hold a News Briefing on the Alliance for Youth Movements Summit at Columbia University at the Foreign Press Center, as Released by the State Department." *Political Transcript Wire*, November 24, 2008.

van Dick, R., P. A. Tissington, and G. Hertel. "Do Many Hands Make Light Work?" *European Business Review* 21, no. 3 (2009): 233–245.

Voelpel, S. C., R. A. Eckhoff, and J. Forster. "David Against Goliath? Group Size and Bystander Effects in Virtual Knowledge Sharing." *Human Relations* 61, no. 2 (2008): 271.

Wagner, J. A., III. "Studies of Individualism-Collectivism: Effects on Cooperation in Groups." *Academy of Management Journal* 38, no. 1 (1995): 152–172.

Williams, K. D. "Social Loafing on Difficult Tasks: Working Collectively Can Improve Performance." *Journal of Personality and Social Psychology* 49, no. 4 (1985): 937–942.

Williams, K. D., S. Harkins, and B. Latané. "Identifiability as a Deterrent to Social Loafing: Two Cheering Experiments." *Journal of Personality and Social Psychology* 40, no. 2 (1981): 303–311.

Witte, E. H. "Köhler Rediscovered: The Anti-Ringelmann Effect." *European Journal of Social Psychology* 19, no. 2 (1989): 147–154.

Worth, Robert F. "Opposition in Iran Meets a Crossroads on Strategy." *New York Times*, February 14, 2010.

CHAPTER 8

Abidin Besleney, Zeynel. "Circassian Nationalism and the Internet." *openDemocracy*, May 21, 2010. www.opendemocracy.net/od-russia/zeynel-abidin-besleney/circassian-nationalism-and-internet.

Arrington, Michael. "Ok You Luddites, Time to Chill Out on Facebook over Privacy." *TechCrunch*, January 12, 2010. techcrunch.com/2010/01/12/ok-you-luddites-time-to-chill-on-facebook-over-privacy/.

———. "Reputation Is Dead: It's Time to Overlook Our Indiscretions." *TechCrunch*, March 28, 2010. techcrunch.com/2010/03/28/reputation-is-dead-its-time-to-overlook-our-indiscretions/.

Baker, L. "The Unintended Consequences of US Export Restrictions on Software and Online Services for American Foreign Policy and Human Rights." *Harvard Journal of Law & Technology* 23, no. 2 (2010).

Barber, B. R. "The Ambiguous Effects of Digital Technology on Democracy in a Globalizing World." In *Innovations for an E-Society: Challenges for Technology Assessment*, edited by Gerhard Banse, Armin Grunwald, and Michael Rader, 43–56. Berlin: Edition Sigma, 2002.

Bartow, A. "A Portrait of the Internet as a Young Man." *Michigan Law Review* 108, no. 6 (2010).

Barlow, John Perry. "Leaving the Physical World." *EFF.org*, 1993. w2.eff.org/Misc/Publications/John_Perry_Barlow/HTML/leaving_the_physical_world.html.

Baumgartner, J. C., and J. S. Morris. "MyFaceTube Politics: Social Networking Web Sites and Political Engagement of Young Adults." *Social Science Computer Review* 28, no. 1 (2010): 24.

Beniger, J. R. "Personalization of Mass Media and the Growth of Pseudo-Community." *Communication Research* 14, no. 3 (1987): 352.

Billing, Soren. "Saudi Campaign to Clean Up YouTube." *ITP.net*, August 13, 2009. www.itp.net/564689-its-just-boredom.

Bimber, Bruce. *Information and American Democracy: Technology in the Evolution of Political Power*. New York: Cambridge University Press, 2003.

Boudreau, John. "Activists Aim to Punch Holes in Online Shields of Authoritarian Regimes." *San Jose Mercury News* (San Jose, California), February 17, 2010.

Brenkert, G. G. "Corporate Control of Information: Business and the Freedom of Expression." *Business and Society Review* 115, no. 1 (2010): 121–145.

———. "Google, Human Rights, and Moral Compromise." *Journal of Business Ethics* 85, no. 4 (2009): 453–478.

Buchstein, H. "Bytes That Bite: The Internet and Deliberative Democracy." *Constellations* 4, no. 2 (1997): 248–263.

Burrell, J. "Problematic Empowerment: West African Internet Scams as Strategic Misrepresentation." *Information Technologies and International Development* 4, no. 4 (2008): 15–30.

Burton, Matthew. "On the Weaponization of the Collaborative Web." Personal Democracy Forum, June 16, 2009. personaldemocracy.com/blog-entry/weaponization-collaborative-web.

Carr, M. "Slouching Towards Dystopia: The New Military Futurism." *Race & Class* 51, no. 3 (2010): 13.

Carr, Nicholas. "Is Google Making Us Stupid?" *Atlantic*, August 2008.

Cavelty, M. D. "Cyber-Terror—Looming Threat or Phantom Menace? The Framing of the US Cyber-Threat Debate." *Journal of Information Technology & Politics* 4, no. 1 (2007): 19–36.

Clarke, Richard A., and Robert Knake. *Cyber War: The Next Threat to National Security and What to Do About It.* New York: HarperCollins, 2010.

Clinton, Hillary. "Remarks on Internet Freedom." The Newseum, Washington, DC, January 21, 2010.

———. "Speech to Kaiser Family Foundation." March 8, 2005.

Cowie, James. "The Proxy Fight for Iranian Democracy." *Renesys Blog,* June 22, 2009. www.renesys.com/blog/2009/06/the-proxy-fight-for-iranian -de.shtml.

Curtin, M. "Beyond the Vast Wasteland: The Policy Discourse of Global Television and the Politics of American Empire." *Journal of Broadcasting & Electronic Media* 37, no. 2 (1993): 127–145.

Dahlberg, L. "Democracy via Cyberspace: Mapping the Rhetorics and Practices of Three Prominent Camps." *New Media & Society* 3, no. 2 (2001): 157.

Damm, J. "The Internet and the Fragmentation of Chinese Society." *Critical Asian Studies* 39, no. 2 (2007): 273–294.

Deibert, R. J., and R. Rohozinski. "Risking Security: Policies and Paradoxes of Cyberspace Security." *International Political Sociology* 4, no. 1 (2010): 15–32.

Dobson, William J. "Computer Programmer Takes On the World's Despots." *Newsweek*, August 6, 2010.

El-Khairy, O. "'Freedom's a Lifestyle Choice': US Cultural Diplomacy, Empire's Soundtrack, and Middle Eastern 'Youth' in our Contemporary Global Infowar." *Middle East Journal of Culture and Communication* 2, no. 1 (2009): 115–135.

Elmusa, S. S. "Faust Without the Devil? The Interplay of Technology and Culture in Saudi Arabia." *Middle East Journal* 51, no. 3 (1997): 345–357.

"Facebook Deletes Hong Kong Groups That Oppose Pro-Beijing Party." *BBC Monitoring International Reports*, February 5, 2010.

Falvey, Christian. "Minister's Web Monitoring Tool Off to Rocky Start." Radio Prague, February 12, 2010.

Fandy, M. "Information Technology, Trust, and Social Change in the Arab World." *Middle East Journal* (2000): 378–394.

Fletcher, Owen. "Apple Censors Dalai Lama iPhone Apps in China." IDG News Service, December 30, 2009.

Fox, J. "The Uncertain Relationship Between Transparency and Accountability." *Development in Practice* 17, no. 4 (2007): 663–671.

Franzese, P. W. "Sovereignty in Cyberspace: Can It Exist?" *Air Force Law Review* 64 (2009): 1.

Fuchs, C. "Some Reflections on Manuel Castells' Book 'Communication Power.'" *Triple C: Cognition, Communication, Co-operation* 7, no. 1 (2009): 94.

Garnham, N. "The Mass Media, Cultural Identity, and the Public Sphere in the Modern World." *Public Culture* 5, no. 2 (1993): 251.

Gilboa, E. "The CNN Effect: The Search for a Communication Theory of International Relations." *Political Communication* 22, no. 1 (2005): 27–44.

———. "Global Communication and Foreign Policy." *Journal of Communication* 52, no. 4 (2002): 731–748.

Glassman, James K., and Michael Doran. "How to Help Iran's Green Revolution." *Wall Street Journal*, January 21, 2010.

Glenny, Misha. "BlackBerry Is but a Skirmish in the Battle for the Web." *Financial Times*, August 6, 2010.

Goldsmith, Jack L., and Tim Wu. "Digital Borders." *Legal Affairs* (2006): 40.

———. *Who Controls the Internet?: Illusions of a Borderless World*. New York: Oxford University Press, 2006.

Gunaratne, S. A. "De-Westernizing Communication/Social Science Research: Opportunities and Limitations." *Media, Culture & Society* 32, no. 3 (2010): 473.

Guynn, Jessica. "Twitter Hires Obama Administration's Katie Stanton." *Los Angeles Times*, July 10, 2010.

Hardy, Michael. "In-Q-Tel, Google Invest in Recorded Future." *Government Computer News*, July 29, 2010. gcn.com/articles/2010/07/29/inqtel-google-fund-web-analysis-firm.aspx.

Hawkins, V. "The Other Side of the CNN Factor: The Media and Conflict." *Journalism Studies* 3, no. 2 (2002): 225–240.

Himma, K. E. "Hacking as Politically Motivated Digital Civil Disobedience: Is Hacktivism Morally Justified?" *Internet Security: Hacking, Counterhacking, and Society* (2007): 73.

Hindman, Matthew. *The Myth of Digital Democracy*. Princeton, NJ: Princeton University Press, 2009.

Hofmann, J. "The Libertarian Origins of Cybercrime: Unintended Side-Effects of a Political Utopia." ESRC Research Centre Discussion Paper no. 62, 2010. w.lse.ac.uk/collections/CARR/pdf/DPs/Disspaper62.pdf.

Holmes, Allan. "Defining Transparency." *Nextgov*, September 3, 2009. www.nextgov.com/nextgov/ng_20090903_7217.php.

———. "The Risks of Open Government." *Nextgov*, September 14, 2009. www.nextgov.com/nextgov/ng_20090914_3118.php.

Howe, Jeff P. "Obama and Crowdsourcing: A Failed Relationship?" *Wired Epicenter Blog*, April 1, 2009. www.wired.com/epicenter/2009/04/obama-and-crowd/.

Ibahrine, M. "Mobile Communication and Sociopolitical Change in the Arab World." *Quaderns de la Mediterránia* 11 (2009): 51–60.

"Interview with Indira Lakshmanan of Bloomberg TV." U.S. Department of State, March 19, 2010.

Issa, Antoun. "Palestine: Twitter Accused of Silencing Gaza Tribute." *Global Voices*, December 29, 2009. globalvoicesonline.org/2009/12/29/palestine-twitter-accused-of-silencing-gaza-tribute/.

Jaeger, P. T., J. Lin, J. M. Grimes, and S. N. Simmons. "Where Is the Cloud? Geography, Economics, Environment, and Jurisdiction in Cloud Computing." *First Monday* 14, no. 5 (2009).

Jakobsen, P. V. "Focus on the CNN Effect Misses the Point: The Real Media Impact on Conflict Management Is Invisible and Indirect." *Journal of Peace Research* 37, no. 2 (2000): 131.

Jenkins, Henry. "The Chinese Columbine." *Technology Review*, August 2, 2002. www.technologyreview.com/read_article.aspx?id=12913&ch=infotech.

Johnson, D. G. "Is the Global Information Infrastructure a Democratic Technology?" *Readings in Cyberethics* 18 (2004): 121.

Katz, J. E., and C. H. Lai. "News Blogging in Cross-Cultural Contexts: A Report on the Struggle for Voice." *Knowledge, Technology & Policy* 22, no. 2 (2009): 95–107.

Kenner, David. "Useless Democracy Promotion Efforts? There's an App for That." *FP Passport, Foreign Policy*, December 31, 2009. blog.foreignpolicy.com/posts/2009/12/31/useless_democracy_promotion_efforts_theres_an_app_for_that.

Khouri, Rami G. "When Arabs Tweet." *International Herald Tribune*, July 22, 2010.

Kingsbury, P., and J. P. Jones III. "Walter Benjamin's Dionysian Adventures on Google Earth." *Geoforum* 40, no. 4 (2009): 502–513.

Kirkpatrick, Marshall. "Jordan Says It Will Begin Censoring Websites." *Read-WriteWeb*, January 14, 2010. www.readwriteweb.com/archives/jordan_to_censor_websites.php.

Klang, M. "Civil Disobedience Online." *Journal of Information, Communication & Ethics in Society* 2, no. 2 (2008): 2.

Kleine, D., and T. Unwin. "Technological Revolution, Evolution and New Dependencies: What's New About ict4d?" *Third World Quarterly* 30, no. 5 (2009): 1045–1067.

Kleinz, Torsten, and Craig Morris. "Higher Regional Court Says Online Demonstration Is Not Force." *Heise Online*, June 2, 2006. www.heise.de/english/newsticker/news/73827.

Kluver, R. "US and Chinese Policy Expectations of the Internet." *China Information* 19, no. 2 (2005): 299.

Kluver, R., and P. H. Cheong. "Technological Modernization, the Internet, and Religion in Singapore." *Journal of Computer-Mediated Communication* 12, no. 3 (2007): 1122–1142.

Lagerkvist, J. "Global Media for Global Citizenship in India and China." *Peace Review* 21, no. 3 (2009): 367–375.

Land, M. B. "Peer Producing Human Rights." *Alberta Law Review* 46, no. 4 (2009).

Landler, Mark. "U.S. Hopes Exports Will Help Open Closed Societies." *New York Times*, March 7, 2010.

Lee, Tae-hoon. "Lawmaker Calls for Stricter Access to NK Sites." *Korea Times*, October 6, 2009.

Lessig, L. "Against Transparency." *New Republic* 9 (2010).

Lewis, James. "Sovereignty and the Role of Government in Cyberspace." *The Brown Journal of World Affairs* 16, no. 2 (2010).

Lichtenstein, Jesse. "Digital Diplomacy." *New York Times Magazine*, July 16, 2010.

Loftus, Meghan. "People Use Social Networking to Fight Violence, Extremism." U.S. Department of Defense, December 2, 2008.

Lonkila, M., and B. Gladarev. "Social Networks and Cellphone Use in Russia: Local Consequences of Global Communication Technology." *New Media & Society* 10, no. 2 (2008): 273.

Luhr, N. L. "Iran, Social Media, and US Trade Sanctions: The First Amendment Implications of US Foreign Policy." *First Amendment Law Review* 8 (2010): 500–533.

Lynch, M. "Blogging the New Arab Public." *Arab Media & Society* 1, no. 1 (2007).

———. "The Internet Freedom Agenda." *Abu Aardvark's Middle East Blog*, January 22, 2010. lynch.foreignpolicy.com/posts/2010/01/22/the_internet _freedom_agenda.

MacKinnon, R. "China's Censorship 2.0: How companies censor bloggers." *First Monday* 14, no. 2-2 (2009).

———. "The Great Chinese Censorship Hoax." *RConversation*, March 14, 2006. rconversation.blogs.com/rconversation/2006/03/the_great_chine .html.

——— "Liberty or Safety? Both—or Neither." *IEEE Spectrum*, May 2010. spectrum.ieee.org/telecom/internet/liberty-or-safety-bothor-neither.

Markoff, John. "U.S. and Russian Accord on Display at Internet Meeting." *New York Times*, April 15, 2010.

Marosi, Richard. "UC San Diego Professor Who Studies Disobedience Gains Followers—and Investigators." *Los Angeles Times*, May 7, 2010.

Martin, K. E. "Internet Technologies in China: Insights on the Morally Important Influence of Managers." *Journal of Business Ethics* 83, no. 3 (2008): 489–501.

McCarthy, Caroline. "Philly Targets Facebook, Twitter After Snowball Fight Turns Ugly." *CNET*, February 17, 2010. news.cnet.com/8301-13577_3 -10455254-36.html.

McConnell, Mike. "Mike McConnell on How to Win the Cyber-War We're Losing." *Washington Post*, February 28, 2010.

McMillan, Robert. "Citing Cybercrime, FBI Director Doesn't Bank Online." IDG News Service, October 7, 2009.

Mearsheimer, John J. *The Tragedy of Great Power Politics*. New York: W. W. Norton, 2003.

Merelman, R. M. "Technological Cultures and Liberal Democracy in the United States." *Science, Technology & Human Values* 25, no. 2 (2000): 167.

Metzl, J. F. "Information Intervention: When Switching Channels Isn't Enough." *Foreign Affairs* 76, no. 6 (1997): 15–20.

———. "Rwandan Genocide and the International Law of Radio Jamming." *American Journal of International Law* 91, no. 4 (1997): 628–651.

Miller, J. "Soft Power and State–Firm Diplomacy: Congress and IT Corporate Activity in China." *International Studies Perspectives* 10, no. 3 (2009): 285–302.

Mite, Valentinas. "Estonia: Attacks Seen as First Case of 'Cyberwar.'" Radio Free Europe / Radio Liberty, May 30, 2007.

Morozov, Evgeny. "More on the Unintended Consequences of DDoS Attacks on Pro-Ahmadinejad Web-sites." *Net Effect, Foreign Policy,* June 18, 2009. neteffect.foreignpolicy.com/posts/2009/06/18/more_on_the_unintended _consequences_of_ddos_attacks_on_pro_ahmadinejad_web_sites.

———. "U.S. Web Firms Practice Self-Censorship." *Newsweek International,* March 7, 2009.

Moses, Asher. "Facebook Bans Doll Nipples." *Sydney Morning Herald,* July 5, 2010.

Mynihan, Colin. "Arrest of Queens Man Puts Focus on Texting to Rally Protesters." *New York Times,* October 9, 2009.

Nakashima, Ellen. "Dismantling of Saudi-CIA Web Site Illustrates Need for Clearer Cyberwar Policies." *Washington Post,* March 19, 2010.

Noam, Eli M. "An Unfettered Internet? Keep Dreaming." *New York Times,* July 11, 1997.

Norris, P. "A Virtuous Circle? The Impact of Political Communications in Post-Industrial Democracies." In *Papers for the Annual Meeting of the Political Studies Association of the UK, London School of Economics and Political Science.* 2000.

Nye, J. S., Jr. *Cyber Power.* Cambridge, MA: Belfer Center for Science and International Affairs, Harvard University, 2010.

"Obama Bemoans 'Diversions' of iPod, Xbox Era." Agence France-Presse, May 9, 2010.

"Obama Pushes China to Stop Censoring Internet." National Public Radio, November 16, 2010.

Oboler, A. "The Rise and Fall of a Facebook Hate Group." *First Monday* 13, no. 11-3 (2008).

O'Reilly, Tim. "My Contrarian Stance on Facebook and Privacy." *O'Reilly Radar,* May 21, 2010. radar.oreilly.com/2010/05/my-contrarian-stance-on -facebook-privacy.html.

Orlowski, Andrew. "Google Buys CIA-Backed Mapping Startup." *Register,* October 28, 2004. www.theregister.co.uk/2004/10/28/google_buys_keyhole/.

Parks, L. "Digging into Google Earth: An Analysis of 'Crisis in Darfur.'" *Geoforum* 40, no. 4 (2009): 535–545.

Peterson, Chris. "In Praise of [Some] DDoSs?" *Chris Peterson's blog,* July 21, 2009. www.cpeterson.org/2009/07/21/in-praise-of-some-ddoss/.

Prior, Markus. "Liberated Viewers, Polarized Voters: The Implications of In-
    creased Media Choice for Democratic Politics." *Good Society* 11, no. 3
    (2002): 10–16.

———. *Post-Broadcast Democracy: How Media Choice Increases Inequality in
    Political Involvement and Polarizes Elections.* New York: Cambridge Univer-
    sity Press, 2007.

Radsch, C. "Core to Commonplace: The Evolution of Egypt's Blogosphere."
    Arab Media & Society 6 (2008).

Rajadhyaksha, M. "Genocide on the Airwaves: An Analysis of the Interna-
    tional Law Concerning Radio Jamming." *Journal of Hate Studies* 5, no. 1
    (2010): 99.

Robinson, P. "The CNN Effect: Can the News Media Drive Foreign Policy?"
    Review of International Studies 25, no. 2 (1999): 301–309.

Salmanov, Oleg, and Anastasia Golitsyna. "Reiman Confirms New Search
    Project." *Moscow Times,* July 8, 2010.

"Security Fears over Map Site." *Daily Record* (Glasgow), December 21, 2005.

"Senators Announce Formation of Global Internet Freedom Caucus." John
    McCain's Press Office, March 24, 2010.

"Sex, Social Mores, and Keyword Filtering: Microsoft Bing in the 'Arabian
    Countries.'" OpenNet Initiative, March 4, 2010. opennet.net/sex-social
    -mores-and-keyword-filtering-microsoft-bing-arabian-countries.

Shachtman, Noah. "Exclusive: U.S. Spies Buy Stake in Firm That Monitors
    Blogs, Tweets." *Danger Room, Wired.com,* October 19, 2009. www.wired
    .com/dangerroom/2009/10/exclusive-us-spies-buy-stake-in-twitter-blog
    -monitoring-firm/.

Sharma, Amol, and Jessica E. Vascellaro. "Google and India Test the Limits
    of Liberty." *Wall Street Journal,* January 4, 2010.

Sheridan, Barrett. "The Internet Helps Build Democracies." *Newsweek,* April
    30, 2010.

Stahl, R. "Becoming Bombs: 3D Animated Satellite Imagery and the
    Weaponization of the Civic Eye." *MediaTropes* 2, no. 2 (2010): 65.

Stanek, Steven. "Egyptian Bloggers Expose Horror of Police Torture." *San
    Francisco Chronicle,* October 9, 2007.

Talbot, David. "Bing Dinged on Arab Sex Censorship." *Technology Review Ed-
    itors' Blog,* March 4, 2010. www.technologyreview.com/blog/editors/24891/
    ?utm_source=twitterfeed&utm_medium=twitter.

"Turkish Engineers Developing Internet Search Engine." *World Bulletin,* No-
    vember 28, 2009. www.worldbulletin.net/news_detail.php?id=50543.

"24 Hours of Video Uploaded to YouTube Every Minute." Agence France-Presse, March 17, 2010.

Vedel, T. "The Idea of Electronic Democracy: Origins, Visions and Questions." *Parliamentary Affairs* 59, no. 2 (2006): 226.

Weintraub, Seth. "Google to Open 'Google Ideas' Global Technology Think Tank." *Google 24/7* blog, *Fortune*, August 15, 2010. tech.fortune.cnn.com/2010/08/15/google-to-open-google-ideas-global-technology-think-tank/.

Wilson, Paul. "IPv6 Answers to Common Questions from Policy Makers, Executives and Other Non-Technical Readers." *CircleID*, November 8, 2009. www.circleid.com/posts/ipv6_answers_to_most_common_questions_f or_non_technical/.

Wong, Albert, and Fanny W. Y. Fung. "Facebook Questioned as Political Pages Shut Down." *South China Morning Post*, February 6, 2010.

Wood, B. D., and J. S. Peake. "The Dynamics of Foreign Policy Agenda Setting." *American Political Science Review* 92, no. 1 (1998): 173–184.

Worthen, Ben. "Internet Strategy: China's Next Generation Internet." *CIO.com*, July 15, 2006. www.cio.com/article/22985/Internet_Strategy_China_s_Next_Generation_Internet.

Wright, Robert. "The Internet vs. Obama." *Opinionator Blog, New York Times*, February 2, 2010. opinionator.blogs.nytimes.com/2010/02/02/obamas-modern-predicament/.

York, Jillian. "Facebook Removes Moroccan Secularist Group and Its Founder." *Global Voices Advocacy*, March 14, 2010. advocacy.globalvoices online.org/2010/03/14/facebook-removes-moroccan-atheist-group-and-its-founder/.

CHAPTER 9

Alexseev, M. A. "Majority and Minority Xenophobia in Russia: The Importance of Being Titulars." *Post-Soviet Affairs* 26, no. 2 (2010): 89–120.

Allen-Mills, Tony. "Mexican Drug Gangs Take Their Turf Wars onto YouTube." *Times of London*, April 15, 2007.

Allnutt, Luke. "Twitter Doesn't Start a Revolution, People Do." *Christian Science Monitor*, February 8, 2010.

Amer, Pakinam. "Muslim Brotherhood Use New Media to Document History." *Al-Masry Al-Youm*, February 23, 2010. www.almasryalyoum.com/en/news/muslim-brotherhood-use-new-media-document-history.

Anderson, Benedict. *Imagined Communities: Reflections on the Origin and Spread of Nationalism*. London: Verso, 1991.

Apodaca, C. "The Whole World Could Be Watching: Human Rights and the Media." *Journal of Human Rights* 6, no. 2 (2007): 147–164.

Armony, Ariel C. *The Dubious Link: Civic Engagement And Democratization.* Stanford, CA: Stanford University Press, 2004.

Arshad, Arlina. "Desperate Indonesians Sell Organs Online." Agence France-Presse, December 17, 2009.

Auten, B. J. "Political Diasporas and Exiles as Instruments of Statecraft." *Comparative Strategy* 25, no. 4 (2006): 329–341.

Bäck, H., and A. Hadenius. "Democracy and State Capacity: Exploring a J-Shaped Relationship." *Governance* 21, no. 1 (2008): 1–24.

Baldauf, Scott. "Can Kenya Stop Violence After Vote?" *Christian Science Monitor*, January 2, 2008.

Balkin, Jack M. "Information Power: The Information Society from an Anti-humanist Perspective." *Social Science Research Network eLibrary* (2010). papers.ssrn.com/sol3/papers.cfm?abstract_id=1648624.

Bangre, Habibou. "Kenya: SMS Text Messages the New Guns of War?" *Afrik-News*, February 20, 2008. www.afrik-news.com/article12629.html.

Bezlova, Antoaneta. "China: Battle with Tradition Spills into Cyberspace." Inter Press Service, April 7, 2002.

Billig, Michael. *Banal Nationalism.* Thousand Oaks, CA: Sage, 1995.

Blum, B. S., and A. Goldfarb. "Does the Internet Defy the Law of Gravity?" *Journal of International Economics* 70, no. 2 (2006): 384–405.

Brenner, N. "Beyond State-Centrism? Space, Territoriality, and Geographical Scale in Globalization Studies." *Theory and Society* 28, no. 1 (1999): 39–78.

Bunt, Gary R. *iMuslims: Rewiring the House of Islam.* Chapel Hill: University of North Carolina Press, 2009.

Burstein, A. "Jefferson's Rationalizations." *William and Mary Quarterly* 57, no. 1 (2000): 183–197.

"Cambodia Lambast Google Earth for Locating Temple in Thai Soil." *Nation/Asia News Network*, February 10, 2010. www.asiaone.com/News/Latest +News/Asia/Story/A1Story20100210-197804.html.

Castells, Manuel. *The Internet Galaxy: Reflections on the Internet, Business, and Society.* Oxford: Oxford University Press, 2003.

Chan, B. "Imagining the Homeland: The Internet and Diasporic Discourse of Nationalism." *Journal of Communication Inquiry* 29, no. 4 (2005): 336.

Craig, G. A. "The Professional Diplomat and His Problems, 1919–1939." *World Politics: A Quarterly Journal of International Relations* 4, no. 2 (1952): 145–158.

Currion, Paul. "Better the Devil We Know: Obstacles and Opportunities in Humanitarian GIS." *Humanitarian.info*, January 25, 2006. www.humanitarian.info/humanitarian-gis/.

———. "Correcting Crowdsourcing in a Crisis." *Humanitarian.info*, March 30, 2009. www.humanitarian.info/2009/03/30/correcting-crowdsourcing-in-a-crisis/.

"Cyber-Nationalism: The Brave New World of E-hatred." *Economist*, July 24, 2008.

Dahlberg, L. "Rethinking the Fragmentation of the Cyberpublic: From Consensus to Contestation." *New Media & Society* 9, no. 5 (2007): 827.

Dewan, Shaila. "Chinese Student in U.S. Is Caught in Confrontation." *New York Times*, April 17, 2008.

Doppelt, G. "What Sort of Ethics Does Technology Require?" *Journal of Ethics* 5, no. 2 (2001): 155–175.

Edmunds, A., and A. Morris. "The Problem of Information Overload in Business Organisations: A Review of the Literature." *International Journal of Information Management* 20, no. 1 (2000): 17–28.

Eriksen, T. H. "Nations in Cyberspace." Short version of the 2006 Ernest Gellner Lecture, delivered to the ASEN conference, London School of Economics, March 27, 2006.

Eriksson, J., and G. Giacomello. "Who Controls the Internet? Beyond the Obstinacy or Obsolescence of the State." *International Studies Review* 11, no. 1 (2009): 205–230.

———. "The Information Revolution, Security, and International Relations: (IR) Relevant Theory?" *International Political Science Review/ Revue internationale de science politique* 27, no. 3 (2006): 221.

Eriksson, J., and M. Rhinard. "The Internal-External Security Nexus: Notes on an Emerging Research Agenda." *Cooperation and Conflict* 44, no. 3 (2009): 243.

Evers, C. "The Cronulla Race Riots: Safety Maps on an Australian Beach." *South Atlantic Quarterly* 107, no. 2 (2008): 411.

Feenberg, A. "Democratizing Technology: Interests, Codes, Rights." *Journal of Ethics* 5, no. 2 (2001): 177–195.

———. "Subversive Rationalization: Technology, Power, and Democracy." *Inquiry* 35, no. 3 (1992): 301–322.

Finel, B. I., and K. M. Lord. "The Surprising Logic of Transparency." *International Studies Quarterly* 43, no. 2 (1999): 325–339.

Fukuyama, F. "Social Capital and Development: The Coming Agenda." *SAIS review* 22, no. 1 (2002): 23–38.

Giridharadas, Anand. "Ushahidi—Africa's Gift to Silicon Valley: How to Track a Crisis." *New York Times*, March 12, 2010.

Glionna, John M. "Korea Activists Target Foreign English Teachers." *Los Angeles Times*, January 31, 2010.

Goble, Paul. "Circassians Using Internet to End Soviet-Imposed Divisions in Advance of 2010 Census." *Georgiandaily.com*, January 8, 2010. georgian daily.com/index.php?option=com_content&task=view&id=16365& Itemid=134.

———. "Russian Nationalists Now Mapping Location of Immigrants in Major Cities." *Window on Eurasia*, June 5, 2008. windowoneurasia.blogspot.com/ 2008/06/window-on-eurasia-russian-nationalists.html.

———. "Will the Internet Integrate the Russian Federation—or Tear It Apart?" *Moscow Times*, April 16, 2009.

Goggin, G. "SMS Riot: Transmitting Race on a Sydney Beach, December 2005." *M/C Journal* 9, no. 2206 (2008): 28.

"Google Admits 'Mistake' of Wrong Depiction of Arunachal." *The Times of India*, August 8, 2009.

Guillén, M. F. "Is Globalization Civilizing, Destructive or Feeble? A Critique of Five Key Debates in the Social Science Literature." *Annual Review of Sociology* 27 (2001): 235–260.

Hancocks, Paula. "Facebook Gets Caught in Golan Heights Dispute." *CNN.com*, September 21, 2009. edition.cnn.com/2009/TECH/09/21/israel.syria .facebook/index.html.

Hanson, Elizabeth C. *The Information Revolution and World Politics*. Lanham, MD: Rowman & Littlefield, 2008.

Harley, Jonathan. "Race Riots Erupt in Sydney." *The 7:30 Report*. Australian Broadcasting Corporation, December 12, 2005.

Herrera, G. L. "Technology and International Systems." *Millennium: Journal of International Studies* 32, no. 3 (2003): 559.

Herold, D. K. "Development of a Civic Society Online? Internet Vigilantism and State Control in Chinese Cyberspace." *Asian Journal of Global Studies* 2, no. 1 (2008): 26–37.

Hess, S. "Dividing and Conquering the Shop Floor: Uyghur Labour Export and Labour Segmentation in China's Industrial East." *Central Asian Survey* 28, no. 4 (2009): 403–416.

Holmes, S. "What Russia Teaches Us Now: How Weak States Threaten Freedom." *American Prospect* (1997): 30–39.

Huang, Annie. "Taiwanese Offer Ancestors Paper Ferraris, iPhones." Associated Press, April 2, 2010.

"India's Youth Hit the Web to Worship." *BBC News*, February 8, 2007.

"Internet Fuels Philippine Election Smear Campaigns." Agence France-Presse, April 14, 2010.

Kaplan, C. "The Biopolitics of Technoculture in the Mumbai Attacks." *Theory, Culture & Society* 26, nos. 7–8 (2009): 301.

Kapor, Mitchell. "Where Is the Digital Highway Really Heading?" *Wired*, August 1993.

Keenan, T. "Mobilizing Shame." *South Atlantic Quarterly* 103, nos. 2–3 (2004): 435.

———. "Publicity and Indifference (Sarajevo on Television)." *Publications of the Modern Language Association of America* 117, no. 1 (2002): 104–116.

Kennan, George F. "Somalia, Through a Glass Darkly." *New York Times*, September 30, 1993.

"Kenya Election Violence Witnesses Get Death Threats." *BBC News*, January 6, 2010.

Kerr, O. S. "Enforcing Law Online." *University of Chicago Law Review* 74, no. 2 (2007): 745–760.

Kimmelman, Michael. "New Weapons in Europe's Culture Wars." *New York Times*, January 17, 2010.

"Koreans Cyber Attack Japanese Site for Anti-Korean Posts." Yonhap News Agency, March 1, 2010.

Kurlantzick, Josh. "China's Next-Generation Nationalists." *Los Angeles Times*, May 6, 2008.

Lee, Jiyeon. "'Witch Hunting on the Web: The Latest Korean Fad?" *Global Post*, January 6, 2010. www.globalpost.com/dispatch/south-korea/091230/witch-hunting-web-trend.

Lee, Tae-hoon. "Lawmaker Calls for Stricter Access to NK Sites." *Korea Times*, October 6, 2009.

Lewis, Leo. "Google Earth Maps Out Discrimination Against Burakumin Caste in Japan." *Times of London*, May 22, 2009.

Linde, Steve. "Israel's Newest PR Weapon: The Internet Megaphone." *Jerusalem Post*, November 28, 2006.

Lord, Kristin M. *The Perils and Promise of Global Transparency: Why the Information Revolution May Not Lead to Security, Democracy, or Peace*. Albany: State University of New York Press, 2006.

Loveless, M. "The Theory of International Media Diffusion: Political Socialization and International Media in Transitional Democracies." *Studies in Comparative International Development* 44, no. 2 (2009): 118–136.

Mann, M. "The Autonomous Power of the State: Its Origins, Mechanisms and Results." *The State: Critical Concepts* 25 (1994): 331.

———. "Has Globalization Ended the Rise and Rise of the Nation-State?" *Review of International Political Economy* 4, no. 3 (1997): 472–496.

———. "Infrastructural Power Revisited." *Studies in Comparative International Development* 43, no. 3 (2008): 355–365.

March, Stephanie. "South Korea Tries to Curb Internet Addiction." Radio Australia, April 5, 2010.

Mathiason, J. "Internet Governance Wars: The Realists Strike Back." *International Studies Review* 9, no. 1 (2007): 152–155.

McCarthy, Michael, and Kevin Rawlinson. "Internet Trade Driving Rare Salamander to Extinction." *Independent*, March 17, 2010.

McLuhan, Marshall. *Understanding Media: The Extensions of Man*. New York: McGraw-Hill, 1964.

Melleuish, G., K. Sheiko, and S. Brown. "Pseudo History/Weird History: Nationalism and the Internet." *History Compass* 7, no. 6 (2009): 1484–1495.

Miller, Michael E. "Mexico Considers Clamping Down on Twitter." *Global Post*, February 2, 2010. www.globalpost.com/dispatch/mexico/100128/twitter-crackdown.

Morozov, Evgeny. "Citizen War-Reporter? The Caucasus Test." *OpenDemocracy*, August 18, 2008. www.opendemocracy.net/article/citizen-war-reporter.

Negroponte, Nicholas. *Being Digital*. New York: Knopf, 1995.

Nicholson, Sophie. "Internet Spreads Mexico Drug Gang Fears." Agence France-Presse, April 26, 2010.

"'No Rapes' in Riot Town." Radio Free Asia, June 29, 2009.

Nossiter, Adam. "Nigerians Recount Night of Their Bloody Revenge." *New York Times*, March 10, 2010.

Nyiri, P., J. Zhang, and M. Varrall. "China's Cosmopolitan Nationalists: 'Heroes' and 'Traitors' of the 2008 Olympics." *China Journal* 63 (2010): 25.

O'Hara, K., and D. Stevens. "The Devil's Long Tail: Religious Moderation and Extremism on the Web." *IEEE Intelligent Systems* 24, no. 6 (2009): 37–43.

Osnos, E. "Angry Youth: The New Generation's Neocon Nationalists." *New Yorker* 28 (2008).

Pallaris, C., S. S. Costigan, and W. B. I. Calcutta. "Shared Knowledge, Joint Pursuits: International Relations Beyond the Age of Information." Working Paper, May 24, 2010.

Peoples, C. "Technology, Philosophy and International Relations." *Cambridge Review of International Affairs* 22, no. 4 (2009): 559–561.

Perritt, H. H., Jr. "The Internet as a Threat to Sovereignty: Thoughts on the Internet's Role in Strengthening National and Global Governance." *Indiana Journal of Global Legal Studies* 5 (1997): 423.

"Police Fail to Protect Victims of Neo-Nazi Threats." *Prague Monitor*, September 14, 2009.

Price, M. E. "End of Television and Foreign Policy." *Annals of the American Academy of Political and Social Science* 625, no. 1 (2009): 196.

Putnam, R. D. *Bowling Alone: The Collapse and Revival of American Community*. New York: Simon & Schuster, 2001.

Quarantelli, E. L. "Problematical Aspects of the Information / Communication Revolution for Disaster Planning and Research: Ten Non-technical Issues and Questions." *Disaster Prevention and Management* 6, no. 2 (1997): 94–106.

Querengesser, Tim. "Cellphones Spread Kenyans' Messages of Hate." *Globe and Mail* (Toronto), February 29, 2008.

Rafael, V. L. "The Cell Phone and the Crowd: Messianic Politics in the Contemporary Philippines." *Public Culture* 15, no. 3 (2003): 399.

Rose, N., and P. Miller. "Political Power Beyond the State: Problematics of Government." *British Journal of Sociology* 43, no. 2 (1992): 173–205.

Rubio, M. "Perverse Social Capital: Some Evidence from Colombia." *Journal of Economic Issues* 31, no. 3 (1997): 805–816.

Saunders, R. A. "Denationalized Digerati in the Virtual Near Abroad: The Internet's Paradoxical Impact on National Identity Among Minority Russians." *Global Media and Communication* 2, no. 1 (2006): 43.

———. "Nationality: Cyber-Russian." *Russia in Global Affairs* 2, no. 4 (2004): 156.

Saunders, R. A., and S. Ding. "Digital Dragons and Cybernetic Bears: Comparing the Overseas Chinese and Near Abroad Russian Web Communities." *Nationalism and Ethnic Politics* 12, no. 2 (2006): 255–290.

Scheuerman, William E. "Liberal Democracy and the Empire of Speed." *Polity* 34, no. 1 (2001): 41–67.

———. *Liberal Democracy and the Social Acceleration of Time*. Baltimore: John Hopkins University Press, 2004.

———. "Realism and the Critique of Technology." *Cambridge Review of International Affairs* 22, no. 4 (2009): 563–584.

Schleifer, Yigal. "Turkey: The Internet Helps Some Rural Men Practice Polygamy." *EurasiaNet.org*, October 22, 2009. www.eurasianet.org/departments/insightb/articles/eav102309a.shtml.

Schlesinger, A., Jr. "Has Democracy a Future?" *Foreign Affairs* 76, no. 5 (1997): 2–12.

Schuler, I. "SMS as a Tool in Election Observation." *Innovations* (2008).

Selinger, E. "Towards a Reflexive Framework for Development: Technology Transfer After the Empirical Turn." *Synthese* 168, no. 3 (2009): 377–403.

Sisci, Francesco. "Who Is Hitting at Hu?" *Asia Times Online*, July 24, 2009. www.atimes.com/atimes/China/KG24Ad01.html.

Soifer, H., and M. vom Hau. "Unpacking the Strength of the State: The Utility of State Infrastructural Power." *Studies in Comparative International Development* 43, no. 3 (2008): 219–230.

"Somalia's Text Message Insurgency." *BBC News*, March 16, 2009.

Spitulnik, D. "Anthropology and Mass Media." *Annual Review of Anthropology* 22, no. 1 (1993): 293–315.

Streeten, P. "Reflections on Social and Antisocial Capital." *Journal of Human Development and Capabilities* 3, no. 1 (2002): 7–22.

Suleymanova, Dilyara. "Tatar Groups in Vkontakte." *Digital Icons* 1, no. 2 (2009).

Tamir, Y. "The Enigma of Nationalism." *World Politics* 47, no. 3 (1995): 418–440.

Tehranian, M. "Communication and Revolution in Iran: The Passing of a Paradigm." *Iranian Studies* 13, no. 1 (1980): 5–30.

"Traffic Bribery Goes Underground." *Daily Nation* (Nairobi), October 29, 2009.

Trifonov, Vladislav. "FSB Razygrala 'Bol'shuju Igru.'" *Kommersant*, August 11, 2009. www.kommersant.ru/doc.aspx?DocsID=1219146.

Trofimov, Yaroslav. "Taliban Force Cellphone Shutdown in Afghanistan." *Wall Street Journal*, March 22, 2010.

Trouillot, M. R. "The Anthropology of the State in the Age of Globalization." *Current Anthropology* 42, no. 1 (2001).

Waisbord, S. "Democratic Journalism and 'Statelessness.'" *Political Communication* 24, no. 2 (2007): 115–129.

Walby, S. "The Myth of the Nation-State: Theorizing Society and Polities in a Global Era." *Sociology* 37, no. 3 (2003): 529.

Warf, B., and J. Grimes. "Counterhegemonic Discourses and the Internet." *Geographical Review* 87, no. 2 (1997): 259–274.

Warren, M. E. "Social Capital and Corruption." Presentation at Social Capital: Interdisciplinary Perspectives, EURESCO Conference on Social Capital, September 15–20, 2001.

Watts, Jonathan. "Old Suspicions Magnified Mistrust into Ethnic Riots in Urumqi." *Guardian*, July 10, 2009.

Weiss, L. "Globalization and National Governance: Antinomy or Interdependence?" *Review of International Studies* 25 (1999): 59–88.

———. "Globalization and the Myth of the Powerless State." *New Left Review*, no. 225 (1997): 3–27.

———. *The Myth of the Powerless State*. Ithaca, NY: Cornell University Press, 1998.

Zuev, D. "The Movement Against Illegal Immigration: Analysis of the Central Node in the Russian Extreme-Right Movement." *Nations and Nationalism* 16, no. 2 (2010): 261–284.

CHAPTER 10

Achterhuis, Hans, ed. *American Philosophy of Technology: The Empirical Turn*. Translated by Robert P. Crease. Bloomington: Indiana University Press, 2001.

Adas, Michael. *Machines as the Measure of Men: Science, Technology, and Ideologies of Western Dominance*. Ithaca, NY: Cornell University Press, 1990.

Alexander, J. "The Sacred and Profane Information Machine: Discourse About the Computer as Ideology." *Archives de sciences sociales des religions* 35, no. 69 (1990): 161–171.

Alvarez, M. R. "Modern Technology and Technological Determinism: The Empire Strikes Again." *Bulletin of Science, Technology & Society* 19, no. 5 (1999): 403.

Armitage, J. "Resisting the Neoliberal Discourse of Technology: The Politics of Cyberculture in the Age of the Virtual Class." *CTHEORY* 1 (1999).

Balabanian, N. "On the Presumed Neutrality of Technology." *IEEE Technology and Society Magazine* 25, no. 4 (2006): 15–25.

Barbrook, R., and A. Cameron. "The Californian Ideology." *Science as Culture* 6, no. 1 (1996): 44–72.

Barley, S. R. "What Can We Learn from the History of Technology?" *Journal of Engineering and Technology Management* 15, no. 4 (1998): 237–255.

Behringer, W. "Introduction: Communication in Historiography." *German History* 24, no. 3 (2006): 325.

Beniger, James R. *Control Revolution: Technological and Economic Origins of the Information Society*. Cambridge, MA: Harvard University Press, 1989.

Bijker, Wiebe E., Thomas P. Hughes, and Trevor J. Pinch, eds. *The Social Construction of Technological Systems: New Directions in the Sociology and History of Technology*. Cambridge, MA: MIT Press, 1989.

Bimber, B. "Karl Marx and the Three Faces of Technological Determinism." *Social Studies of Science* (1990): 333–351.

Blondheim, Menahem. *News over the Wires: The Telegraph and the Flow of Public Information in America, 1844–1897,* Cambridge, MA: Harvard University Press, 1994.

Boccaccio, Giovanni. *The Decameron.* Vol. 1. New York: Modern Library, 1955.

Boorstin, Daniel Joseph. *The Republic of Technology.* New York: HarperCollins, 1979.

Briggs, Asa, and Peter Burke. *A Social History of the Media: From Gutenberg to the Internet.* 2nd ed. Malden, MA: Polity, 2005.

Cardwell, Donald. *Wheels, Clocks, and Rockets: A History of Technology.* New York: W. W. Norton, 1995.

Carey, J., and J. J. Quirk. "The Mythos of the Electronic Revolution." *American Scholar* 39, no. 1 (1970).

Carnes, Mark Christopher. *The Columbia History of Post-World War II America.* New York: Columbia University Press, 2007.

Ceruzzi, P. E. "Moore's Law and Technological Determinism." *Technology and Culture* 46, no. 3 (2005): 584–593.

Comor, E. "Harold Innis and 'the Bias of Communication.'" *Information, Communication and Society* 4, no. 2 (2001): 274–294.

Corn, Joseph J. *The Winged Gospel: America's Romance with Aviation.* Baltimore: Johns Hopkins University Press, 2002.

Cortada, J. W. "Do We Live in the Information Age? Insights from Historiographical Methods." *Historical Methods: A Journal of Quantitative and Interdisciplinary History* 40, no. 3 (2007): 107–116.

Cowan, Ruth S. *More Work for Mother: The Ironies of Household Technology from the Open Hearth to the Microwave.* New York: Basic Books, 1983.

Craig, Douglas B. *Fireside Politics: Radio and Political Culture in the United States, 1920–1940.* Baltimore: Johns Hopkins University Press, 2005.

Czitrom, Daniel J. *Media and the American Mind: From Morse to McLuhan.* Chapel Hill: University of North Carolina Press, 1982.

David, P. A. "The Dynamo and the Computer: An Historical Perspective on the Modern Productivity Paradox." *American Economic Review* 80, no. 2 (1990): 355–361.

de la Peña, Carolyn. "'Slow and Low Progress,' or Why American Studies Should Do Technology." *American Quarterly* 58 (2006): 915–941.

de la Peña, Carolyn, and Siva Vaidhyanathan, eds. *Rewiring the "Nation": The Place of Technology in American Studies.* Baltimore: Johns Hopkins University Press, 2007.

Diamond, L. "Liberation Technology." *Journal of Democracy* 21, no. 3 (2010): 69–83.

Douglas, Susan J. *Inventing American Broadcasting, 1899–1922*. Baltimore: Johns Hopkins University Press, 1987.

———. *Listening In: Radio and the American Imagination*. Minneapolis: University of Minnesota Press, 2004.

———. "The Turn Within: The Irony of Technology in a Globalized World." *American Quarterly* 58, no. 3 (2006): 619–638.

Dunlap, Orrin E., Jr. *The Outlook for Television*. New York: Harper & Brothers, 1932.

Durbin, P. T. "Technology and Political Philosophy." *Technology in Society* 6, no. 4 (1984): 315–327.

Elliott, E. D. "Against Ludditism: An Essay on the Perils of the (Mis) Use of Historical Analogies in Technology Assessment." *Southern California Law Review* 65, no. 1 (1991): 279.

Ezrahi, Yaron, Everett Mendelsohn, and Howard Segal. *Technology, Pessimism, and Postmodernism*. Boston: Kluwer Academic, 1994.

Feenberg, Andrew. *Alternative Modernity: The Technical Turn in Philosophy and Social Theory*. Berkeley: University of California Press, 1995.

———. "Marcuse or Habermas: Two Critiques of Technology." *Inquiry* 39, no. 1 (1996): 45–70.

———. *Questioning Technology*. New York: Routledge, 1999.

Ferkiss, V. C. "Man's Tools and Man's Choices: The Confrontation of Technology and Political Science." *American Political Science Review* 67, no. 3 (1973): 973–980.

——— "Technology and American Political Thought: The Hidden Variable and the Coming Crisis." *Review of Politics* 42, no. 3 (1980): 349–387.

Fischer, Claude S. *America Calling: A Social History of the Telephone to 1940*. Berkeley: University of California Press, 1994.

Fischer, E. "Contemporary Technology Discourse and the Legitimation of Capitalism." *European Journal of Social Theory* 13, no. 2 (2010).

Forest, Lee De. *Television, Today and Tomorrow*. New York: Dial, 1942.

Foster, T. "The Rhetoric of Cyberspace: Ideology or Utopia?" *Contemporary Literature* 40, no. 1 (1999): 144–160.

Friedel, Robert Douglas. *A Culture of Improvement: Technology and the Western Millennium*. Cambridge, MA: MIT Press, 2007.

Galston, W. A. "Does the Internet Strengthen Community?" *National Civic Review* 89, no. 3 (2000): 193–202.

Gane, N. "Speed Up or Slow Down? Social Theory in the Information Age." *Information, Communication & Society* 9, no. 1 (2006): 20–38.

Gladney, G. A. "Technologizing of the Word: Toward a Theoretical and Ethical Understanding." *Journal of Mass Media Ethics* 6, no. 2 (1991): 93–105.

Graham, S., and S. Marvin. "Planning Cybercities? Integrating Telecommunications into Urban Planning." *Town Planning Review* 70, no. 1 (1999): 89–114.

Grier, David Alan. *When Computers Were Human*. Princeton, NJ: Princeton University Press, 2005.

Grint, K., and S. Woolgar. "On Some Failures of Nerve in Constructivist and Feminist Analyses of Technology." *Science, Technology & Human Values* 20, no. 3 (1995): 286.

Halleck, DeeDee. *Hand-Held Visions: The Impossible Possibilities of Community Media*. New York: Fordham University Press, 2002.

Hand, M., and B. Sandywell. "E-topia as Cosmopolis or Citadel: On the Democratizing and De-Democratizing Logics of the Internet, or, Toward a Critique of the New Technological Fetishism." *Theory, Culture & Society* 19, no. 1-2 (2002): 197.

Hannay, N. B., and R. E. McGinn. "The Anatomy of Modern Technology: Prolegomenon to an Improved Public Policy for the Social Management of Technology." *Daedalus* 109, no. 1 (1980): 25–53.

Headrick, Daniel R. *The Invisible Weapon: Telecommunications and International Politics, 1851–1945*. New York: Oxford University Press, 1991.

———. *The Tools of Empire: Technology and European Imperialism in the Nineteenth Century*. New York: Oxford University Press, 1981.

———. *When Information Came of Age: Technologies of Knowledge in the Age of Reason and Revolution, 1700–1850*. New York: Oxford University Press, 2000.

Heidegger, Martin. *The Question Concerning Technology, and Other Essays*. New York: Harper Perennial, 1982.

Henderson, Peter. "Coal Fuels Much of Internet Cloud, Says Greenpeace." Reuters, March 30, 2010.

Herf, J. "Technology, Reification, and Romanticism." *New German Critique* (1977): 175–191.

Hine, C. "Internet Research and the Sociology of Cyber-Social-Scientific Knowledge." *Information Society* 21, no. 4 (2005): 239–248.

Hughes, Thomas P. *Human-Built World: How to Think About Technology and Culture*. Chicago: University of Chicago Press, 2004.

———. "Lusting for the Gratifications of Technology." *Reviews in American History* 14, no. 2 (1986): 265–269.

————. "The Seamless Web: Technology, Science, Etcetera, Etcetera." *Social Studies of Science* 16, no. 2 (1986): 281–292.

Ihde, Don. *Ironic Technics*. Copenhagen: Automatic Press, 2008.

Introna, L. D. "Maintaining the Reversibility of Foldings: Making the Ethics (Politics) of Information Technology Visible." *Ethics and Information Technology* 9, no. 1 (2007): 11–25.

Jacobs, Meg, William J. Novak, and Julian E. Zelizer. *The Democratic Experiment: New Directions in American Political History*. Princeton, NJ: Princeton University Press, 2003.

Jones, S. "Fizz in the Field: Toward a Basis for an Emergent Internet Studies." *Information Society* 21, no. 4 (2005): 233–237.

Katz-Kimchi, M. "Historicizing Utopian Popular Discourse on the Internet in America in the 1990s: Positions, Comparison, and Contextualization." Presentation at The Long History of New Media conference, Montreal, May 22, 2008.

Kenny, Charles. "Revolution in a Box." *Foreign Policy* (November 2009).

Khiabany, G. "Globalization and the Internet: Myths and Realities." *Trends in Communication* 11, no. 2 (2003): 137–153.

Kling, R. "Reading 'All About' Computerization: How Genre Conventions Shape Nonfiction Social Analysis." *Information Society* 10, no. 3 (1994): 147–172.

Latour, Bruno, and Peter Weibel. *Making Things Public: Atmospheres of Democracy*. Cambridge, MA: MIT Press, 2005.

Layton, E. T., Jr. "Technology as Knowledge." *Technology and Culture* (1974): 31–41.

Mackay, H., and G. Gillespie. "Extending the Social Shaping of Technology Approach: Ideology and Appropriation." *Social Studies of Science* 22, no. 4 (1992): 685–716.

MacKenzie, Donald. *Knowing Machines: Essays on Technical Change*. Cambridge, MA: MIT Press, 1998.

Mander, Jerry. *Four Arguments for the Elimination of Television*. Goa, India: Other India Press, 1998.

————. *In the Absence of the Sacred: The Failure of Technology and the Survival of the Indian Nations*. San Francisco: Sierra Club Books, 1991.

Manjoo, Farhad. "The Co-Founders of Twitter Say It Will Change the World. They Should Remind People That It's Also Fun." *Slate*, April 15, 2010. www.slate.com/id/2250991.

Marx, Leo. *The Pilot and the Passenger: Essays on Literature, Technology, and Culture in the United States*. New York: Oxford University Press, 1988.

———. "Technology: The Emergence of a Hazardous Concept." *Social Research* 64, no. 3 (1997): 965–988.

McFarland, F. B. "Clarence Ayres and His Gospel of Technology." *History of Political Economy* 18, no. 4 (1986): 617.

McLoughlin, I., R. Badham, and P. Couchman. "Rethinking Political Process in Technological Change: Socio-Technical Configurations and Frames." *Technology Analysis & Strategic Management* 12, no. 1 (2000): 17–37.

Melzer, Arthur M., Jerry Weinberger, and M. Richard Zinman. *Technology in the Western Political Tradition*. Ithaca, NY: Cornell University Press, 1993.

Merrin, W. "Media Studies 2.0: Upgrading and Open-Sourcing the Discipline." *Interactions: Studies in Communication and Culture* 1 (2009): 17–34.

Michael, D. N. "Too Much of a Good Thing? Dilemmas of an Information Society." *Technological Forecasting and Social Change* 25, no. 4 (1984): 347–354.

Mickelson, Sig. *From Whistle Stop to Sound Bite: Four Decades of Politics and Television*. New York: Praeger, 1989.

Misa, Thomas J. "How Machines Make History, and How Historians (and Others) Help Them to Do So." *Science, Technology, and Human Values* (1988): 308–331.

———. *Leonardo to the Internet: Technology and Culture from the Renaissance to the Present*. Baltimore: Johns Hopkins University Press, 2004.

———. "Theories of Technological Change: Parameters and Purposes." *Science, Technology, and Human Values* 17, no. 1 (1992): 3–12.

Misa, Thomas J., Philip Brey, and Andrew Feenberg. *Modernity and Technology*. Cambridge, MA: MIT Press, 2004.

Mosco, V. "From Here to Banality: Myths About New Media and Communication Policy." *Seeking Convergence in Policy and Practice* (2004): 23.

———. "Review Essay: Approaching Digital Democracy." *New Media & Society* (2009).

Mosco, V., and D. Foster. "Cyberspace and the End of Politics." *Journal of Communication Inquiry* 25, no. 3 (2001): 218.

Mowshowitz, A. "Computers and the Myth of Neutrality." In *Proceedings of the ACM 12th Annual Computer Science Conference on SIGCSE Symposium*, 92. 1984.

Munir, K. A., and M. Jones. "Discontinuity and After: The Social Dynamics of Technology Evolution and Dominance." *Organization Studies* 25, no. 4 (2004): 561.

Nerone, J. "The Future of Communication History." *Critical Studies in Media Communication* 23, no. 3 (2006): 254–262.

Neuman, J. "The Media's Impact on International Affairs, Then and Now." *SAIS Review* 16 (1996): 109–124.

Nguyen, A. "The Interaction Between Technologies and Society: Lessons Learned from 160 Evolutionary Years of Online News Services." *First Monday* 12, no. 3-5 (2007).

Nissenbaum, H. "How Computer Systems Embody Values." *Computer* 34, no. 3 (2001): 120.

Noble, David F. *America by Design: Science, Technology, and the Rise of Corporate Capitalism*. New York: Oxford University Press, 1979.

———. *Forces of Production: A Social History of Industrial Automation*. New York: Oxford University Press, 1986.

Nye, David E. *American Technological Sublime*. Cambridge: MIT Press, 1996.

Olsen, Jan-Kyrre Berg, and Evan Selinger. *Philosophy of Technology: Five Questions*. Copenhagen: Automatic Press/VIP, 2007.

Olsen, Jan-Kyrre Berg, Evan Selinger, and Søren Riis. *New Waves in Philosophy of Technology*. New York: Palgrave Macmillan, 2009.

Ornatowski, C. M. "Techne and Politeia: Langdon Winner's Political Theory of Technology and Its Implications for Technical Communication." *Technical Communication Quarterly* 11, no. 2 (2002): 230–234.

Pärna, K. "Believing in the Net: Implicit Religion and the Internet Hype, 1994–2001." PhD diss., Leiden University, 2010.

Patnode, R. "Path Not Taken: Wired Wireless and Broadcasting in the 1920s." *Journal of Broadcasting & Electronic Media* 49, no. 4 (2005): 383–401.

Pease, Edward C., and Everette E. Dennis, eds. *Radio: The Forgotten Medium*. New Brunswick, NJ: Transaction Publishers, 1995.

Pfaffenberger, B. "Fetishised Objects and Humanised Nature: Towards an Anthropology of Technology." *Man* 23, no. 2 (1988): 236–252.

———. "Social Anthropology of Technology." *Annual Review of Anthropology* 21, no. 1 (1992): 491–516.

———. "The Social Meaning of the Personal Computer; or, Why the Personal Computer Revolution Was No Revolution." *Anthropological Quarterly* (1988): 39–47.

———. "Symbols Do Not Create Meaning—Activities Do; or, Why Symbolic Anthropology Needs the Anthropology of Technology." *Anthropological Perspectives on Technology* (2001): 77–86.

———. "Technological Dramas." *Science, Technology & Human Values* 17, no. 3 (1992): 282.

Pool, Ithiel de Sola. *Technologies of Freedom*. Cambridge, MA: Harvard University Press, 1983.

Post, Robert C. "Missionary: An Interview with Melvin Kranzberg." *American Heritage of Invention & Technology* 4, no. 3 (1989).

———. "No Mere Technicalities: How Things Work and Why It Matters." *Technology and Culture* 40, no. 3 (1999): 607–622.

Postman, Neil. "Informing Ourselves to Death." Speech at the German Informatics Society, October 11, 1990.

Pursell, C. W., Jr. "Government and Technology in the Great Depression." *Technology and Culture* 20, no. 1 (1979): 162–174.

Radder, H. "Normative Reflexions on Constructivist Approaches to Science and Technology." *Social Studies of Science* 22, no. 1 (1992): 141–173.

Rürup, R. "Historians and Modern Technology: Reflections on the Development and Current Problems of the History of Technology." *Technology and Culture* (1974): 161–193.

Scannell, P. "The Dialectic of Time and Television." *Annals of the American Academy of Political and Social Science* 625, no. 1 (2009): 219.

Schaniel, W. C. "New Technology and Culture Change in Traditional Societies." *Journal of Economic Issues* 22, no. 2 (1988): 493–498.

Segal, Howard P. *Technological Utopianism in American Culture*. Syracuse, NY: Syracuse University Press, 2005.

Shen, X. *The Chinese Road to High Technology: A Study of Telecommunications Switching Technology in the Economic Transition*. New York: St. Martin's, 1999.

Sibley, M. Q. "Utopian Thought and Technology." *American Journal of Political Science* (1973): 255–281.

Smith, Merritt Roe, and Leo Marx, eds. *Does Technology Drive History? The Dilemma of Technological Determinism*. Cambridge, MA: MIT Press, 1994.

Spar, Debora L. *Ruling the Waves: Cycles of Discovery, Chaos, and Wealth from Compass to the Internet*. New York: Harcourt, 2001.

Standage, Tom. *The Victorian Internet: The Remarkable Story of the Telegraph and the Nineteenth Century's On-Line Pioneers*. New York: Walker, 1998.

Staudenmaier, John M. "Rationality, Agency, Contingency: Recent Trends in the History of Technology." *Reviews in American History* (2002): 168–181.

———. *Technology's Storytellers: Reweaving the Human Fabric*. Cambridge, MA: Society for the History of Technology and the MIT Press, 1989.

Stump, D. J. "Socially Constructed Technology." *Inquiry* 43, no. 2 (2000): 217–224.

Sturken, Marita, and Douglas Thomas. *Technological Visions: The Hopes and Fears That Shape New Technologies.* Philadelphia: Temple University Press, 2004.

Tedre, M., E. Sutinen, E. Konen, and P. Kommers. "Ethnocomputing: ICT in Cultural and Social Context." *Communications of the ACM* 49, no. 1 (2006): 130.

Teich, Albert H., ed. *Technology and the Future.* 9th ed. Belmont, CA: Wadsworth/Thomson, 2003.

Thorne, K., and A. Kouzmin. "Cyberpunk-Web 1.0 'Egoism' Greets Group-Web 2.0 'Narcissism': Convergence, Consumption, and Surveillance in the Digital Divide." *Administrative Theory & Praxis* 30, no. 3 (2008): 299–323.

Thrift, N. "New Urban Eras and Old Technological Fears: Reconfiguring the Goodwill of Electronic Things." *Urban Studies* 33, no. 8 (1996): 1463.

Van Dijck, J., and D. Nieborg. "Wikinomics and Its Discontents: A Critical Analysis of Web 2.0 Business Manifestos." *New Media & Society* 11, no. 5 (2009): 855.

Verheul, Jaap, ed. *Dreams of Paradise, Visions of Apocalypse: Utopia and Dystopia in American Culture.* Amsterdam: VU University Press, 2004.

Warf, B., and J. Grimes. "Counterhegemonic Discourses and the Internet." *Geographical Review* 87, no. 2 (1997): 259–274.

Weightman, Gavin. *Signor Marconi's Magic Box: The Most Remarkable Invention of the 19th Century and the Amateur Inventor Whose Genius Sparked a Revolution.* Cambridge, MA: Da Capo Press, 2003.

Wellman, B., and B. Hogan. "The Immanent Internet." *Netting Citizens: Exploring Citizenship in a Digital Age* (2004): 54–80.

White, David Manning, ed. *Popular Culture.* New York: New York Times, 1975.

Williams, Raymond, and Ederyn Williams. *Television: Technology and Cultural Form.* New York: Routledge, 2003.

Winner, Langdon. *Autonomous Technology: Technics-Out-of-Control as a Theme in Political Thought.* Cambridge, MA: MIT Press, 1978.

———. "Social Constructivism: Opening the Black Box and Finding It Empty." *Science as Culture* 3, no. 3 (1993): 427–452.

———. *The Whale and the Reactor: A Search for Limits in an Age of High Technology.* Chicago: University of Chicago Press, 1988.

Winseck, Dwayne R., and Robert M. Pike. *Communication and Empire: Media, Markets, and Globalization, 1860–1930*. Durham, NC: Duke University Press, 2007.

Wise, George. "Technological Prediction, 1890–1940." PhD diss., Boston University, 1976.

Woolgar, S., and G. Cooper. "Do Artefacts Have Ambivalence? Moses' Bridges, Winner's Bridges and Other Urban Legends in S&TS." *Social Studies of Science* 29, no. 3 (1999): 433.

Wyatt, S. "Technological Determinism Is Dead; Long Live Technological Determinism." In *The Handbook of Science and Technology Studies*, edited by Edward J. Hackett et al., 165. Cambridge, MA: MIT Press, 2008.

CHAPTER 11

Arendt, Hannah. *On Violence*. New York: Harcourt, Brace, Jovanovich, 1970.

Austin, E. K., and J. C. Callen. "Reexamining the Role of Digital Technology in Public Administration: From Devastation to Disclosure." *Administrative Theory & Praxis* 30, no. 3 (2008): 324–341.

Bostrom, N. "Technological Revolutions: Ethics and Policy in the Dark." *Nanotechnology and Society* (2007).

Brown, M. B. "Can Technologies Represent Their Publics?" *Technology in Society* 29, no. 3 (2007): 327–338.

Carey, J. W. "Historical Pragmatism and the Internet." *New Media & Society* 7, no. 4 (2005): 443.

Coyne, R. "Wicked Problems Revisited." *Design Studies* 26, no. 1 (2005): 5–17.

David, E. E., Jr. "On the Dimensions of the Technology Controversy." *Daedalus* 109, no. 1 (1980): 169–177.

Dörner, Dietrich. *The Logic of Failure: Recognizing and Avoiding Error in Complex Situations*. Translated by Rita and Robert Kimber. New York; Metropolitan Books, 1996.

Duff, A. S. "Social Engineering in the Information Age." *Information Society* 21, no. 1 (2005): 67–71.

Freeman, M. "Sociology and Utopia: Some Reflections on the Social Philosophy of Karl Popper." *British Journal of Sociology* 26, no. 1 (1975): 20–34.

Grunwald, A. "Converging Technologies: Visions, Increased Contingencies of the Conditio Humana, and Search for Orientation." *Futures* 39, no. 4 (2007): 380–392.

Hamlett, P. W. "Technology Theory and Deliberative Democracy." *Science, Technology & Human Values* 28, no. 1 (2003): 112.

Horner, D. S. "Digital Futures: Promising Ethics and the Ethics of Promising." *ACM SIGCAS Computers and Society* 37, no. 2 (2007): 64–77.

Jopson, Barney. "Hope Founders Where Ministers Lack E-mail." *Financial Times*, February 17, 2010.

Kakabadse, N. K., A. P. Kakabadse, and A. Kouzmin. "Designing Balance into the Democratic Project: Contrasting Jeffersonian Democracy Against Bentham's Panopticon Centralisation in Determining ICT Adoption." *Problems and Perspectives in Management* 1 (2007).

Karlsson, R. "Why the Far-Future Matters to Democracy Today." *Futures* 37, no. 10 (2005): 1095–1103.

Keulartz, J., M. Schermer, M. Korthals, and T. Swierstra. "Ethics in Technological Culture: A Programmatic Proposal for a Pragmatist Approach." *Science, Technology & Human Values* 29, no. 1 (2004): 3.

Klosterman, Chuck. *Eating the Dinosaur*. New York: Scribner, 2009.

Krotoski, Alex. "MediaGuardian Innovation Awards: Austin Heap v Iran's censors." *Guardian*, March 29, 2010.

Lanki, J. "Why Would Information and Communications Technology Contribute to Development at All? An Ethical Inquiry into the Possibilities of ICT in Development." *E-Learning and Digital Media* 3, no. 3 (2006): 448–461.

Layne, L. L. "The Cultural Fix: An Anthropological Contribution to Science and Technology Studies." *Science, Technology & Human Values* 25, no. 3 (2000): 352.

Lazarus, R. J. "Super Wicked Problems and Climate Change: Restraining the Present to Liberate the Future." *Cornell Law Review* 94, no. 5 (2009).

Lessnoff, M. "The Political Philosophy of Karl Popper." *British Journal of Political Science* 10, no. 1 (1980): 99–120.

Morrison, A. H. "An Impossible Future: John Perry Barlow's 'Declaration of the Independence of Cyberspace.'" *New Media & Society* 11, no. 1-2 (2009): 53.

Norman, D. A. "Affordance, Conventions, and Design." *Interactions* 6, no. 3 (1999): 38–43.

Oliver, M. "The Problem with Affordance." *E-Learning and Digital Media* 2, no. 4 (2005): 402–413.

O'Loughlin, B. "The Political Implications of Digital Innovations: Trade-offs of Democracy and Liberty in the Developed World." *Information, Communication & Society* 4, no. 4 (2001): 595–614.

Petrina, S. "Questioning the Language That We Use: A Reaction to Pannabecker's Critique of the Technological Impact Metaphor." *Journal of Technology Education* 4, no. 1 (1992).

Pitkin, B. "A Historical Perspective of Technology and Planning." *Berkeley Planning Journal* 15 (2001): 34–59.

Popper, K. "The Poverty of Historicism, I.." *Economica* 11, no. 42 (1944): 86–103.

———. "The Poverty of Historicism, II. A Criticism of Historicist Methods." *Economica* 11, no. 43 (1944): 119–137.

Rittel, H. W. J., and M. M. Webber. "Dilemmas in a General Theory of Planning." *Policy Sciences* 4, no. 2 (1973): 155–169.

Rosner, Lisa. *The Technological Fix: How People Use Technology to Create and Solve Problems.* New York: Routledge, 2004.

Searle, J. "I Married a Computer." *New York Review of Books*, April 8, 1999.

Stahl, B. C. "Democracy, Responsibility, and Information Technology." In *Proceedings of the European Conference on e-Government*, 429–439. 2001.

Tenner, Edward. *Why Things Bite Back: Technology and the Revenge of Unintended Consequences.* New York: Vintage Books, 1997.

Weinberg, Alvin Martin. "Can Technology Replace Social Engineering." *Bulletin of the Atomic Scientists* 22, no. 10 (1966): 4–8.

———. *Nuclear Reactions: Science And Trans-science.* New York: American Institute of Physics, 1992.

Wexler, M. N. "Exploring the Moral Dimension of Wicked Problems." *International Journal of Sociology and Social Policy* 29 (2009).

Winner, Langdon. "Citizen Virtues in a Technological Order." *Inquiry* 35, no. 3 (1992): 341–361.

———. "Mythinformation in the High-Tech Era." *IEEE Spectrum* 21, no. 6 (1984): 90–96.

———. "Myth Information: Romantic Politics in the Computer Revolution." *Philosophy and Technology II: Information Technology and Computers in Theory and Practice* (1986): 269.

———. "Technology Today: Utopia or Dystopia? Technology and the Rest of Culture." *Social Research* 64, no. 3 (1997): 989–1017.

Zuckerman, Ethan. "Internet Freedom: Beyond Circumvention." *My Heart's in Accra*, February 22, 2010. ethanzuckerman.com/blog/2010/02/22/internet-freedom-beyond-circumvention/.

INDEX

ALLEN LANE
an imprint of
PENGUIN BOOKS

Recently Published

Jesse J. Prinz, *Beyond Human Nature: How Culture and Experience Shape Our Lives*

Robert Holland, *Blue-Water Empire: The British in the Mediterranean since 1800*

Jodi Kantor, *The Obamas: A Mission, A Marriage*

Philip Coggan, *Paper Promises: Money, Debt and the New World Order*

Charles Nicholl, *Traces Remain: Essays and Explorations*

Daniel Kahneman, *Thinking, Fast and Slow*

Hunter S. Thompson, *Fear and Loathing at Rolling Stone: The Essential Writing of Hunter S. Thompson*

Duncan Campbell-Smith, *Masters of the Post: The Authorized History of the Royal Mail*

Colin McEvedy, *Cities of the Classical World: An Atlas and Gazetteer of 120 Centres of Ancient Civilization*

Heike B. Görtemaker, *Eva Braun: Life with Hitler*

Brian Cox and Jeff Forshaw, *The Quantum Universe: Everything that Can Happen Does Happen*

Nathan D. Wolfe, *The Viral Storm: The Dawn of a New Pandemic Age*

Norman Davies, *Vanished Kingdoms: The History of Half-Forgotten Europe*

Michael Lewis, *Boomerang: The Meltdown Tour*

Steven Pinker, *The Better Angels of Our Nature: The Decline of Violence in History and Its Causes*

Robert Trivers, *Deceit and Self-Deception: Fooling Yourself the Better to Fool Others*

Thomas Penn, *Winter King: The Dawn of Tudor England*

Daniel Yergin, *The Quest: Energy, Security and the Remaking of the Modern World*

Michael Moore, *Here Comes Trouble: Stories from My Life*

Ali Soufan, *The Black Banners: Inside the Hunt for Al Qaeda*

Jason Burke, *The 9/11 Wars*

Timothy D. Wilson, *Redirect: The Surprising New Science of Psychological Change*

Ian Kershaw, *The End: Hitler's Germany, 1944-45*

T M Devine, *To the Ends of the Earth: Scotland's Global Diaspora, 1750-2010*

Catherine Hakim, *Honey Money: The Power of Erotic Capital*

Douglas Edwards, *I'm Feeling Lucky: The Confessions of Google Employee Number 59*

John Bradshaw, *In Defence of Dogs*

Chris Stringer, *The Origin of Our Species*

Lila Azam Zanganeh, *The Enchanter: Nabokov and Happiness*

David Stevenson, *With Our Backs to the Wall: Victory and Defeat in 1918*

Evelyn Juers, *House of Exile: War, Love and Literature, from Berlin to Los Angeles*

Henry Kissinger, *On China*

Michio Kaku, *Physics of the Future: How Science Will Shape Human Destiny and Our Daily Lives by the Year 2100*

David Abulafia, *The Great Sea: A Human History of the Mediterranean*

John Gribbin, *The Reason Why: The Miracle of Life on Earth*

Anatol Lieven, *Pakistan: A Hard Country*

William Cohen, *Money and Power: How Goldman Sachs Came to Rule the World*

Joshua Foer, *Moonwalking with Einstein: The Art and Science of Remembering Everything*

Simon Baron-Cohen, *Zero Degrees of Empathy: A New Theory of Human Cruelty*

Manning Marable, *Malcolm X: A Life of Reinvention*

David Deutsch, *The Beginning of Infinity: Explanations that Transform the World*

David Edgerton, *Britain's War Machine: Weapons, Resources and Experts in the Second World War*

John Kasarda and Greg Lindsay, *Aerotropolis: The Way We'll Live Next*

David Gilmour, *The Pursuit of Italy: A History of a Land, Its Regions and Their Peoples*

Niall Ferguson, *Civilization: The West and the Rest*

Tim Flannery, *Here on Earth: A New Beginning*

Robert Bickers, *The Scramble for China: Foreign Devils in the Qing Empire, 1832-1914*

Mark Malloch-Brown, *The Unfinished Global Revolution: The Limits of Nations and the Pursuit of a New Politics*

King Abdullah of Jordan, *Our Last Best Chance: The Pursuit of Peace in a Time of Peril*

Eliza Griswold, *The Tenth Parallel: Dispatches from the Faultline between Christianity and Islam*

Brian Greene, *The Hidden Reality: Parallel Universes and the Deep Laws of the Cosmos*

John Gray, *The Immortalization Commission: The Strange Quest to Cheat Death,*

Patrick French, *India: A Portrait*

Lizzie Collingham, *The Taste of War: World War Two and the Battle for Food*

Hooman Majd, *The Ayatollahs' Democracy: An Iranian Challenge*

Dambisa Moyo, *How The West Was Lost: Fifty Years of Economic Folly - and the Stark Choices Ahead*

Evgeny Morozov, *The Net Delusion: How Not to Liberate the World*

Ron Chernow, *Washington: A Life*

Nassim Nicholas Taleb, *The Bed of Procrustes: Philosophical and Practical Aphorisms*

Hugh Thomas, *The Golden Age: The Spanish Empire of Charles V*

Amanda Foreman, *A World on Fire: An Epic History of Two Nations Divided*

Nicholas Ostler, *The Last Lingua Franca: English until the Return of Babel*

Richard Miles, *Ancient Worlds: The Search for the Origins of Western Civilization*

Neil MacGregor, *A History of the World in 100 Objects*

Steven Johnson, *Where Good Ideas Come From: The Natural History of Innovation*

Dominic Sandbrook, *State of Emergency: The Way We Were: Britain, 1970-1974*

Jim Al-Khalili, *Pathfinders: The Golden Age of Arabic Science*

Ha-Joon Chang, *23 Things They Don't Tell You About Capitalism*